The Cattle, Sheep and Pigs of Great Britain
Various Breeds of the United Kingdom, Their History, Management, Etc

by Charles John Darent Blake Coleman

with an introduction by Jackson Chambers

This work contains material that was originally published in 1887.

This publication is within the Public Domain.

This edition is reprinted for educational purposes and in accordance with all applicable Federal Laws.

Introduction Copyright 2018 by Jackson Chambers

Self Reliance Books

Get more historic titles on animal and stock breeding, gardening and old fashioned skills by visiting us at:

http://selfreliancebooks.blogspot.com/

Introduction

I am pleased to present another title in the "Cattle" series.

The work is in the Public Domain and is re-printed here in accordance with Federal Laws.

As with all reprinted books of this age that are intended to perfectly reproduce the original edition, considerable pains and effort had to be undertaken to correct fading and sometimes outright damage to existing proofs of this title. At times, this task is quite monumental, requiring an almost total "rebuilding" of some pages from digital proofs of multiple copies. Despite this, imperfections still sometimes exist in the final proof and may detract from the visual appearance of the text.

I hope you enjoy reading this book as much as I enjoyed making it available to readers again.

Jackson Chambers

PREFACE.

IN issuing a Second Edition of THE CATTLE, SHEEP, AND PIGS OF GREAT BRITAIN, it has been thought desirable to combine the whole in one volume, which, much handier in form and reasonable in price, will it is hoped prove useful to both the student and those about embarking in a business which, to conduct with success, requires much more thorough study and training than has as a rule been given to it. Most of the original articles reappear in amended form, and it is believed that the information as to the various breeds accurately describes their present position and influence.

The original drawings, by Mr. Harrison Weir, have been reduced in size by a photographic process in order to adapt them to the smaller pages of the present volume.

As in the original books, it has been thought desirable to introduce the articles on the different breeds of cattle, sheep, and pigs, by observations on management, which are based upon the Editor's practical experience extending over a period of more than forty years.

The Editor desires to express his sense of obligation to those who have so willingly aided him in the work of reproduction.

JOHN COLEMAN.

YORK, *March*, 1887.

CONTENTS.

THE CATTLE OF GREAT BRITAIN.

CHAPTER I.
Introductory ... ...*page* 1

CHAPTER II.
Breeding and General Management 12

CHAPTER III.
Principles of Feeding, and Value of different kinds of
    Food 31

CHAPTER IV.
Buildings, and the Management of Manure 59

CHAPTER V.
Dairy Management, the Milk Trade, &c. 82

CHAPTER VI.
Shorthorns. By John Thornton 104

CONTENTS.

Chapter VII.
The Hereford Breed of Cattle. By T. Duckham ...*page* 112

Chapter VIII.
Devon Breed of Cattle. By Lieut.-Col. J. T. Davy.. .. 125

Chapter IX.
The Longhorns. By Gilbert Murray... 132

Chapter X.
The Sussex Breed of Cattle. By A. Heasman... 139

Chapter XI.
Norfolk and Suffolk Red-Polled Cattle. By Thomas Fulcher 144

Chapter XII.
Galloway Cattle. By Gilbert Murray 151

Chapter XIII.
The Angus-Aberdeen Cattle... 158

Chapter XIV.
The Ayrshire Breed of Cattle. By Gilbert Murray ... 168

Chapter XV.
West Highland Cattle. By John Robertson 176

CONTENTS.

Chapter XVI.
The Glamorgan Breed of Cattle. By Morgan Evans...*page* 188

Chapter XVII.
Pembrokeshire or Castlemartin Cattle. By Morgan Evans 195

Chapter XVIII.
The Anglesea Cattle. By Morgan Evans... 206

Chapter XIX.
The Kerry Breed of Cattle. By the late R. O. Pringle ... 213

Chapter XX.
The Jersey Breed of Cattle. By John M. Hall... 219

Chapter XXI.
The Guernsey Breed of Cattle. By "A Native" 227

THE SHEEP OF GREAT BRITAIN.

Chapter I.
Introductory ... ...*page* 241

Chapter II.
The Management of Ewes up to Lambing... 247

Chapter III.
Preparations for and Attention during Lambing 257

Chapter IV.
Management from Birth to Weaning 265

Chapter V.
From Weaning to Market 269

Chapter VI.
On Wool 276

Chapter VII.
Leicester Sheep 283

Chapter VIII.
Border Leicesters. By John Usher 289

CONTENTS.

CHAPTER IX.
Cotswold Sheep ... *page* 297

CHAPTER X.
Long-Woolled Lincoln Sheep ... 304

CHAPTER XI.
The Devon Longwools. By Joseph Darby ... 310

CHAPTER XII.
Romney Marsh Sheep ... 324

CHAPTER XIII.
Southdown Sheep ... 327

CHAPTER XIV.
The Hampshire or West Country Down Sheep. By E. P. Squarey ... 335

CHAPTER XV.
Shropshire Sheep ... 341

CHAPTER XVI.
Oxfordshire Down Sheep. By Messrs. A. F. M. Druce and C. Hobbs ... 351

CHAPTER XVII.
The Roscommon Sheep. By the late R. O. Pringle ... 357

CONTENTS.

CHAPTER XVIII.
Negrette Merino Sheep... ... ...*page* 363

CHAPTER XIX.
Exmoor Sheep 369

CHAPTER XX.
The Black-faced or Scotch Mountain Sheep 376

CHAPTER XXI.
Cheviot Sheep. By John Usher... 387

CHAPTER XXII.
Dorset Horned Sheep. By Joseph Darby... 395

CHAPTER XXIII.
Welsh Mountain Sheep. By Morgan Evans 408

CHAPTER XXIV.
The Radnor Sheep. By Morgan Evans 414

CHAPTER XXV.
Herdwick Sheep. By H. A. Spedding 417

CONTENTS.

THE PIGS OF GREAT BRITAIN.

Chapter I.
Introductory ... ...*page* 425

Chapter II.
The Berkshire Pig 446

Chapter III.
Black Suffolk Pigs 455

Chapter IV.
The Large White Breed of Pigs 459

Chapter V.
Small White Pigs 465

Chapter VI.
Middle Bred White Pigs 470

Chapter VII.
The Black Dorset Pig 473

CONTENTS.

Chapter VIII.

The Tamworth Pig... ... ...*page* 476

Index 479

The Cattle, Sheep, and Pigs
OF
Great Britain.

THE CATTLE OF GREAT BRITAIN.

CHAPTER I.

INTRODUCTORY.

HORNED CATTLE have been cultivated in this country from the earliest times. We are told that the Britons neglected the art of cultivation, so well known to the Romans, and contented themselves with looking after cattle, living on their flesh and milk. No doubt breeds were kept up by a process of natural selection. Little care would be bestowed upon the selection of sires, but the stronger animals would be reserved as males, and, running out with the cows, lived in a condition of semi-wildness, of which we see instances now in the white cattle of Chillingham Park. It is probable that the descendants of the original cattle are those which we see in Sussex, Devonshire, Wales, and Scotland, and it would not be difficult to trace a certain likeness. They are all middle-horned type, but climate and food have doubtless caused great changes. The Longhorns, which were originally derived from Ireland, first took root in Lancashire, and from thence spread to the midland counties, where for a time they formed the prevailing breed, being superseded by the Hereford, which probably were derived from the same stock as the Devon.

The Shorthorns are evidently of very mixed origin, possibly owing some of their merits to foreign blood. They were at first restricted to Yorkshire and Durham, but have, from their superior qualities as rent-paying cattle, pushed their way in all directions, and become established wherever climate and soil are sufficiently good.

Different breeds differ considerably in aptitude to feed, the more cultivated sorts—those which differ most from the original types—are as a rule the quickest feeders. Thick-skinned animals are proverbially slow, although the quality of flesh is good because more compact. The less highly cultivated sorts thrive upon harder keep, and can be kept with a profit when the better sorts would starve. And this to a certain extent limits the utility, and points out the localities best suited for the different breeds. The Shorthorn above all other breeds exercises an influence on quality. Ireland affords the most telling example of this fact. Only a few years ago Irish was a term of reproach as applied to cattle, and the hard-skinned, big-boned mongrels exhibited in our fairs were distinguished at a glance from home-breds. Now, thanks to the prepotency of Shorthorn sires, the best lots of Irish are quite equal in appearance, and often higher in price, than good stock bred here. The marvellous change that has been made proves how great the value of good sires, and how important it is to use well-bred and well-shaped animals. But it is not necessary to go to Ireland for examples of Shorthorn influence. Compare the character of the stock in any of our markets now with what it was thirty years since. The change is entirely attributable to the influence of good bulls.

Whoever travels through the length and breadth of this country cannot but be struck with the general quality of the cattle which everywhere meet his eye, and, if he can look back for thirty or forty years, the progress that has been made will appear in a very favourable light. North, south, east, and west, with few exceptions, it is the same, some counties being better off in this respect than others, those in which the holdings are the largest being the best. Whilst we may well be proud of the progress that has marked the present century, and which has given us up to the present time a leading position as breeders of the most valuable and purest animals in the world, thereby giving us a market for a large portion of the cattle-breeding countries of the world; we must not rest upon our laurels, and imagine that, because our leading breeders have done such great things in the past, therefore we can afford to leave things as they are. The watchword should be progress. We must take care that nothing is wanted on our part—the great meat and milk producing community—to make the most of our opportunities, and take care to maintain, and, if possible, increase, the leading position that has been won for us by the energy, enterprise, and skill of our leading breeders. In considering to whom merit is chiefly due for the work that has been done, it would be base ingratitude to overlook the great encouragement that has been afforded by those landlords who, truly realising the duties as well as the privileges of their position, have devoted time and money to cultivate the best animals of their particular kinds. The influence of such centres has been most marked upon the cattle of surrounding districts. Recent troubles have (for a time only, we trust) checked such enterprise; but the good seed has been sown, and a more correct appreciation generally exists as to the importance of using good sires—not only animals of undeniable lineage, by which alone we can hope for the transmission of hereditary qualities, but animals that exhibit in themselves all the best characteristics of their respective lines of blood. There was a time, during what may now be styled the breeding mania, when, both here and in America, pedigree, almost regardless of personal qualities, carried the day, and what is known as line-breeding was regarded as of vital importance. If only there was no bar sinister in the scutcheon, personal defects were overlooked. The closer an animal's ancestry had been allied, the greater its value, although constitution and real utility had been seriously injured in the process. Experience has now happily dissipated erroneous notions, and, whilst the value of blood is justly appreciated, it is felt that the best test of that value is the possession of personal excellence, and the necessity for an occasional outcross of new, though it may be distantly-allied, blood, by which (if judiciously used) vigour and good points are maintained and increased, is fully recognised. We remember the time when a single cross of blood, introduced, it may be, several generations back, most seriously affected market value, although the possessor had every personal merit to justify the innovation. Fortunately for our future prosperity as cattle breeders, more sensible views are now held, and it is generally allowed that, in order to maintain and improve our best breeds, there must be from time to time a dash of fresh blood brought in. Although much has been done in the way of improving our breeds of cattle, there still remains a very great deal to do. For example, in several of our principal dairy countries the quality of the cows is not as good as it might be; and, as it is from these districts that much of our store stock comes, it is through their medium that we must look for an improvement in the general quality of the stock spread over the surface of the country. Too often the cows kept have no tendency whatever to make flesh, even when dry and well fed. No amount of good food could render such animals fat, and their offspring must inherit their bad properties. An idea prevails that, by improving the quality and meat-producing properties of dairy cows, their value as milkers will be lessened; but this need not be the case if proper care be taken in selecting bulls of well-known milking families to cross with the existing stock. There are, even among the very highest-bred Shorthorns, cows which give as much milk and butter as common ones; and, where milking powers have not been disturbed by unnatural feeding, and due care has been exercised in the choice of bulls, Shorthorns yield more milk than any other pure breed. It should also be kept in mind that we breed oxen as well as cows.

If a dairy farmer buys a bull simply because it has a pedigree and is a bull, without any reference to the milking properties of the family from which he springs, such a man does not deserve to be lucky in his purchase. Farmers should take more pains to inquire into these matters before purchasing, and then breeders of bulls would find it conduce to their benefit to pay more attention than they do now to the milking properties of their stock. The results of the use of high-bred bulls with rough cows are sometimes astonishing, the calves partaking so much of their sire's quality, being smarter-looking and having far better coats than their mothers.

The great loss of cattle from the first attack of cattle-plague from 1864 to 1866 led gradually to a great increase of rearing, and, although this desirable feature was seriously checked from time to time by the prevalence of contagious diseases, especially pleura and foot-and-mouth disease, yet of late years there has been a gradual and steady increase in numbers, until at the present time (1886) our stock of cattle has nearly reached the numbers previous to 1864; how large the increase has been since 1866 will be seen by the following figures, extracted from the statistics published by the Agricultural Department of the Privy Council:

	England.	Wales.	Scotland.	Ireland.	Islands.	Total.
1866	3,307,034	541,401	937,401	3,746,157	37,700	8,569,693
1885	4,713,101	708,859	1,176,004	4,228,751	42,045	10,868,760

A few years since, the high rates for store cattle and the remunerative sales of dairy produce encouraged those whose circumstances were favourable to go in more for breeding and rearing. Production was greatly stimulated, and, as a consequence, supply has overtaken demand, and during the last year or two, and especially in 1885, reaction occurred. Store cattle fell considerably, though commanding prices which a few years ago would have been considered fairly remunerative. It may be that for a time the business of rearing may not prove so remunerative as buying store animals that others have bred. Yet, wherever the conditions are favourable, we strongly advise breeders and rearers to stick to their system, believing that, with increasing population and more spending capacity by the wage-earning class, there will be room for profitable management. And it is quite certain that the use of good bulls on dairy stock will so improve the progeny that, whether reared at home or sold at birth, the outlay in a good male animal will be amply repaid. Everyone who has had any experience in rearing must know that a mongrel-bred calf is much dearer at 1*l*. than a good one at 2*l*. When they are both a year old, the one will be a very different animal from the other.

We must impress upon farmers generally, and young ones in particular, the importance to be attached to colour in choosing animals to breed from. More stress should be laid on this point than would at first appear, so-called coloury stock finding a much readier sale than when the predominating colour is white. It may be prejudice, but graziers do not like white stock, and will not have them if possible. They are not considered so hardy as reds and roans, and as store stock, especially if it is intended to do them roughly, they have many disadvantages. It is a fact that white animals seem more liable to parasites than their richer-coloured companions. How this is to be accounted for we cannot vouch, but there can be no doubt as to its truth. Cattle will be found to detoriate, so to speak, in colour, i.e., they will, generation after generation, become lighter. In order to obviate this, it is necessary to keep the colour up to the mark by the introduction of fresh blood of the desired colour, and *descended from stock of a similar colour*. As regards shorthorns, blood-red sires are now very much more in demand than roans, though the blending may be rich, and it is evident that this results from experience. Reds are more likely to beget stock like themselves in colour, or at least of a roan colour; whereas roans will often produce white animals. Animals exercise influence on their offspring in proportion to the purity and *length* of their descent; therefore a pure white bull will be likely to beget more white calves than coloured ones, although all the cows put to him were red ones, always supposing the cows not to be of equal purity of blood. If we start with red as a foundation colour, we can easily get the stock lighter, and therefore this colour will be found the best. The use of red bulls bred from red stock will, in a short time, influence very greatly the colour of a herd. As farther evidence in the same direction, we may remind our readers that the breeders of Charolais cattle in the centre of France, which should be of a creamy white colour, have introduced English Shorthorn bulls from time to time with great advantage to quality of flesh and early maturity and little risk as to colour, so long as they used only white animals descended from white parents.

It is a mistake to suppose that well-bred animals require more food, and that of better quality, than rough ones. No doubt a Welsh runt will do well where an average-bred beast would comparatively starve; but between the common black-nose, hard-skinned, light-fleshed cow so often seen and one with two or three crosses of pure blood, there is no difference whatever in this respect. We will not say that the better-bred cow will do *better* on *very poor* food, because we believe such not to be the case; but, if the quality of the food be improved, then the well-bred animal will soon show a marked superiority, and, when store animals are brought to market, the difference in the value between good and bad stock, either barren or in calf, cannot but be of vital importance to the farmer. In the case of poor-bred, barren cows there might be very considerable difficulty in selling them at all, whereas in the other they will command a very good price indeed, a really good cow being often worth almost as much barren as in calf.

We have endeavoured to explain how it is possible easily and speedily to enhance the value of a herd without being at any great expense. Shorthorn bulls may be bought at a moderately low price as yearlings, or at a more mature age, either at sales or by private contract. At sales the purchaser, unless he have a previous knowledge of the herd, must remain more or less in the dark in regard to their qualities and their probable utility for his purpose. Against this he must set the chance of securing a cheap bargain. Bulls of from four to six years old may often be bought at a butcher's price, and, providing they will get stock, will often answer the dairyman's purpose better than a younger one. For this reason, young animals of fashionable blood will always command good prices, whereas older ones will be sold cheap. It must not be inferred, because a bull has failed to get show animals from high-bred cows, that he will not answer when used upon rough ones. The better his quality, the more marked his effects.

The Shorthorn, with its capacity for early development, combining, when properly selected and carefully bred, milking and feeding properties, yielding for a given quantity of food a larger return than any other breed, is the animal that seems to offer the greatest advantage to the breeder, and it is not exceeding the truth to say that, when circumstances are suitable, Shorthorns will be patronised. The very fact that, originating a century since on the banks of the Tees, in a comparatively small district, they have become distributed through the length and breadth of the land, and penetrated to the far distant shores of America and Australia, speaks volumes for the merits of this breed. Ireland, once remarkable for its mongrel stock, owes the vast improvement of its herds to Shorthorn influence. In parts of Scotland the Shorthorn is to be found, and, if he does not displace the aborigines, he blends with them, and produces magnificent cross-breeds. The Shorthorn, however, requires good and abundant food, and is not suited to exposure in a severe or very moist climate. The young animals must be well cared for, as after-growth greatly depends upon a supply of nutritious food during early life. Another remarkable illustration of the spread of these cosmopolitan cattle is to be found in Cumberland and Westmoreland. Half a century since they were scarcely known, and we were told by a farmer still living that, forty years since, he had to go to Darlington market to buy calving heifers; now and for a long period Penrith has been one of the best marts in the kingdom for well-bred stock. The influence of good bulls in our dairy districts has been very marked. We have heard it said by a successful breeder in Gloucestershire that thirty-five years since, when he commenced breeding, there was not a Shorthorn within many miles; now it would be an exception to find a herd that is not three-parts pure, and many of the farmers possess pedigree stock. The animals that have been displaced were great at filling the cheese-tub, but so coarse and slow-feeding that, although the produce may have been to some extent sacrificed, the total return is better and quicker. The draft cows, which formerly were sold poor at low rates, are now finished off with a little cake and roots, and com: mand good prices. The young stock also sell well, calving-down heifers making 25l. to 28l. at three years, and steers coming out fat at the same figure when between two and three years old. Shorthorns are more subject to sterility than less-cultivated breeds. This arises from the unnatural condition in which high-bred animals are too frequently reared, and may be guarded against by giving our young stock plenty of exercise and keeping them from fat-producing food. The researches of physiologists have demonstrated that excessive fatness of the carcase is accompanied with deposition of fat in the tissues, and when this is the case the breeding tendencies in both sexes are seriously compromised. Wherever the stock is treated naturally and the bulls selected with care, barrenness is not present in the Shorthorn to a greater extent than in other cultivated breeds.

The Herefords, a much older breed, have never become so thoroughly scattered although they have pushed their way towards the circumference of a wide circle, and have displaced or become mixed with the native breeds. Thus we find them in Shropshire, Warwick, Stafford, Monmouth, and several of the Welsh counties. Great attention has been paid to this breed of late years, and in many points of general utility the Herefords are unrivalled. The quality of the beef, and the capacity of the beast to lay it on rapidly and in beautiful proportions of fat and lean, are remarkable. The springy firm touch of a well-fed Hereford is due to the distribution of the fatty globules. It is this combination of hardy and grazing properties that has of recent years created such a demand for well-bred animals from American breeders, and has given a great stimulus to breeding and great care as to the registration of pedigrees. So important is the breed considered in the States, where many breeding stations have been founded, that at the present time English cattle are subject to a duty, and this and the bad times have greatly affected the foreign trade. The Hereford is very hardy, capable of doing upon poorer fare than the Shorthorn, not requiring such careful attention whilst young, and thriving in exposed situations. The pasture lands of the old red sandstone suit them well; here they rear their produce, but are not, as a rule, large milkers; perhaps the tendency to make beef is too prominent to allow of large dairy produce. A well-bred heifer or young cow will, on good food, fatten whilst milking, and no feeding stock will pay so well as Hereford cows about five years old. The great object in the district is to rear the produce and bring it forward, so as to sell either as yearlings or two-year-olds; they are bought for grazing principally by the farmers of the midland and eastern districts. Formerly a good trade existed for working bullocks; but latterly the great value of young beef has caused everything good to go that way, and it is only the inferior lots, and very few of them, that go into harness. The Hereford is a hardy healthy animal, with many valuable qualities.

Since the first edition of this work was published the Polled Angus Aberdeen cattle have come prominently into notice, principally on account of their great success either as pure stock or as crossed with Shorthorns in our fat shows, whereas they were up to a recent period confined mainly to the counties in Scotland from which they are named; there are at the present time a good many herds in England, and every year they are acquiring greater publicity. Like the Herefords they have been found to exercise a great and rapid influence on ranche cattle, and they have been largely exported of late years. They are noted for hardy character, ability to winter in exposed situations, extraordinary aptitude for feeding, combined with better dairy properties than the Herefords. The champion heifer at both Birmingham and London in December, 1885—Mr Clement Stephenson's Luxury—yielded the extraordinary and hitherto unexampled proportion of $78\frac{3}{4}$ per cent. of dead meat, and the young class at Smithfield showed a greater weight for age than any other breed exhibited, not excluding the Shorthorn.

Wherever the climate or food approaches to the character of our mountainous districts, we shall find that local breeds are the most desirable to keep. We may improve them by care, or even by judicious crossing; but we cannot dispossess them without certain loss. Thus the North Devon cattle are admirably adapted, from their active habits and hard nature, to feed over the exposed ranges of Exmoor and similar districts; whilst both in Wales and Scotland we find distinct breeds modelled, as it were, by the force of circumstances, into forms best suited to withstand the climate. Still in many of these cases, as our system of management improves, we may be in a position to make use of a more cultivated animal, and either improve the originals by careful selection and good treatment, or try the effect of crossing.

We have said enough to show that the subject requires general rather than particular treatment. At the same time, as it is necessary to have some type of animal before our eyes, we shall describe the management suitable to any of the more cultivated breeds, and especially to animals of the Shorthorn type.

INTRODUCTORY.

We do not propose to treat of high-bred animals, as such are not generally desirable for the rent-paying farmer; or, at any rate it is not wise to commence by a heavy outlay in cows with long pedigrees, unless we have time, taste, pluck, and money to go in for breeding prize stock—a more interesting than profitable affair with most. Good-looking roomy animals, got by a pure-bred bull out of ordinary cows, or animals whose length of pedigree does not materially affect their price, must be sought for. It is bootless to describe those points that indicate the dairy animal—such knowledge can only be acquired by experience and observation; and the young farmer may very reasonably doubt his own judgment, and will do well to commission a respectable dealer, who is generally able to make a more profitable selection and obtain the animals on better terms than the farmer himself; the practice of employing a middle man, both for buying in and selling out, is increasing in our grazing districts.

CHAPTER II.

BREEDING AND GENERAL MANAGEMENT.

 A GOOD lot of young cows having been secured, the next point is to select a bull, and the wiser policy is to obtain a thoroughly good animal with a sound pedigree, even if we pay handsomely for the same. The first male will have a most important influence on the herd, and a few pounds therefore should not be grudged. Generally a good yearling can be had for between thirty and forty guineas.

We are not venturing into an exhaustive treatise on the principles of breeding, or intend more than briefly to touch upon this point; but we must urge upon all young farmers the value of quality, and the improvement of their stock by the use of good bulls. This, be it remarked, is quite apart from keeping highly-bred stock. Pedigree breeding is a business for the few, requiring special conditions to render success even probable; but everyone who breeds, whatever the class of animal he selects, should aim at quality, by which we understand the qualification to mature at the earliest possible period, and accumulate the maximum weight from a given quantity of food. What the difference is in this respect, according to quality, has never been accurately tested, but we believe it is quite sufficient to determine profit and loss. Now, as a rule, the influence of the male preponderates; consequently, whilst careful to select good-looking females, we must spare neither money nor time in finding the right sort of bull. Suppose we require milking stock (and whatever the particular direction in which we farm, milk must always be an important consideration), not only should we select heifers that give promise, but we must seek a sire that comes of a good milking stock, for these qualities are to a great extent hereditary. We must, moreover, take care that the qualities that existed in the ancestry have not been weakened or destroyed by injudicious breeding or feeding. Many an animal with a natural tendency to milk has been ruined by early forcing. Whilst, as we hope to show, in the following pages, generous diet from birth is necessary to quick and healthy development, undue forcing, such as is resorted to in order to develop abnormal growth in show animals, weakens and often completely destroys milk-producing qualities. It is in this way, principally, that discredit has been thrown on certain families of Shorthorns as milk-producers, and thus the race that were originally noticeable for the quantity of their yield are now frequently unable to rear their produce, and choice animals require foster-mothers to supply their wants.

The folly of forcing young animals for show purposes is acknowledged on all sides, and those who possess the most valuable blood will not run the risk of damaging their animals by forcing. It may be allowable to make an animal extraordinarily fat for the butcher, in order to show the public of what a breed is capable. The animal is for the shambles, and, provided he lives till ready for the knife, the end sought for is obtained, and the feeder is the only loser, the extra fat costing more to put on than it will yield; but this extra fat state is not a healthy condition, and animals so fed lose much of their vital energy, as was too evident from the collapse of the beasts at the Smithfield Show of 1873. Lean stock would not have suffered to the same extent. Shorthorns, especially, suffer from the disease known as fatty degeneration of the heart, which may be explained to the unlearned as meaning a deposit of fat between the muscles of the heart, which greatly lessens and sometimes altogether arrests the expansive and contractive powers. The object of the breeder should be to so treat his young animals as to develop frame and flesh by supplying food containing the constituents of bone and muscle, and allowing of sufficient exercise to develop and strengthen the frame and constitution.

Very little is really known on the subject of breeding. Mr. J. K. Fowler, who delivered a lecture on the subject before the Central Farmer's Club some years since, considers that the sire influences form, the dam the internal organisation, and there is some general truth in this, as is proved by the cases of the Mule and the Hinney. The former, which results from connection of the male ass with the mare, has all the external points of the ass, only a rounder barrel to give room for the bowels, which resemble those of the dam. When a stallion is put to the female ass, the result is a modified horse (the hinney), only the barrel is smaller, in form resembling that of the ass. If this is correct, we should be most careful that the dam is sound in wind and of good constitution, being most particular that the sire possesses beauty of form. In the case of Shorthorns, certain sires, such as Hubback, Favourite, and the Earl of Dublin, were noticeably prepotent in their influence on progeny. The last-named bull, in the hands of Mr. Adkin and Sir C. Knightley, impressed deep milking properties in all the animals he was put upon, and this, a result of internal organisation, is due to the influence of Princess, from which he was directly and closely descended. This latter fact rather tells against Mr. Fowler's theory, since the milking properties derived from Princess were transmitted more directly through the male than the female. Prepotency may arise from an intensifying of certain qualities from very close breedings, and this also, to some extent, militates against Mr. Fowler's theory. The influence of the male may be due to his being deeper-bred than the female. Cross-bred cattle offer a good example. It is nearly always possible to find out the sire by the strong likeness to him, and here we have an illustration in support of Mr. Fowler as to external form. Mr. Fowler further illustrated his views by reference to facts noticed in poultry breeding. The Brahma and Dorking fowls were crossed with the following results: When the Brahma cock was used on the Dorking hen, the chickens had four claws generally, feathered legs; the pullets laid white eggs, and the cockerels, though resembling the Brahma, crowed like the Dorkings. When the process was reversed, the produce likewise followed the change closely; the cockerels were like Dorkings, but roared like the Brahmas; indeed the illustration was perfect. The same results were noted when the Rouen and Aylesbury ducks were crossed.

Great importance, undoubtedly, is attached to the first

impregnation, and the imagination has a good deal to do with colour; and it is said that the late Mr. McCombie, the celebrated breeder of black cattle, was most careful to have all his buildings, gates, &c., painted black.

The period of the year at which our cows should calve will depend upon circumstances; if our object is rearing and dairying, the calves should drop from Christmas to March or April; if we are for cheese, March and April; and if we are milk-sellers, the cows must come in at all periods. Early calving is best for the offspring. The difference of wintering between a calf dropped in January and May, both receiving equal care, is very great.

As to the best age to commence breeding, different opinions exist, and everything depends upon the class of animals we possess, and the quality of our land to favour early maturity. As the question is a very important one, it would be well if experiments were carried out to determine at what age heifers will breed. If conditions are favourable, the calf that drops in from December to April may be brought to the bull when fifteen months old, that is during the summer of the succeeding year. The first calf will then be dropped when the heifer is about, or a little over, two years old. Of course, we presuppose careful attention and abundance of food. The heifer may be small at this time, but grows rapidly afterwards, and we bring our animal into a productive state at the earliest period. The milk may not be very abundant, but there is plenty for the calf, which should be allowed to suck, at any rate for some time, as the bag is thereby developed and rendered soft; at six months it is good policy to dry the heifer, as she is thus enabled to lay on flesh and take care of the fœtus. These remarks apply, of course, only to cultivated breeds. We could refer to many excellently managed dairies where a number of the best heifer calves are reared, and are so well cared for that the bull is turned with them about mid-summer, and thus the first calf is dropped at about two years, and it is really extraordinary how much milk these heifers will yield. In breeds that are not so forward, or where circumstances are unfavourable for early development, the heifer calves at from two and a half to three years old. In reference to this point, we could multiply examples of accidentally early breeding which have turned out well. So fully satisfied are we, from our own experience, that with generous feeding heifers may calve down when two years old, without injury either to growth or milking qualities, that we adopt the plan of taking a calf from animals intended for beef. Those who buy in much of their stock might carry out this practice. We can often buy, during summer or autumn, yearling Irish heifers; if these were served at once, and done well to through the winter, the plan would answer. Where breeding is carried on, we are quite certain of the profitable nature of this practice. We ascertain at an early period whether the animals are likely to make valuable milkers, and such as are not promising can be fed off after the calf is weaned. An impression prevails that early breeding affects after growth; but we have not found it so where care was exercised as to food. The late Mr. Edward Bowley, who was well known as an authority on Shorthorn management, alludes to the subject in his prize Essay in the "Journal of the Royal Agricultural Society." His practice was to bull the heifers dropped from December to the end of February, in July or August of the following year—that is when they range from sixteen to eighteen months old—they thus would calve just before going to grass—when they are about two years and four months old. He says: "I allow their calves to run with them during the summer. When four or five months old I take the calves away, and dry the dams, by which means the heifers get a much longer rest than the older cows before they calve again, thereby encouraging their growth; and under this system they can produce calves at an early age without interfering with the full development of their forms." He also mentions a case of very early breeding by a heifer that calved at fifteen months, having been served by a six-months old bull calf whilst both were with their dams. The heifer took the first prize as a two-year-old in-calf heifer, and a second prize the following year as a cow in milk, in a strong class, and was afterwards sold at a high price to go abroad. This is important evidence bearing on the point we are anxious to see elucidated. Our own experience is that, so far from early breeding injuriously affecting future size, the heifer, if generously fed, appears to grow out in consequence, and we are quite certain that for feeding purposes they are the better for having dropped a calf. We have not, however, had so much experience of early breeding for the dairy. We are quite aware that our suggestions do not meet with universal approval, and we believe that a high authority on animal physiology—Professor Brown—takes a directly contrary view, but we can only give our own experience. Before calving, the cow should be placed in a loose box or shed by herself, and not tied up. It is a most unnatural proceeding to allow cows to calve when tied up in their usual places, among, perhaps, twenty others. The calf runs considerable risk of being injured by the other cows in case it is born when no one is in attendance. The plan is simply cruel, and on no account to be followed. Of course, sometimes such a case occurs on the best regulated farms, but in these instances the cow either calves before her time or unexpectedly. After calving, the cow should have warm gruel and a little sweet hay. Chilled water should be used for the first three days, after which, if no unfavourable symptoms occur, all danger ceases. If there is any fear of fever (as is always the case with large milkers), a moderate aperient may be given. Linseed oil is useful for this purpose, and safer than salts and sulphur, which, however, are frequently given. Cleansing drinks should be always at hand, in case the animals do not clean properly.

Great loss is often occasioned by cows slipping their calves. There seems to be no accounting for this; as a rule, the food the animals have been fed on gets the blame, but if anyone will be at the trouble to inquire into the matter, they will find that cows slip calf on all kinds of food, and under all sorts of management. When one cow among a lot of others slips her calf, she should be at once separated from them, and not allowed to be with them again for some time. Unless this is done, and at once, the farmer cannot be sure when it may stop—perhaps not before half his cows have followed the example. Mr. Clement Stephenson, in a valuable paper published in the "R.A.S.E. Journal," 1885, attributes much of our heavy losses from this cause to impure water, and especially water contaminated with sewage and decaying vegetable matters; and he truly says this is not surprising since such waters not only contain injurious organic matter, but swarm with bacteria and other living organisms which play an important part in producing morbid changes and disease; and, he adds, that whilst non-breeding adult animals may drink such water without visible injury, "In pregnant animals the delicate fœtus is like a sensitive barometer, its development and life depending absolutely upon the purity of the maternal blood; it is influenced by variations and agents against which independent life may be proof." Cases are mentioned by Mr. Stephenson in which abortion has been caused in a herd by the use of bulls that are not efficient workers, or are actually diseased. Ergotised grass is very generally considered a frequent cause, and there is some colour for this in the fact that abortion often occurs in wet, cold, ungenial seasons, which also favour the fungus; but Mr. Stephenson thinks it is often innocently blamed, and that, instead of being a frequent cause of abortion, it is an exceedingly rare one. If the natural food is unwholesome—i.e., deficient in nutritive properties, which is the case in ungenial summers, the system becomes deranged, and we may have abortion. We think there is little doubt that the disease may be, and is often, spread by sympathetic action set up by the act of abortion and by the smell given off, and in this sense Mr. Stephenson considers it infectious. We quote his remarks on treatment, as they are concise and comprehensive. "Promptly isolate all cows that have aborted, or in which the premonitory symptoms of it are observed. Destroy the aborted calf, membranes, and discharges; do not bury them in a manure heap or in any place to which cattle have access, but in some place where they cannot possibly do any harm. Disinfect the place where the abortion occurred, and the cow in her seclusion; this must be continued for some time. Have the floor, bedding, manure, and air saturated with the disinfectant, and if no recognised disinfectant is at hand use quicklime freely. If possible find out the cause of the abortion; if successful take immediate measures to prevent further mischief. Keep the patient on light sloppy food, attend to her general health, and the condition of the bowels. If the uterine discharges are copious, offensive, or the membranes are attacked, wash out the vagina with tepid water containing Sanitas or Condy's fluid." This may be done once or twice a day. On no account should the cow be again served until she is perfectly healthy and regular. We may add that carbolic acid, either as McDougall's disinfectant or Dr. Calvert's preparation, will be found probably the most powerful agent known.

The management of the calf is the next point for consideration. In the case of the heifer it is well to let the calf suck, and possibly run with its dam, during the summer. In such cases we advise the farmer to procure a second calf, and when the heifer has become accustomed to it, to turn her out into good pasture with her two attendants, who will make good use of their time and pay well for the heifer's milk. With older cows there are two plans open to us—first, to remove the calf at birth, before the cow has noticed or licked it; this plan is frequently pursued in the north of England with great success. The calf is carried to a warm well-littered house, and thoroughly rubbed with a wisp of straw until dry and warm. The beastlings are then drawn from the cow, supplied to the calf a small quantity at a time, and frequently. The fingers should be introduced and the calf's mouth drawn down to the milk. They will thus readily learn to drink, and the great point is to prevent their drinking too fast. We must imitate as much as possible the process of sucking, by which a good deal of air enters the stomach and assists digestion. Patent feeding-mouths are very useful for this purpose. The calf should be fed three times a day; many people prefer only twice, but it is too long for the stomach to remain without food, and is contrary to the natural habits of the calf. The second plan is to leave the calf with the mother two or three weeks, or at least allow it to suck night and morning; but if we have a good cowman who understands the other plan, it is preferable for some reasons. The cow gives her milk down more freely, does not fret at the separation, and is apt to take the bull sooner than when the calf sucks. Adopting the first plan, we may use new milk for a fortnight; then skimmed milk of the same temperature as the new milk, and thickened with linseed jelly or fine dust oil-cake, which supplies the fatty matters removed in cream, besides enriching the food in other ways. Boiled flour porridge is frequently used in conjunction with skimmed milk. We prefer dust cake, provided it be from fresh genuine linseed;

great care should be exercised to secure a good article. Care should be taken in all cases that the milk be given warm; cold milk produces scouring, and all manner of evils. The cake may be soaked in hot water first, or else added to the milk and gently heated; in either case it produces a rich soup, which is very palatable and nutritious; a handful of finely-ground oatmeal may be added, and a little later a small quantity of fine pollards. Rock-salt and chalk should always be placed within reach. One great advantage in the plan of separation consists in the earlier date at which the calf eats. As soon as this is accomplished, we may by degrees discontinue the liquid; at first supplying it only once a day, and soon leaving it off altogether. And thus a good cow will rear eight to ten calves, provided her produce is entirely used for this purpose.

The greatest trouble in rearing arises from scouring, to which calves are very liable. This may occur from various causes, but it is an evidence that the food is not properly digested. In the farm prize report of 1884, R.A.S.E. Journal, vol. 20, new series, page 539, will be found a valuable recipe for this disease. Mrs. C. Holmes, of Burley Fields, near Stafford, who is very successful in rearing calves, states "that after long experienced and serious losses, she believes she can now cure the scour, which she is convinced arises from the new milk being too rich for the calf's digestion. When the disturbance is first manifest, the food is changed, a dose of castor oil and laudanum is given to allay irritation; boiled skim milk and sago, nearly cold, are substituted for the new milk; and if this food proves too rich, then boiled sago alone, which is also given cold. "The second day, and until the diarrhœa is stopped, one or two tablespoonfuls of the following mixture, viz., 4oz. prepared chalk, 1oz. grains of paradise, 2oz. cummin seed, 2oz. aniseed, well mixed in a pint of starch gruel, to which may be added twenty drops of laudanum."

When a month old, calves will begin to nibble a little sweet hay, finely-sifted chaff, pulped roots, and meal. We cannot begin too soon to teach them to eat, although the longer they get the skim milk and porridge, the better for future growth. Calves weaned too early seldom thrive well afterwards. Oil-cake, crushed, and then boiled to a jelly, and mixed either with the porridge or skim milk, is excellent food for calves when a month or six weeks old. What we here call porridge, should be called more properly gruel, and should not be made too thick. When ten weeks old, the calf should be weaned; this should be done by degrees, the daily allowance being decreased, so as to accustom it to the change. By this age the calf can eat a considerable quantity of hay, chaff, pulp, and corn, and should receive at least half-a-pound of cake and corn. Various preparations of finely prepared meals are sold for calf spice, such as Simpson's, Bibby's, &c. We believe, from our own experience, that such foods are very useful, and may be considered not only healthy but well suited to the digestive powers of the young animals. In another place we shall have more to say on this subject.

The treatment of the calf during its first year is most important. As the spring comes on and the sun gets power, the calves, in small lots, should be allowed the range of a comfortable yard and shed, taking particular care that they are warmly housed at night. If we can depend upon our arable land for a succession of green food, then the calves will do best if kept in yards all the summer, and, indeed, not suffered to go out into the pastures until turned one year old. This is often impracticable, but it is much to be recommended in all cases where it is possible. The advantages of this plan are, that we have the animals more under control, the least malady is at once detected, the supply of food can be regulated with greater exactness, and the manure economised. Considerable variety of food is necessary. Thus, we must have vetches, trifolium, cabbages, and artificial grasses, coming on in regular succession. Chaff, meal, and cake correct the laxative tendency of our green crops. All who are acquainted with rearing know how frequently calves will scour when in the pastures, and how difficult it is to cure this. In many instances we believe the disorder is occasioned by poisonous acrid weeds growing in the grass. Again, as the hot weather comes on, the animals are driven wild by the flies, and while thus irritated cannot make any progress. The yard system relieves us of these difficulties. The green food may be supplied partly long, in racks, and partly cut up and mixed with hay-chaff, a slight fermentation being often desirable. Over this a mixture of several kinds of food may be dusted, amongst which old beans should have a prominent place, being particularly adapted to calves eating green food, counteracting the too laxative effects of the latter, and supplying a large amount of flesh-forming material; we have seen excellent results from the use of beans. A moderate quantity of fine-ground oil-cake, and a little home-grown grain will complete the mixture, unless we can buy bran cheap, in which case it may be added. The quantity of such a mixture to be used depends upon the age and condition of the calf—from 1lb. to 2lb. a day expresses about the range of quantity. With this mixture, continually varied as to the green food, the calves should be fed three times a day at early morning, noon, and night, and clean water always placed within their reach. Occasionally, every two or three weeks, some flowers of sulphur, about half to three-quarters of an ounce per head, may be given with the food; this purifies the blood, and helps to preserve a healthy condition of skin. Judgment and attention are required in supplying the food. Some stockmen fancy that too much cannot be given, and so they fill up the fresh food on what remains from the last meal—a dirty, slovenly practice, that cannot be too severely condemned. An hour or two before feeding time the manger should be thoroughly cleaned out, every particle of dirt removed, and, besides this, every now and then thoroughly washed and scrubbed. Eating off a dirty plate would not tend to improve our appetite, and though the ruminant is not so particular, cleanliness is very desirable. The refuse from the calves' mangers will be readily eaten by the cows, and thus nothing need be lost, and the calves will come to their food with an appetite. Should an animal hang back, we must at once separate it, and carefully watch the symptoms.

There is one point upon which we would specially remark—viz., the importance of keeping the skin healthy; it has certain functions to perform, its excretions relieve the system of waste matters, and there is great sympathy between the skin and the digestive organs; indeed, the lining membrane of the stomach, &c., is but a continuation of the skin. Now, if the pores become choked up with dirt, or if parasites, such as lice, ticks,

&c., make their habitation therein, the action of the skin is impaired; the constant irritation frets the animal, and progress is checked, even if more serious injury is not caused. Occasional washing with soft soap and warm water will be found very beneficial, and if parasites are present we may add a portion of carbolic acid, or we may use one of the preparations of this powerful agent which contains a certain quantity of soap, and makes a milky emulsion with water. We are acquainted with herds where this is used, not only for the calves, but through the dairy, with satisfactory results. Moreover, the well known antiseptic properties of the carbolic acid group render them doubly useful as a possible means of keeping off contagious diseases; and although the evidence is but negative, and must be taken only for what it is worth, we believe that during the cattle plague no disease occurred where this practice of occasionally washing over the animal's body with a properly diluted solution of carbolic acid was persevered with. Whether this be so or not, there can be no question about the importance of keeping the skin of the young animal clean, as we thus insure a good circulation, waste matters are duly carried off through the pores, and the internal organs are not overworked. In an ordinary way two washings in the first year will be sufficient—say, when five or six months, and again at nine or ten months old. During the winter, and especially if there is barley straw about, we must watch for lice, and wash at once if we find them. It is want of attention to little matters of this kind that so frequently prevents success; our animals contract some irritating skin disease, or are plagued with vermin, and are restless and worried, however well they may be fed.

The plan we have advocated as to summer feeding the calves may not always be possible. In that case we must turn out where there is a good bite of fine herbage, avoiding rank strong pastures. By the end or middle of May the oldest calves on a dairy farm will probably be three months old, and may be turned out by day in a well sheltered grass field, with a shed attached—if they have no shelter to run into they will suffer much both from sun and rain. The most convenient shed for calves is one in which they can be shut up at night, thus saving their having to be driven every night to the homestead. When such a shed is available, by the middle of June the calves may be left to go in or out as they like at night. Care must be taken to have some ground to change them on occasionally, and plenty of good water and a shelter hovel are indispensables. Chalk and rock salt must be placed within reach, and a supply given night and morning of artificial food and chaff—the latter composed of a mixture of green hay and whatever forage crop we are cutting at the time. Best linseed cake, at the rate of 1lb. per head daily, will answer admirably, provided the grass is not very watery, in which case we should prefer using either a portion of beans or decorticated cake, say $\frac{1}{2}$lb. of linseed and half a pint of beans, or $\frac{1}{2}$lb. of decorticated cotton cake. Common cotton cake should never be used for this purpose, as it cannot be properly digested. Later on in the autumn, whatever the farmer intends to give them during the coming winter in the way of corn, should be gradually introduced and mixed with the oil cake, &c., so as to accustom them to the change, violent transitions in the way of food are always productive of loss of condition, and therefore, especially with young animals, the change should be gradual.

By such treatment carefully persevered in we shall gain a year in point of bulk and maturity over the old starving system, and have a good chance of ultimate size, which is out of the question when growth has been checked. The advantage of this system of management will be in proportion to the quality of our stock; badly-bred coarse animals do not repay us for our care in the same way as the better sort. Whilst, however, advocating liberal treatment, we must guard against giving an excessive quantity of nutritious food. The young animals should be placed under those conditions that favour rapid growth rather than the accumulation of fat or a great quantity of soft flesh. We like to feel a loose soft hide, with plenty of room under; we must have, in short, a well-covered frame, with a skin that fits on the animal like a loose great coat.

It is not good policy to let calves lie out even during the height of summer, but it is decidedly injurious late in autumn, when heavy dews and frost indicate the chilly nature of the surface. By the 1st of October the calves should be housed for the winter; if allowed to remain out later than this they will not thrive well, the nights becoming cold, and they will also be liable to disease. Of course they may be turned out during the day, but even this we do not recommend for long, as the grass they eat gives them a dislike to the drier food they ought to consume during the hours they are in the yards. They should be gradually brought from grass to dry food and winter management, so as to become accustomed to the change without loss of condition. Sheltered yards, well littered, with good hovels, are the best situations for calves in winter. The smaller the lot in each yard the better. Pulped roots and chaff put together for a short time, to allow of a gentle fermentation are the materials on which we must chiefly rely. The chaff may be a mixture in equal quantities of oat straw and hay; over this a small quantity of artificial food may be dusted, in which the greater the mixture of different materials the better—say, barley, wheat, and bean meal, with finely-ground cotton-cake or linseed-cake, and perhaps a little palm nut meal. Yearlings will eat from 1½lb. to 2lb. per day. The cost of such food, taking 2lb. per diem as the average consumption, will not exceed from 1s. 2d. to 1s. 6d. a week. The gain in the greater growth and better health of the animal will be very manifest. The natural food goes further, and the manure is improved.

As a preventive to quarter ill or blackleg—a very fatal disease, to which calves are subject on certain soils, and which is most prevalent in autumn, when the natural food is often at its richest, we strongly recommend setons being inserted in the dewlap and occasionally moved. Whether the seton acts as a counter irritant, or in some other way, we do not know, but we have had abundant evidence to satisfy as to the efficiency of the preventive.

We should not recommend for calves, the use of any condimental foods which are so loudly advocated by the different makers. As a rule, these foods consist of a mixture of finely ground grain, with a percentage of locust beans, flavoured and seasoned with certain stomachics and stimulants; the latter being commonly fenugreek, aniseed, carraways, gentian, mustard, and possibly nitre. We could not say a word in favour of such compositions as food for calves, or as the seasoning for their food, because we consider them very objectionable. If the young animal is in a healthy state, and under proper management, digestion and appetite will be quite equal to the wants of the body. The use of such food will tend to impair the digestive faculties. We should not consider it desirable to give a child mustard or hot pickles, inasmuch as such stimulating food would injure the stomach; and so the use of cattle condiments must be condemned for young stock, however advantageous in the case of fattening animals. Even if there were not these drawbacks, the condiment, as tending to create an abnormal appetite, would induce the deposition of fatty matter, and thus predispose to disease. To quote from a writer on this subject— "It is now generally admitted by physiologists that an extra development of fat is opposed to the normal growth and health of muscle, and when the accumulation of fat is in connection with any of the vital organs, as the heart, the degree of health is then very low, the slightest influence from without—extra heat or cold, or sudden exercise or excitement of any kind—being sufficient to endanger life. An extra deposition of fat about the liver or kidneys is also attended with similar conditions unfavourable to the general health. High-bred stock too often inherit a tendency to accumulate fat, which all our care is unavailing to counteract. The best antidotes are plenty of exercise, which acts by exciting the excretory organs to free action, with a supply of food suitable for building up the frame; and if, in spite of such treatment, animals will still get fat, we don't know that there is any remedy. Nothing is more opposite to nature than the system of forcing upon stimulating food, and keeping young stock shut up in houses—plans that are too frequently adopted in the case of animals intended for exhibition."

Having wintered well during their first year, after management is comparatively easy. The animal will be able to take care of itself. Our practice must now depend upon the end in view. If we intend fattening the animals, the process must be continuous, and a supply of cake and other artificials must be given at grass or supplied with the forage crops in the yards;

whereas the animals destined for stock may be left to get their own living during summer in the pastures, and, provided they have plenty of food and occasional change, will keep in growing condition, and be quite ready for the bull in the summer and autumn.

Contrast the system we have sketched out, which we know to be based on scientific principles and sound practice, with the absence of management that results in the miserably underfed and underbred animals that may be found at any of our large fairs. Granted that the cost of keep has been only half, are they half as valuable? A good yearling will often make 9*l*. 10*s*. to 10*l*., which is more than a half-starved two-year-old is worth, for the latter are often dear at any price—they have no go in them, no quality that can be developed by keep. It is well sometimes to buy a young animal in poor condition, because we know it will improve rapidly, provided there is quality; but a half-starved mongrel remains a sorry brute to the end of the chapter. Therefore keep a good sort, rear well, and bring to market or into productive condition at as early an age as is consistent with the health and stamina of the animal.

Many years ago Mr. Henry Ruck, of Cricklade, Wilts, described his practice in the management of calves before the Cirencester Farmers' Club, and showed that he had for some years been in the habit of rearing on an average fifty to sixty calves with the produce of four cows. The description embraced three years, and during the first two not an animal was lost. There were three deaths in the third year, but this loss was due to mismanagement. Independent testimony was forthcoming as to the healthy appearance of the stock. The following is Mr. Ruck's description of operations: "I take all the calves (from a neighbour) after about the beginning of March. Every Wednesday I send for what are above ten days old, as up to that time they require their mothers' milk, which is unfit for the dairy. The price I pay is 30*s*. each. They have for the first three or four days two or three quarts of milk at a meal; then gradually some food in the shape of gruel is added, and by degrees water is substituted for milk. Mixing oilcake with gruel is the secret of success. I use half oilcake, the best I can buy. Take a large bucket, capable of holding six gallons; put into it two gallons of scalding water; then add 7lb. of linseed cake, finely ground, which is obtained by collecting the dust that falls through the screen of the crusher, and passing it through one of Turner's mills; well stir the oilcake and water together, and add two gallons of hay tea. The hay tea is made every morning by filling a small tub with sweet hay, pouring on scalding water; use this in the evening; add a sufficient quantity of scalding water to the hay leaves, and cover down for next morning. The hay tea is very sweet, dark in colour, and I think the extract from the different herbs assists digestion. Again the mess is stirred, and 7lb. of mixed flour well worked in; the mixture consists of one-third wheat, one-third barley, and one-third beans; add sufficient cold water to fill the six-gallon bucket, and well stir. Two quarts of this with two quarts of cold water will be sufficient for a calf at a meal, and about the right temperature. The food should be given at regular hours—say six in the morning and six at night. Each bucket of gruel will be a meal for twelve or fifteen calves, and costs about 1s. 6d., or 3d. a day for each calf. We always measure the food with a two-quart cup, and never overload the stomach of young calves. After fifteen days, when the calf chews the cud, some of the difficulty and danger is passed, and when the calf eats well we gradually diminish the gruel." The calves are tied up whilst being served, and Mr. Ruck prefers the old-fashioned plan of letting them suck through the cowman's fingers, as this prevents bolting, and a proper quantity of air is taken in, which assists digestion. As soon as they can eat, crushed corn, sweet hay, and roots are placed within reach; vetches, as soon as ready, and mangolds, of which a supply should always be stored if practicable. The calves live in a cool, well-ventilated house, are kept very clean and quiet, supplied with fresh water daily, and the manure frequently removed. The addition of the decoction of hay is a very sensible practice, supplying some, if not nearly all, the nutriment of good hay, which the calf cannot otherwise obtain. Since the object of the feeder should be to imitate nature as closely as possible, we would suggest the introduction of a small quantity of sugar, just sufficient to give the requisite sweetness of new milk. Sugar plays an important part in the juvenile economy, and we find it present to a large extent in milk. It will also take the place and act the part of the cream which the calf gets in the natural state, although this may be more directly imitated by using a small quantity of palm-nut meal, a substance containing about 13 per cent. of fatty matter indentical in composition with the fatty matter of cream. We must introduce this substance by degrees, for animals do not take it very kindly at first, on account of its gritty texture; but if used in conjunction with sugar, we are satisfied from our own experience that it would do good. Mr. Ruck advocates the use of a little hay cut up with straw, especially when the calf is just deprived of its liquid food. During the first winter the following mixture is recommended: 5cwt. of straw chaff, 5s.; 10cwt. of pulped mangolds, 5s.; 1cwt. of oil cake, 10s.; and 4cwt. of mixed crushed corn at 30s.; put together and allowed to heat moderately. This gives a ton of stuff superior to hay for 50s.; a small quantity of hay might, however, be added with advantage. Mr. Ruck touches upon the diseases of calves, referring especially to murrain, blackleg, husk, scour, and lice, and gives some simple rules which are familiar to stock breeders. Thus murrain, which is a stoppage in the circulation of blood in the extremities, is easily distinguished by the crackling under the skin when the hand is passed over the infected part. It usually occurs when calves that have been badly kept are suddenly put into a luxuriant though watery pasture; blood is made too fast, and the system is unable to get rid of the excess of carbon. Prevention is better than cure, which is seldom possible; therefore always keep stock thriving. With regard to husk, Mr. Ruck believes it arises from a threadlike worm in the windpipe, resulting from ova taken in with lattermath grass, and therefore recommends that calves should, if turned out, be always kept on land that has been fed in the spring. The risk of disease from this cause would be obviated by yard feeding. Scour, which is very common with badly fed calves, may be prevented by generous diet regularly supplied. Of course, it is impossible always to have our animals in health, but nine-tenths of the maladies of young stock arise from mismanagement. Everything depends upon close attention to details, feeding with great regularity, supplying the proper amount of food to each animal in a suitable condition for rapid digestion, taking care that young stock are well housed, yet allowed air and liberty, and keeping the pores of the skin open. Another and not insignificant argument for good treatment is found in the greater strength thereby provided to withstand sudden attacks of disease, and rally from their effects.

CHAPTER III.

PRINCIPLES OF FEEDING—NATURE AND VALUE OF DIFFERENT KINDS OF FOOD.

WE now proceed to consider the question of feeding, which may be divided into several heads; thus we may settle the age at which an animal can be brought out, the nature and value of different kinds of food, and the profit or loss that may be looked for from the operation. Evidently our views upon the first point will depend upon the nature of external circumstances. In the case of very rich feeding land, such as the Leicestershire pastures, the best return will generally result from purchasing well-bred full-grown animals in fresh condition, and finishing them off rapidly. Such land is too valuable to do the earlier work, which may be effected on cheaper material. Wherever sufficient straw is grown to provide litter and some forage, it is often on such farms a good plan to purchase in the autumn, winter well on straw and oilcake, and thus turn out fresh in May. At any rate, it is good policy to secure a portion of stock before winter, and make up our quantity in the spring, buying more or less according to the prospect of keep and the state of trade. On such land, breeding and rearing seldom pays. A cow, living on three acres, would not do as well as three fattening animals, from each of which we may look for £5 or £6 as the return for grass consumed. Summer feeding may be dismissed without further comment, save to remark that we should always try and purchase well-bred animals, such, as a rule, having greater tendency to fatten; that some sort of shelter should be provided during the heat of summer; and that the use of a very moderate quantity of cake is often found, especially during the latter stages of feeding, a very profitable addition to natural food. Cake is most commonly used, because the most easily given, and less liable to waste than meal; a mixture of oil-cake and crushed beans would often be an improvement, especially in wet weather or when the grass grows fast. At such times the grass is laxative, and the linseed cake, instead of correcting, rather increases this tendency. Well-ground cotton-cake might be substituted for linseed cake; for, being obtainable at a reduced price, and being equally feeding, it is, weight for weight, much cheaper food. There are some who altogether condemn the practice of outdoor grazing as extravagant, and tending to a loss of food; and it must be confessed that it is not always easy to regulate the mouths to the growth, and have the food constantly eaten to the greatest advantage. If the keep increases on the beasts, then it is trodden down, soiled on, and, so far as these animals are concerned, wasted; but after the last lot goes to the butcher, and the land has lain three or four weeks to sweeten, our comparatively hungry store cattle gladly gnaw up all that is left, so that very little is wasted. The injury from flies is often very considerable, especially when there is no shade or shelter; but this may to a great extent be prevented. No doubt animals tied up in cool, airy sheds, and supplied with grass *ad libitum*, would, on the whole, fatten faster than animals at large; but the injury to the quality of the food by mowing would be so great as to render the plan unfeasable. Lastly, we might save the trampling of the grass by a system of tethering, which is carried out on some dairy farms with Jersey stock; but our beasts would fret at first, and suffer much from the fly, and, we fear, do but indifferently well. We have long been convinced that the most paying practice, when the nature of the grass will allow it, is to give cake or its equivalent on grass during summer, and supply the market with moderate weights when meat is most scarce—viz., during the summer months. To do this, the animals must be brought forward during the winter and early spring with generous food. They should be placed in well-sheltered yards, to which a roomy shed is attached, and fed with a mixture of pulped roots, chaff, and meal—indeed, much in the same way as if feeding, only the nourishing food in smaller quantities. Those who have good grass and a large proportion, and whose arable land, both from its nature and limited proportions, is not capable of supplying roots to a large extent, will act wisely in not attempting winter feeding, but content themselves with growing and developing the cattle destined for summer grazing, which can be done without roots at all, although a few pulped and distributed through the chaff help the latter down wonderfully. Winter feeding is an expensive necessity for those who have but little grass, and whose land requires liberal supplies of fold-yard manure. In such cases the problem to be solved is, how to obtain the maximum return at the least cost. Few pretend to say that house feeding can be made to pay *per se*; but great will be the advantage if the increase in the animals covers the outlay in food, and we have the manure as our profit, for that represents a very considerable item. We may calculate that during four months (which is about the average time fresh beasts required to be housed) each animal will make from eight to ten yards of manure, which at 6*s.* a yard—a fair price for such manures—gives a return of from 48*s.* to 60*s.* per head. Now, we believe that this result can be attained by careful management, and we proceed to sketch out a programme which we have adopted for some years, and which has proved satisfactory. The forwardest animals are drawn out from the rest, at the period when the lattermath is ready for pasturing. They are supplied, by means of cattle cribs, with a mixture of equal parts of decorticated cotton cake and palm-nut meal, two pounds of each per head, given in the mornings. At first they do not eat this well, picking out the cotton cake, but a few days suffice to bring them to their food, and they soon let one know if it is not given punctually. Not only does such a mixture supply a large proportion of feeding material, but the rather binding nature of the cotton cake tends to correct the too laxative influence of the grass. According to the state of the weather, these cattle are brought into the yards about the middle of October, and have pulped turnips, chaff composed of a mixture of hay and straw, and 6lb. of meal, *i.e.*, 2lb. of barley meal added to their previous food. The proportion of pulp depends upon our resources; about 70lb. to 80lb. per day is amply sufficient, which is about half the quantity that would be necessary if the animals were fed on the old plan. The reason for placing them in yards is to bring them by degrees to a life of confinement; if at once placed in the boxes or tied up, they sweat so much as to loose flesh. They should remain in the yard about three weeks, after which they will settle in the boxes. Many complain with justice of the expense of boxes, which is quite double that of byres; but it must be allowed that animals thrive much faster in them than when tied up, and the manure is most excellent. We do not believe in stale fermented or cooked food; hence the pulp is made each day, and consumed within twenty-four hours. The animals are fed three times a day with the mixture, and the cotton cake being in lumps is given and readily eaten by itself at noon; the meal, that is, the palm-nut and the barley, being scattered over the pulp and chaff. Rock salt should always be supplied, and the animals require water; it is best when this is laid on, and always at command, but, as this is often not possible, we supply it once a day. Cattle eating 80lb. of roots with 20lb. of chaff will drink from four to six gallons a day. Some people recommend the use of the brush and currycomb, and no doubt the circulation improves when the pores are kept open; but in boxes cattle can rest themselves and keep themselves clean better than when tied up—moreover, the expense is considerable. If cattle are tied up in byres, we believe it would aid the process if they were turned out daily to stretch their legs. At first they would probably run about a good deal. Young growing animals especially require a little exercise, and any loss of force would be made up by improved appetite. We have found animals feed quite as rapidly in a sheltered yard, with plenty of shed and crib room, as when tied up. When within about a month of the market, we add 2lb. of the best linseed cake, making eight pounds a day of artificial food. This is expensive feeding, as we cannot calculate such food at much less than 1$d.$ a lb. We calculate the cost of house feeding as follows: $\frac{3}{4}$ of a cwt. of roots, $4\frac{1}{2}d.$; 6lb. of artificial food, say at 5$d.$; hay and straw used as chaff, say 20lb. a day, at 3$l.$ a ton, 6$d.$; attendance 2$d.$; total, 1$s.$ 5$\frac{1}{2}d.$ per day, or 10$s.$ 2$\frac{1}{2}d.$

per week, charging a full price for the hay and straw. It will be found good work—more, we suspect, than an average result—if the animals increase 14lb. a week of dead weight; and as the market value even at present rates hardly reaches the cost, it will be evident that stall feeding can seldom be made to pay expenses and leave the manure clear. The farmer must consider himself well-off if he gets his manure at half the price it would cost him to buy; and such manure as will result from box feeding on the food described is well worth the price we have estimated, viz., 6s. a ton. Feeding cattle is a necessity on light thin soils. On strong land artificial manure and thorough cultivation will produce remunerative crops, at any rate, for a number of years; but we cannot do without fold-yard manure on weak sandy soils.

Where we rear and feed—which with good management will on mixed farms pay, as a rule, better than buying in animals to feed—there can be no doubt that we should feed from birth, never allow the animal to stand still, much less go back, and thus bring our animal out at an early period. How this can be done will depend upon circumstances. Our practice in this respect has undergone a wonderful revolution in the last ten or twelve years. By pursuing the system sketched out, it is quite possible to bring out two-year-old steers averaging from nine to ten score a quarter, and equally practicable to have our heifers producing their first calves at the same age. Unless our soil is rich, and the produce very nourishing, we may not be able to make beef so economically at two as at three years. In the latter case we force less, but always should have the animal gently thriving. Where practicable, we recommend turning the animals off at two and a half years old, as a quick return is a great point in these days; but this requires a continuation of good feeding during the second year. If in the field, the pasture must be abundant, and night and morning trough food must be given. Comparatively early in autumn the beast should be housed. A well-ventilated box, which allows of more freedom of motion than the stall, is preferable for young beasts. The taste of turnips should have been previously acquired by a few roots scattered about the pasture. With the chaff and corn they are familiar enough, and so we get them to settle down after a day or two without a stand-still period, so common to sudden changes of diet. During the second summer at grass we recommend a mixture of bean meal, oil-cake, or cotton-cake, and barley meal, always given with chaff, the quantity of artificial being increased from 2lb. at the spring to 4lb. in the autumn; and it may be as well to pulp a few turnips in August to add to the mixture, and thus more thoroughly accustom the animals to their winter food. For a few days after first coming in, a portion of green food might be given, so that the change should be as gradual as possible. By such means we cannot fail, under careful superintendence, to keep our stock thriving, and by May or June, when our roots are on the wane, we may expect some very pretty ripe beasts, weighing from 90 to 100 stones of 8lb. dead weight. In the event of a third year being considered advisable, we should winter in yards the second year precisely as during the first, taking care to run them thinly, and feed moderately well. We have used green German rape with good effect in such cases, breaking it up and steaming, or else macerating it in boiling water, and then pouring it over the chaff. The price of this article has considerably risen of late years; moreover, it is difficult to obtain, and we should consider palm-nut meal as equally cheap and more feeding, on account of its greater richness in fatty matter. This, however, brings us to another part of our subject, viz., the nature and value of different kinds of food.

The system of feeding horned stock, though more rational than of old, is still in many districts costly and extravagant, and must be amended if we hope to make it a profitable operation; or, perhaps it would be more correct to say, escape a heavy loss. In former times, the high price of wheat rendered it desirable that rich manure should be made, even though the expense of its manufacture was great; especially was this the case at a time when the farmer could not, as now, supplement his home-made manure by artificials; consequently we find that large quantities of costly food were given, the greater part of which remains in the manure. We know that even now, in some cases, as much as 10lb. to 14lb. of linseed cake are given daily, a third of which only is assimilated. Now, such treatment is wasteful and unscientific. We cannot afford to keep animals merely as expensive machines for the manufacture of manure; and unless it be possible so to feed cattle as to cover our expenses, we had better abandon the practice altogether, and obtain our manure from other sources. We believe that it is possible so to feed as to make both ends meet. Let us examine the progress that has been made. In Scotland, where feeding was largely pursued in the days before linseed cake was known, the custom was to give sliced turnips *ad libitum*, and long hay. The quality of the swedes being good, excellent beef was made, though the process was rather slow, and decidedly expensive. This barbarous practice is still the rule, though we are glad to know that the pulper is slowly forcing its way into favour. In vain has Dr. Lyon Playfair shown that heat is an equivalent for food, and that every manifestation of force is accompanied by loss. Scotchmen, as a rule, still persist in pouring gallons of cold water, in the form of watery turnips, into their bullocks, thereby lowering the temperature of the body to such an extent that the animals may be seen to shiver after a hearty meal, and much extra fuel (*i.e.*, food) is consumed in raising the temperature thus needlessly lowered. A large ox will eat as much as 1½ to 2cwt. of sliced roots a day; 90 per cent. being water, we have more than twenty gallons of fluid: how much heat must be absorbed in raising this to the temperature of the animal's body! The effect is similar to pumping too much cold water into the boiler of an engine. The temperature falls rapidly, and extra fuel is required to regain the original condition. By the introduction of linseed cake a reduction was effected in the quantity of roots and hay, but the system of feeding was still defective. The mixture of different materials into a compound suitable for digestion was rendered possible only when the chaff-cutter, and more lately the pulper, were introduced, and we must regard these inventions as of the utmost importance to agricultural success. Every farmer who keeps stock, whether he fattens them out or only grows them, should have a chaff-cutter and pulping machine. The mere fact of reducing the food into a form which renders it more easily digested, though important, does not half express the value of these inventions. It is the means they afford of economising the more costly part of our food

—viz., the roots—and enabling us to substitute cheap straw for expensive hay, that gives them such importance, and marks their introduction as inaugurating a new era in cattle management.

Pulping-machines are constructed on two distinct principles, viz., first, such as have the cutters fixed on a barrel, differing only from Gardener's turnip slicers in the size of the blades and the spaces in the barrel, through which the cut food escapes; and, secondly, such as have the knives fixed in a vertical disc. The latter have important advantages, and principally that the roots are not rolled round and round, and bruised, the form of the hopper being such as to allow of their remaining stationary whilst being cut; whereas when the root comes in contact with the revolving barrel there is a tendency to fly off, and thus a rolling motion is communicated, the root is bruised, and the juice more or less extracted, which is to be avoided. Of all disc pulpers yet brought out, that of Messrs. Hornsby and Sons stands first; not for the quantity of work done in a given time, but for the perfect way in which the roots are brought into the required condition with the minimum loss of juice. The difference in this respect in different machines is really remarkable. At the Oxford trials in 1870 samples of the pulp were taken and examined a few hours after being cut; the difference in colour and freshness was very great; in some cases, notably the produce of barrel cutters, the pulp of mangold was already nearly black. The disc cutters cut to the last piece without the root being squeezed. The price of a pulper is so inconsiderable, the largest power size being only 6*l*. 10*s*., and the saving is so great that, as we said before, it is only ignorance of its value that prevents universal use. In order to have the greatest advantage from pulp, it should be fresh. Advocates for fermentation are not wanting, but we have always noticed that if kept to the third day the cattle do not eat it so well. Whether in case of necessity the pulp might not be compressed and kept from the air, especially if slightly salted, we cannot say, but think it very probable and worth testing. With steam power, which is only at command one or two days a week, the preservation of the pulp becomes an important question; but with horse-power always at hand we like using the pulp fresh, never keeping it more than from twenty-four to thirty hours.

The improved one or two horse gear is particularly adapted for driving chaff-cutter and pulper. We are acquainted with a farmer who has about 100 head of cattle of all ages in his boxes, stalls, and yards, during the winter. A strong Galloway pony does the work, the operations of pulping and chaff-cutting being performed separately; the chaff, as cut, falls down on to the floor where the pulper stands. The largest feeding beast gets 80lb. a day; the younger animals, which are not feeding, but growing, from 20lb. to 40lb. The store animals have more straw chaff than the fatting beasts, the rule being to allow them as much as they can eat up clean. The mixture of roots and chaff offers an excellent medium for the distribution of artificial food in the form of meal.

Animals thus fed should have water offered to them once a day; the cattle in the yards can drink when they like. The fatting beasts will usually take about two or three gallons a day. Bearing in mind Liebig's views as to the loss sustained by each motion of the body, there must be a decided gain in presenting the food in a state requiring so little labour in mastication. The work of filling the belly is effected in a much shorter time than formerly, consequently there is more time for rest, which is a condition favourable to the deposition of fat. The number of animals we can feed out is usually determined by the supply of roots. If, then, by pulping, we can economise the roots by at least one-third, it follows that the system allows of extra stock to that amount being made out. Taking the time of feeding to be six months, and the quantity of pulped roots to be 80lb. against 120lb. sliced, it follows that we shall save fully 3 tons, worth at least 30s. a head. Winter feeding of cattle is often held to be an expensive necessity, and no wonder, with the costly method too frequently employed. We believe that, under judicious management, the winter feeding of cattle is the cheapest method of maintaining and increasing the fertility of our land.

We have said that in many districts the system of feeding is still costly and extravagant. This is very apparent in the wasteful practice with regard to straw. Now, if properly made and taken due care of, a large portion of the straw should be used as food, as it bears a considerably higher value for this purpose than as mere manure. The quality varies considerably, according to nature of soil, climate, and, above all, method of harvesting, the difference in this latter respect being very remarkable and worthy of most serious attention. Some years since we experimented on a crop of white Poland oats, cutting one portion very unripe, another green but still ripening, and leaving a third portion until dead ripe. We may dismiss the first, as the analysis proved that material was left in the straw which should have gone to the grain. The second—cut green, but sufficiently forward to allow of a full and fine sample—contained as much as 10 per cent. of sugar and gum, 30 per cent of digestible fibre, and $2\frac{1}{2}$ per cent. of soluble (i.e., available) protein compounds. In the over-ripe the percentage of sugar, gum, &c., was reduced to 3·19, the digestible fibre to 27·75, whilst indigestible woody fibre, useless matter, was increased from 31·78 to 41·82 per cent., the soluble protein being reduced to 1·29. The effect of ripening straw is in all cases to increase the insoluble material at the expense of that which is available as food, and therefore the practical question for each to arrive at is the exact period of time when the greatest amount of nutriment remains in the straw, and the grain is properly matured. It will be seen, by a careful study of the analysis of these straws, made by the late Dr. Vœlcker, and recorded in the twenty-second volume of the "Royal Agricultural Society's Journal," that the actual amount of solid matter is identical in the two samples; therefore the grain is not robbed, and moreover early cutting secures a finer quality, and is economical in many respects. Now let us compare this oat straw with a sample of well-made clover hay. We shall find that the difference of nutritive matter is not so considerable as we might suppose—

	Clover Hay.	Oat Straw.
Water	20·50	16·00
Oil	3·59	1·05
Soluble protein compound	5·00	2·62
Insoluble ditto	8·75	1·46
Sugar, gum, &c.	13·07	10·57
Digestible fibre	16·42	30·17
Indigestible fibre	25·62	31·78
Soluble mineral matter	4·43	3·64
Insoluble ditto	2·62	2·71

The hay, though much richer in oil and protein compounds, does not contain so large a proportion of non-nitrogenous elements as the straw, and we may fairly calculate that the straw is worth two-thirds as much as the hay. Now hay, such as this, would be valued at about 50s. a ton for consumption, and therefore the straw may be estimated at from 30s. to 35s., whereas the actual manuring value of the same would probably not exceed 10s. to 12s. The loss from treading such material into manure, instead of passing it through the animals, must be apparent. Let us remember, moreover, that other straw is even more valuable than the oat. Pea haulm, cut early and well harvested, is most valuable, quite equal to much of the artifical hay from poor land, and such material should be taken the utmost care of. This fact is not recognised, and in valuations between tenants we find that pea haulm is put in some counties actually lower than wheat straw.

Bean straw, especially if the pods are attached, is more valuable than is generally supposed; and though, from its hard woody nature, it is not so adapted for food as softer straws, still a portion may be used with advantage; while, as we are now in the habit of cutting it in a much greener condition than formerly, it should not be overlooked as a source of food. Bean straw is increased in value by steaming, principally by being made softer, and therefore more easy of digestion, but to a limited extent its solubility is increased. Enough has been advanced to show that straw is a valuable part of our natural food, and when properly made may be economically substituted for hay in the feeding of cattle. The cost of chaff-cutting, on the most economical plan, will not exceed about 6s. a ton. Now, the use of chaff enables us to effect a great saving in the consumption of roots; and as roots hauled off the land are very expensive food, costing probably from 7s. 6d. to 10s. a ton, this is one great feature of modern farming. The pulper, by mashing up the roots, causes the juices to mix with and saturate the chaff, and thus we produce a composition that is readily eaten, and obtain the same effects as formerly with half the quantity of roots. A fatting animal does not require more than about 70lb. to 80lb. of roots, which may be distributed through some 20lb. of straw.

Since the first edition of this work, published in 1875, we have had a new form of green food added to our resources, viz., silage, which promises to be of great service, especially on farms where tillage land is limited, where the nature of the soil does not allow of the growth of roots, or when the climate renders hay-making difficult. Although so recently introduced into England, and hitherto principally in the hands of landlords or amateur farmers, the system of preserving fodder crops by burying, and excluding the air, has been handed down from ancient times. Over forty years ago the German practice of making sour hay was described in the "Journal of the Highland Agricultural Society." Professor Wrightson, reporting on Austro-Hungary in 1874, noticed, as well worth attention, this "sour hay;" and again, in 1875, he described the system in the *Times*; and another writer, Mr. T. Schwann, wrote on the subject in the *Field* in 1870, and again in 1876; yet little notice was taken of these communications, and it was not till the value of silage had been tested in America, and written of in very glowing terms, that the subject became thoroughly ventilated in this country, and was experimented on by many leading agriculturists. Indeed, little had been done prior to 1882, when the Viscount de Chezelles visited this country and explained his operations. Credit must also be given to M. Goffart, who not only was very successful himself, but issued a manual, which was translated into English, and attracted much attention in America. Numbers at once made silos, and the result has been decidedly successful. At first it was held to be indispensable that the fodder should be stored in an air-tight chamber, and hence great expense was incurred in building the receptacle called the silo, or in converting existing buildings into silos. It was also considered desirable that the material should at once be submitted to great pressure, so that the fermentation which takes place, whilst it has the tendency to break down and render the ingredients more soluble and digestible than in the fresh material, should not degenerate into the stage of acetic and alcoholic fermentation, but should be arrested in what is known as the lactic acid stage of fermentation, and this may be considered as the perfect condition of sour silage. But it has been pointed out that, as our fodder crops contain, as a rule, very little sugar, the lactic fermentation does not take place to any extent, but as soon as fermentation begins, alcohol is formed, which means a loss of nutritive matter, and this is intensified by the almost immediate conversion of the alcohol into aldehyd, and when this exists a very considerable loss of feeding material occurs. Our readers will recognise this material as the same produced in over-heated hay. Lastly, if the fermentation goes farther, the aldehyd is converted into acetic acid. It is probable that the most important effect of fermentation is on the woody fibre, which is rendered more digestible by the process. Now it is quite evident that great care is required to ensure proper fermentation, and that any mismanagement must result in serious loss, and Mr. Jenkins, in his valuable paper, vol. 20 R. A. S. E., new series, says: "In the case of our own fodder crops, the evidence given in the foregoing pages tends to show that crops cut before they begin to get woody (in other words, while they are still full of sap), pitted after having been chopped, then carefully trodden layer by layer, then covered with boards, and moderately weighted, stand the best chance of not going much beyond the alcoholic fermentation. On the other hand, crops put in unchopped, dripping with wet, imperfectly trodden, and no matter how heavily weighted, will rapidly go through all the processes, and even beyond those I have spoken of, namely, into the putrefaction stage." Mr. Jenkins suggests that the chemists will, by their investigations, be able to give us a Dr. and Cr. account as between the loss of feeding material by the conversion of carbohydrates into alcohol and acetic acid on the one hand, and the gain of feeding material by the conversion of indigestible into digestible woody fibre on the other. In the meantime the practical testimony as to the wholesome and nutritious character of well-made silage is simply overwhelming, and especially as regards milk production; there is every reason to believe that it compares very favourably with the same material made into hay. We rather agree with Mr. Jenkins in regarding the system of ensilage as a valuable addition to the resources of the English farmer, rather than a complete substitute for haymaking. Within the last two or three years much light has been thrown on the subject by the experiments of Mr. Geo. Fry,

F.L.S. This gentleman has demonstrated that the nature of the product, whether sweet or sour, entirely depends upon management. If the material is placed in the silo, free from rain water, and the silo is slowly filled without undue pressure, a fermentation, accompanied with a high temperature, say 120° to 130°, ensues, which is sufficient to destroy bacteria, which are the cause of the acid stages of fermentation. And when the fermentation has gone on for three or four days, it can be arrested at the sweet stage by applying pressure. In this way he has succeeded in producing a very good silage, in which the loss of carbohydrates is not serious, and provided it is used at once, and not left exposed to the air after being cut out, it supplies excellent food. Another very important stage of our knowledge, only recently reached, is that good silage can be made in stacks without the intervention of costly silos at all. The Council of the R. A. S. E. very wisely supplemented their last president's prize of 100 guineas for the best silo in operation during the winter of 1885-86 by a prize of 25l. "for the best stack or other system of obtaining silage without a silo." There were nine competitors, and the prize was awarded to Mr. C. G. Johnson, of Croft, near Darlington. We subjoin Mr. C. G. Roberts's Report: "In this design the stack is built in the ordinary form, the pressure is supplied by flexible galvanised iron wire rope passing over the top of the stack, and looped on to a crutch at one end of crossheads, which move loosely up and down on the screwed bars made fast to logs of wood which are held down by the weight of the stack built upon them. The rope is passed over the top of the stack and hitched on to a corresponding crutch at the other side, and returning at short intervals, is laced over the whole series of crutches on all the crossheads, and then made fast by hitching it round the last crutch. The crossheads are then tightened down, one at a time, by screwing the nut on the upper side of each of them; the screws are four feet long, to allow amply for the settling of the stack between one day's stacking and the next. The screws are so adjusted that the spanners, pulled with a force of 40lb., exert a pressure of $1\frac{1}{2}$ cwt. per square foot on the stack. When the silage is used the wire ropes are thrown off from one crosshead at a time, and the pressure continues undisturbed on the remainder." The judges, Messrs. Fowler and Baker, found a stack 19ft. long, 17ft. wide, and 11ft. to the eaves and 17ft. high to the peak of the ridge, composed of silage from autumn-sown tares, seeds, and old grass—supposed to contain 130 tons of silage. The cost of the apparatus for a stack of these dimensions is 18*l*., and Mr. Johnson estimates the cost of cutting the forage, carrying the same, and pressing the stack, at about 18*s*. an acre. The silage was sweet and good; about 8in. of the ends and sides was damaged, but this was calculated to represent only 1½ per cent. of the whole, and though dark and inferior it was readily eaten by the cattle. Whilst the stack was in process of building the temperature reached 160 to 177 degrees. Above the wires the stack was topped up with rough hay and straw, and thatched. It is, of course, probable that equal pressure will be obtained at less cost, but we have here a most important advance in our practice, and the process of ensilage is now brought well within reach of the tenant-farmer, who may well be excused if he pauses before undertaking costly structural works in making air-tight silos. Now let us suppose a case. The grass farmer provides himself with the necessary pressing apparatus. Weather is unfavourable for hay-making, he puts the grass into a stack, secures the proper heat for fermentation, and then straps it down. We foresee great results. Sour silage, if used in large quantities, may have an injurious effect on breeding animals; sweet silage, on the contrary, will be conducive to health. Much has yet to be learnt before we arrive at the scientific treatment of silage; but a great deal has been done, and we may now justly regard silage as a most important and available winter food, within reach of nearly everyone who takes the trouble to understand the process of manufacture.

On the use of artificial food, in which we include home-grown grain as well as purchased materials, we find a great diversity of practice, partly arising from varying conditions, and partly from ignorance of the digestive processes. We shall not arrive at uniformity until the feeding of the cattle has been made the subject of carefully conducted experiments. The commercial mind, accustomed to subject every process to experiment and test, will be surprised to find that questions of such importance as the exact proportions of natural and artificial food most suitable at different ages and stages of feeding, are still matters only of opinion, but there are great difficulties in the way of demonstration. The comparison between living subjects is always influenced by the variation that exists in constitution, temperament, and feeding quality; and it is a nice point, and opens a large field for error, to allow for these disturbing causes. It would be important to ascertain exactly what proportions of natural and artificial food prove most economical at different ages and stages of feeding. We are quite certain that money is often lost from over-feeding with rich nitrogenous foods, especially in the early stages of the process. The proportion which the system can take in and convert into flesh is so small, that much the larger part goes into the manure; and, remembering the vast complicated stomach of the ox, it is clear that a bulky food is required which contains a considerable portion of indigestible fibre. The proportion between the stomach and intestines affords an indication as to the nature of the food required by different animals. Thus for each 100lb. live weight the ox has $11\frac{1}{2}$lb. of stomach, and only $2\frac{1}{2}$lb. of intestines; the sheep has much less stomach and more intestines, and altogether a smaller percentage of digestive apparatus, indicating the necessity for more concentrated food; whilst the pig has only $1\frac{1}{3}$lb. of stomach and 6lb. of intestines to each 100lb. live weight, demonstrating that concentrated and generous food is required. These are important points, indicating the kind of food most suitable to each. Thus bullocks have been fattened entirely on good straw and oil-cake. Such a food would not fatten sheep, as their digestive apparatus needs a considerable proportion of starchy food; whilst pigs do best when fed entirely on meal. In the absence of direct experiments on the subject, we venture to suggest a dietary suitable for animals growing and feeding at the same time, and which would be applicable to the animal from the period of weaning until death, the quantity to be given depending on age. It is well to remember that there is virtue in a mixture of different substances, even though we might supply the same constituents in a simple form. Variety suits the digestion: linseed cake, or cotton cake, according to market, 2lb.; barley, or wheat meal, or palm nut meal, or, better still, an equal mixture of all three, 4lb.; beans, peas, or lentils, according to market, 2lb.; locust beans or malt, 1lb.—the whole being reduced to a fine powder, and thoroughly mixed. In the above we have a due admixture of the flesh-forming and fat-producing elements. During the later stages of fattening we might discontinue the beans, and increase the proportion of either linseed cake or palm-nut meal. Commencing with the calf when weaned, ¼lb. of such a mixture daily, distributed over the chaff and pulped roots, would be ample, to be increased to 1lb. for the first winter, 2lb. during the second; and when the animal is put up for feeding, 4lb. to commence, and 7lb. to 8lb. to finish, will be found quite sufficient with a due proportion of roots and chaff. If we are anxious to over-feed animals—as for show, for example—it is necessary to stimulate the appetite by the use of tonics and carminatives, just as the East Indian with a damaged liver and bad digestion craves for pickles; but the use of such food in early life must be injurious, and cannot be recommended.

We need hardly insist upon the advantage of using articles good of their kind, well grown, properly matured, and free from adulteration; though more costly, they are in their effects cheaper than inferior articles. Linseed oil-cake ranks at the head of all purchased foods, being from its complex nature admirably suited to feeding purposes; the chief bar to its use is the comparatively high price of a genuine article, and the great difficulty of finding such. The proportion of really pure cake made is small compared with the quantity of inferior, and that, again, differs according to the variety of the ingredients of which it is composed. In Yorkshire this has been so much felt that at Driffield the farmers have made a company for crushing pure linseed, and thus have secured a first-rate article. Manufacturers have been blamed for the low quality of their cake, when too often the fault rests with the consumer, who is not willing to pay the higher price. Great progress has been made towards sound views on this matter, and cheap rubbish is not now, as formerly, eagerly bought up. The publicity given by the Royal Agricultural Society to adulterated and inferior makes has been most serviceable. It would have been a great misfortune to the farming community if, from prudential consideration, the Society had ceased to pursue the course they have adopted, and which has done more to gain them respect than any other of their useful efforts. The quality of a cake may be ascertained to some extent by a careful examination under a microscope, and by maceration in hot water; but it is wise before making a large purchase to have an opinion from a good authority.

We may here pause to notice the reasons why linseed cake possesses such high feeding qualities, and is so generally esteemed. In order to arrive at a satisfactory answer, we must first inquire what the animal system requires. First and foremost, matter that will supply the daily waste of muscle, or which will build up new muscular structure; secondly, that which provides fuel for keeping up the heat of the body, and which when supplied in excess of these requirements can be deposited as fat; and, lastly, mineral substances suitable for the construction of bone. The following analysis by the late Dr. Vœlcker may be relied on as giving the contents of a good average cake:

Moisture	12·44
Oil	12·79
Nitrogenized, or flesh forming principles	27·28
Heat-giving substances	41·36
Mineral matters (ash)	6·13
	100·00

The above requires but little consideration. Linseed oil-cake contains a large proportion of two most valuable ingredients—ready-made oil and ready-made flesh. Mr. Lawes has proved by experiments that vegetable oil is two and a half times more valuable than the class of starch compounds which exist to so large an extent in most vegetable substances: hence one explanation of the value of linseed cake. The oil being so valuable, the use of the seed has been advocated. Indeed, we believe that the system of feeding animals in boxes was introduced by Mr. Warnes, in order to feed on linseed. There are two good reasons against the use of the seed. First, the value of the oil for commercial purposes is too great to allow of its economical use; and, secondly, we may have too much of a good thing. When properly mixed with other food, linseed is very feeding, but its action in excess is purgative; and cheaper material can be found, which is equally effective. We can well understand the value of linseed cake for animals eating nothing but hard dry straw; thousands of beasts are so wintered, and thrive. Linseed cake has special advantages for growing animals, owing to the large proportion of ready-made flesh formers. We may supply fattening food from other sources, at less cost, but we question if any food save decorticated cotton cake can be found so valuable for growing stock. With such qualities, it is not surprising to find a lively demand, and consequently a price which is excessive in comparison with other feeding materials. The late Mr. Hope, of Fenton Barns, as illustrating the change that had taken place in the comparative value of corn and horn, mentioned, at a discussion before the Edinburgh Chamber of Agriculture, that forty-five years since, when a child, he remembered one of his father's men, coming from Leith with two carts of linseed cake, describe how the people of Preston Pans came out of their houses wondering what he had in his carts, and that he had told them it was a kind of " bannocks," but he did not think it would suit those who drank tea.

When linseed cake is from 9*l*. to 10*l*. a ton, we question whether it may not be replaced with advantage by cotton-seed cake, green German rape, or palm-nut meal. The first of these is now largely used, and, when free from coarse shell, is a safe food with a considerable amount of feeding property, rich in nitrogenous elements, it is best when mixed with such a material as palm-nut meal, which contains a higher per-centage of fatty matter: the prices of these articles are at least 50 per cent. lower than linseed cake. We would especially caution our readers against the use of any but the best samples of cotton-seed cake, viz., those that are yellow in colour, and finely ground. Several deaths have been clearly traceable to coarsely ground cotton cake, the indigestible fibre of which has accumulated in the intestines, and caused inflammation. Thirty years ago this article had no existence; then the cotton-seed was thrown on one side as useless. At first the cake was made from the whole seed, afterwards machinery was invented for removing the black outer skin, and as this formed a considerable proportion of the whole, and consisted entirely of indigestible fibre, the superior value of the decorticated cake, as a feeding material, was soon established, and has since maintained and increased its position. The late Dr. Vœlcker wrote a valuable paper in the Royal Agricultural Society's Journal for 1868, from which the following figures, being the mean of seven analyses, are taken:

DECORTICATED COTTON SEED CAKE.

Water	9·28
Oil	16·05
Albuminous compounds (flesh-forming matters)*	41·25
Gums, mucilage, sugars, &c. (heat-producing materials)	16·45
Indigestible fibre	8·92
Mineral matter (ash)	8·05
	100·00

* Containing nitrogen 6·58 per cent.

If we compare the above with the analysis of linseed cake, we shall find the cotton-cake superior in the proportion of oil and flesh-formers. The former is not like linseed, purgative in its action, but is sweet and agreeable to the taste. Cattle eat it readily. Dr. Vœlcker considers it quite equal to linseed cake. We think it very superior to much of the rubbish sold under the title of linseed cake, although it can be bought at one-third less price. The cake we are describing is the thin cake, imported chiefly from New Orleans, and which finds its way into Liverpool. It should be of a bright yellow colour, close texture, and, when cut with a knife, the surface exhibits a shining appearance, caused by the oil. Only occasional specks of black should be visible. Any larger portion of shell indicates imperfect decortication. This cake is admirably adapted for cattle on grass, as its slightly astringent property tends to prevent scouring, which the luxuriance of the grass would otherwise produce. The large percentage of flesh formers renders it peculiarly valuable for growing animals. Decorticated cake is generally about a quarter less than the best linseed cake, and for many purposes we consider it decidedly superior. It must not be forgotten that Mr. Lawes places it at the head of his list as a manure producer, and we cannot imagine any more immediate or certain way of restoring fertility to land than by applying manure made from decorticated cotton-seed cake. Complaints have been justly made that recent brands are so dry and hard that they are imperfectly digested; indeed, so serious is the evil from this cause that when it can be done we recommend that the cake should be ground and used as meal.

Very much the larger portion of the cotton-cake used in this country is undecorticated, the English make nearly entirely so. When made from clean seed it is a wholesome food, although much less nutritious than the decorticated cake. The following is an analysis by Dr. Vœlcker:

Water	10·53
Oil	6·10
Albuminous compounds (flesh forming)*	22·62
Gums, mucilage, &c.	26·48
Indigestible fibre	26·96
Mineral matters	7·31
	100·00

* Containing nitrogen 3·62 per cent.

A comparison of the two analyses proves that decorticated cake is nearly twice as valuable, although seldom costing above one-third more. Hence, at present rates, we think it decidedly the cheaper food. Another point in its favour is being so much more digestible. Common cotton-cake cannot be safely used for young stock; it is likely to cause indigestion and stoppage. Good cake should be yellow, but much darker than the decorticated cake, and speckled with black shell. It is, however, easily distinguished from a very worthless article made by pressing the refuse from decorticated cake with a small portion of common seed. A good deal of such rubbish has been sold at different times, and especially in early days, and several fatal cases have followed its use. The late Dr. Vœlcker described the death of a bullock belonging to a Mr. John Fryer, of Chatteris, which was fed upon such cakes with mangolds, barley-meal, and clover hay. The post-mortem showed the paunch enormously distended with food, the lower stomach quite empty, the duodenum for twenty-four inches in length entirely blocked up with two or more pounds of the irregular-shaped concave and comminuted husks, which were found to be identical with the husks in the cake. No wonder such was the result, for the cake contained more than half its weight of husks. A slight acquaintance with the appearance of cotton cake will enable the farmer to distinguish good, bad, and indifferent. Although we have never suffered from the use of the common cake, further than with calves—for which it is not suitable—we strongly advise our readers to buy the best decorticated. It must be remembered that cotton-cake is a binding, rather than a relaxing, food; hence it is not suitable for animals feeding entirely on dry food. Under such circumstances the addition of linseed cake is desirable, but as a mixture for growing stock that get turnips, and especially for cattle on grass, we have a very high opinion of the decorticated cotton-cake.

Rape cake has not acquired the importance it deserves, for two reasons: animals do not eat it readily, owing to its hot, bitter taste, though when once accustomed there is no difficulty on this score; secondly, inferior samples, especially English-made, frequently contain the seeds of wild mustard, the oil of which is a violent poison. Fatal cases have occurred from this cause. The oil, being volatile, is easily destroyed by subjecting the cake to a certain temperature: hence if we are at all in doubt as to the quality of the cake, boiling or steaming is a safeguard. The presence of mustard can be readily detected by treating a portion of finely-powdered cake with boiling water. The oil, which is very volatile, escapes, and may be recognised by its pungent odour and powerful irritating action, making the operator's eyes water. According to the late Dr. Vœlcker's analysis, good rape cake should contain—

Moisture	10·68
Oil	11·10
Nitrogenised, or flesh-forming matters	29·53
Heat-giving substances	40·90
Mineral matter (ash)	7·79
	100·00

It is evident from the above that rape cake is not deficient in nutritive properties, and the late Mr. T. Horsfall stated, in his valuable "Articles on Dairy Husbandry," that he used it with marked success. His mixture for milch cows comprised rape cake, 5lb., and bran, 2lb., for each cow, mixed with bean straw, oat straw, and shells of oats, in equal proportions, supplying this food three times a day *ad libitum*. The materials were moistened, well steamed, and given in a warm state. Mr. Horsfall further states that he had no difficulty in inducing his animals to eat the cake, and never found the butter affected by it. We have known others equally successful. Few people steam now, even when there are facilities. The practice is not on the increase, and we must continue to regard the hot, unpleasant taste of rape cake as a serious impediment to its extensive use.

Smith's *Palm-nut Meal*, which results from the grinding and pressing of palm kernels, possesses considerable fattening properties, owing to the large percentage of vegetable oil which it contains. This oil is very similar in appearance to lard, and appears to be readily convertible into animal fat. Dr. J. Vœlcker's analysis is as follows:

Moisture..	9·12
Oil and fatty matter...	12·50
Albuminous compounds (flesh-forming matters)*	15·50 } 63·45
Mucilage, gum, sugar, &c.	35·45
Woody fibre (cellulose)......................................	23·69
Mineral matter (ash)...	3·74
	100·00

* Containing nitrogen 2·48 per cent.

We have connected the meal with the name of Messrs A. M. Smith and Co., of Kent Street Oil Mill, Liverpool, because they were the original importers of palm kernels, and we believe they are the only crushers who sell a meal with this high percentage of fatty matter. As it is perfectly certain that the value of this food depends upon the oil, it is evident that the sample containing 12 or 13 per cent. must be much more valuable than having only two-thirds that quantity. The meal is dry and harsh looking, palm oil being solid at ordinary temperatures. It is sweet, and will keep good for any length of time. Some patience is required in accustoming animals to its rather gritty taste; when once they take to it, however, they eat it freely. For cattle on grass, when liable to scour, we consider its use in conjunction with decorticated cotton-cake is of great value; the mixture possesses high feeding properties, and affects the manure much in the same degree as linseed cake. The price of this meal is from 5*l*. 10*s*. to 6*l*. a ton. About the year 1876 this firm commenced to crush the flesh of the cocoanut, which they imported from Singapore and the neighbouring islands. The result is a very feeding cake, containing oil and a fair proportion of flesh-forming ingredients, which has proved a highly valuable food for dairy purposes. A recent analysis by Dr. J. A. Vœlcker gives the following results:

Moisture	8·40
Oil	11·36
Albuminous compounds (flesh-forming matters)*	20·37
Mucilage, sugars, digestible fibre	40·61
Woody fibre (cellulose)	12·97
Mineral matter (ash)	6·29
	100·00

* Containing nitrogen 3·26 per cent.

Indian corn or maize is an important feeding material, and when either soaked or ground into meal, forms a useful mixture with other substances. It is largely grown in the United States, and, with improved railway communication in the future, we may anticipate increased supplies. We have not been able to find a detailed analysis, and therefore quote from Mr. Horsfall's comparison of different foods for dairy cows. He gives—

Oil	7·00	per cent.
Starch, sugar, &c.	60·00	,,
Nitrogen	2·25	,,
Mineral matters (phosphoric ash, ·19; potash, ·17)	·36	,,

The albuminous compounds required to furnish 2·25 of nitrogen would amount to nearly 14 per cent., which leaves about 18 per cent. for water and indigestible matters. The large proportion of starchy matters, the moderate percentage of flesh formers, and the deficiency in minerals, all indicate that Indian corn is more adapted for feeding than growing animals, also that is not a food to be used alone. Thanks to Mr. Lawes, this is not a mere speculative opinion. In his pig-feeding experiments, the result of using Indian corn alone are given in the following table :—

No. of Pigs.	1st Period of 14 Days.	2nd Ditto.	3rd Ditto.	4th Ditto.	Total Period of 8 Weeks.
1	31lb.	6lb.	40lb.	19lb.	96lb.
2	15lb.	13lb.	13lb.	13lb.	54lb.
3	12lb.	17lb.	19lb.	23lb.	71lb.
	58lb.	36lb.	72lb.	55lb.	221lb.

The following, in explanation of the above, is extracted from the report which will be found in the fourteenth volume of the "Royal Agricultural Society's Journal," page 472:—"One of the pigs gained more than 2lb. a day during the first fortnight of the experiment, but the other two only about half as much. Before the end of the first period it was observed, however, that this fast-gaining pig, and one of the others (No. 3), had large swellings on the side of their necks, and that at the same time their breathing had become much laboured. It was obvious that the Indian corn meal was in some way defective diet, and it occurred to us that it was comparatively poor both in nitrogen and mineral matter, though we were inclined to suspect that it was a deficency of the latter rather than of the former, that was the cause of the ill effects produced. We accordingly determined to continue the food as before, but at least to try the effect of putting before the pigs a trough of some mineral substances, of which they could take if they were disposed. The mixture which we prepared was as follows: 20lb. of finely-sifted coal ashes, 4lb. of common salt, and 1lb. of superphosphate of lime. A trough, containing this mineral mixture, was put into the pen at the commencement of the second period, and the pigs soon began to lick it with evident relish. From this time the swellings or tumours, as well as the difficulty of breathing, which probably arose from the swellings, began to diminish rapidly. Indeed, at the end of the second period, the swellings were very much reduced; and at the end of the third they had disappeared entirely. Notwithstanding this serious drawback, it was found that the animals were satisfied with less of this food, though so poor in nitrogen, in proportion to their weight, than, with one exception, of any of the others; and it will be found that the increase is satisfactory when compared with the food consumed." Indian corn, at any rate for pigs, possesses considerable feeding properties, and there is no reason to doubt its value for cattle when judiciously mixed with other food. As a general rule, we may consider it cheap when it rules 4s. to 5s. a quarter below grinding barleys.

Barley is largely used for feeding. Either the coarser samples of home-growth, which are not suitable for malting, or the hinder ends—that is the tail corn which comes out of finer samples—is available for meal, and when well harvested, has a marked effect in the production of flesh. Foreign barley makes excellent meal, being generally drier and harder. The advocates for the remission of the malt tax argued that better results would be obtained on malt. Sir J. Lawes, however, is of a different opinion, and as his conclusions are the results of direct experiments, we place most confidence in them. The increase in weight of sheep on a certain quantity of barley was considerably greater than on the same after being subjected to the malting process; and he says, "Not only is the weight of the malt considerably less than that of the barley from which it was produced, but that weight for weight, independently of loss and cost of process (estimated at 2s. per quarter), the feeding qualities of malt are not superior to barley. At the same time, he admits that, as a mixture with other food, or as an occasional stimulant to digestion, malt may be usefully employed. The composition of barley is as follows:—

Water	14
Flesh-formers	14
Starch, &c.	68
Fatty matter	2
Ash	2
	100

As a confirmation of Sir J. Lawes's researches, it may be noted that, though the tax has been removed for some years, malt is very little used for feeding purposes; and hence we may conclude that, taking into account the cost of the operation and the loss of weight, it is not found advantageous.

Beans, peas, and lentils are so identical in composition, and similar in their effect, that they may be substituted for each other according to our convenience, and may be considered together. These are valuable feeding materials, and, when used with judgment, give satisfactory results. Owing to mechanical condition, and also to the large proportion of flesh-forming elements, all three are more or less indigestible if given without due preparation, are partially wasted, and if largely used, are apt to cause constipation of the bowels. Beans, which are the hardest, should either be broken small or ground into meal. Peas are much softer in their nature, and will be sufficiently prepared by being kibbled or broken small. Lentils, which in their natural state are, owing to a hard skin, very indigestible, should be reduced to a coarse powder. The predominating feature of these substances is the large percentage of flesh-forming material, which points out their peculiar value for working horses, growing animals, or for young stock that are growing and feeding at the same time. We give Dr. Cameron's analysis:—

	Common Beans.	Peas.	Lentils.
Water	13·0	14·0	13·0
Flesh formers	25·5	23·5	24·0
Fat formers	48·5	50·0	50·5
Woody fibre	10·0	10·0	10·0
Mineral matter	3·0	2·5	2·5
	100·0	100·0	100·0

Bean meal is admirably adapted for calves after they are weaned, and when either out at grass or receiving green food in the yards. Peas appear particularly suitable for young sheep. Lentils may be substituted for either, provided they are properly prepared. Egyptian beans are largely imported, and, weight for weight, are, when sound and good, quite as valuable as home-grown corn; indeed, from their drier condition, we should prefer them to new home-grown beans, which, owing to the presence of more water, are not always desirable food.

Locust, or carob bean, when ground, forms a considerable percentage of some of the condimental food, and might with advantage be used in small proportions, as it contains a large quantity of sugar; but, like German rape cake, the supply is uncertain, and always very limited. Either this or malt is desirable to the extent of 10 or 15 per cent., as tending to render the food more palatable, but neither is to be used in large quantities. Sugar is too soluble for the ruminant, and much saccharine food tends to cloy the appetite. Such at least has been our own experience, and we cannot coincide with much that has been written as to the use of malt. As an alterative, especially for sheep that are out of health, we believe it will prove most valuable; but as a constant food in large quantites we do not think it desirable.

In addition to the above, and especially in the latter stages of feeding, something in the way of a condiment may be given, as encouraging an appetite that has become a little delicate. The following mixture has been used with success: Fenugreek seed, 32lb. at 1¼d.; mustard, 8lb. at 2d.; linseed, 8lb. at 1½d.; carraways, 4lb. at 4d.; fennel, 4lb. at 5¾d., making a total cost of 8s. 11d. for the half-hundredweight, or in round numbers 2d. a pound. From 2oz. to 4oz. a day would be sufficient for each animal. The expense is trifling, and the effects very satisfactory.

We come now to the question of profit or loss—and here much will depend upon our skill in buying well, selecting such animals as have aptitude to feed, and not paying more than they are worth. The cost of feeding may be variously estimated, according as we place a higher or lower value on the materials we use. Taking the roots at 10s. a ton, and straw at 1l. a ton, and artificials at 8l. a ton, the cost per week will be as follows: —5cwt. of roots at 6d. a cwt., 140lb. of straw, at 1s. a cwt., 42lb. of artificial at 8l. a ton—total, 8s. 2d. We have to estimate on the other side the progress and the value of the manure. We think it may be fairly assumed that a beast on such food as above should gain 12lb. of dead weight weekly:—12lb. of meat at 6½d., value of manure, say at 2s. a week—total, 8s. 6d.

Nothing but experiments carefully carried out can give us data on these points. We believe that under judicious management fatting may be made, not perhaps to pay in one sense, but to be the means of providing a large supply of valuable manure which it would cost a good deal of money to replace in artificial manures.

CHAPTER IV.

BUILDINGS, AND THE MANUFACTURE OF MANURE.

THE importance of providing shelter for our live-stock is generally acknowledged; the economy of preserving the animal from the depressing influences of a low temperature and exposure to rainfall is no longer questioned. It is not now, as formerly, necessary to explain how food supplies combustible material wherewith the temperature of the body is maintained, or to show that the quantity of such food will bear a direct proportion to the temperature in which the animal lives, and that an equable and rather high temperature is the most favourable for economising the food. Nor is it necessary to dwell upon the injury arising to the health and growth of the animal when, in place of being comfortably housed, it has to face the pitiless blast on a bare pasture, or to stand in an open yard, over fetlocks in sludge, as we too frequently see in our dairy districts. These questions, we repeat, are now fully understood, and it is rather with a view to moderate opinions, so that we may not exaggerate the value of shelter, that we propose to discuss the question of covered *versus* open yards.

It is always difficult to get a new idea entertained, however good it may be, and a long time must elapse and much discussion is required before the agricultural mind is prepared to receive any novelty; but when once the crust of prejudice is broken through, and we are fairly afloat, there is the danger of being carried away by the tide that sets in, and landed far beyond the point we aimed at in starting. Enthusiasts, with more ardour than judgment, are prepared to advocate their hobby under all circumstances. It appears to us that the question of covered yards is in danger of being injured by the intemperate zeal of its advocates. We believe that, under many conditions, the covering of our homesteads will prove an economical investment. The saving of straw and concentration of manure are points of great importance. But, on the other hand, we are perfectly certain that manure may be made in open yards, when properly constructed, without serious loss of valuable materials. "How is this?" we hear someone exclaim. Has it not been stated by Alderman Mechi and others that manure made under cover is worth much more, weight for weight, than that from open yards? And have we not all heard of the losses to which manure is liable when exposed to the action of rain? The graphic picture of the hillside yard, deluged with water collected by the long unspouted roofs of stable and barn on the upper side, with the duck-pond below, which receives the filtrate, has been too often brought forward as the type of the open yard system. But is this fair? As well might we condemn covered yards because failures have occurred through improper construction, causing imperfect ventilation. Divested of exaggeration, we believe the real facts of the case are these. Manure may be made in open yards properly constructed without the loss of valuable materials; but a greatly-increased quantity of litter will be required, and must be supplied according to the greater or less rainfall, in order to soak up the liquid and prevent waste.

This is the most cogent argument against open yards. We cannot, in many cases, afford to throw away so much straw, which, if properly harvested, has a considerable feeding value; and, although not going so far as some who advocate its entire use as food, we believe that the perfection of manure-making consists in avoiding all excess of straw, and using no more than is absolutely required to absorb the solid and liquid excrements. Manure in open yards varies in quality in its different layers according to the state of the weather when those layers were formed. If dry, little or no fresh litter will be required, whereas in a wet time litter must be spread once or perhaps twice a day. The proportion of excrement, therefore, differs greatly, and in order to obtain a uniform bulk we must mix the different layers in a heap and occasionally turn, in doing which we are liable to some loss of valuable ingredients. Then, again, we have extra labour in carting and spreading—serious items, if we consider that thirteen loads of covered dung are equal in effect to twenty loads from open yards. An open yard should have a water-tight floor, either quite level or sloping slightly towards the centre, be of moderate size, capable of holding some six or eight beasts, and the hovel properly spouted. Under such conditions, if we are careful about the litter, good though bulky manure may be made; and this is consolatory for those who for many reasons may be obliged to adhere to the old system.

It is evident, then, from what has been advanced, that the amount of rainfall materially influences this question; that it will be much easier to make good manure on the eastern side of England, where the rainfall does not exceed twenty inches, than in the west, where we have frequently double that quantity. Consequently, we should expect to find covered yards most in use on the western side of our island, especially as the larger proportion of grass land renders bedding a much more valuable commodity. This, however, is not so, as, with some notable exceptions, the movement has not progressed nearly so rapidly in the breeding districts of the west as in the feeding farms on the eastern coast. Now, why is this? Some may say *this* is an evidence of the superior intelligence and enterprise of eastern farmers, and that after a time their example will be followed. To such we would remark that in Norfolk covered yards have existed for many years, and in these railway days farmers are very soon alive to the improvements in other districts. We think, therefore, the more natural conclusion would be that the covered yard has not been considered to suit the circumstances of the case. Now, what are these reasons? We are going to hazard a rather strong assertion.

We believe *that covered yards have not been more generally adopted under an idea that they are not suitable for breeding animals,* and that as breeding is pursued generally in the districts we have named, therefore the system of covered yards does not find favour. If this can be shown to be prejudice then the difficulty would be overcome, and these very farms, often consisting of two-thirds pasture, present the exact conditions under which the greatest advantage would follow the adoption of covered yards. In such cases, straw is always a scarce article. To economise the straw, and at the same time make our manure to the greatest advantage, are objects for which much might be sacrificed. Breeding and growing animals require plenty of ventilation. The earlier covered yards were made too confined, and, consequently, young animals lost their hair, were made tender, and when sent to market, did not sell so well as those that had been in an open yard. Modern experience has modified our own convictions on this important question, and we are now satisfied that with ventilation without draft, and a temperature only 2° to 3° above that of the external air, animals can be wintered under cover without losing hair or being made less able to withstand sharp changes of weather, especially if a little common sense is used, and care is exercised as to gradually introducing changes. Of course we contemplate that the young growing animals shall have just the same opportunities for exercise in the covered as the open folds. We do not say but what the open yard is the more natural system, but, in order to make farming profitable, we must economise food and preserve manure from waste, and we are quite satisfied, from a large experience, that the objection against covered yards, as regards breeding animals, arises from prejudice, and defects of construction which prevailed in earlier days. There is one very strong evidence in favour of the views we are advocating, and that is, that we have never met with any one, who, having had the opportunity of practically testing the merits of well constructed covered yards, ever wished to go back to the original arrangements.

No doubt covered yards are specially suitable for animals that are kept entirely under cover from birth to death, and on arable farms where breeding is attempted, the produce being destined for the butcher at the earliest possible age, this system can be successfully carried out, and great weights attained at an early period. Again, when animals are bought in for winter grazing, the use of boxes, or even tying up under cover, is an excellent system, provided we have good ventilation. The farmers in the eastern counties, with farms chiefly arable, require stock in large quantities to feed. They produce a great bulk of straw, which must be made into manure, either passed through or put under the animal. Animals placed in roomy compartments, with plenty of air, at an even temperature, must thrive faster and be in a more healthy condition than when crowded up in a feeding house, with little head room, and often exposed to draughts. This is almost self evident. Neither do we agree with those who see an objection to covered yards, on the score that when straw is abundant, it cannot be consumed. Such an argument might have had weight in olden times; but now, with the pulper and chaff-cutter, a considerable portion may be advantageously passed through the animal. And, indeed, this economising of straw we regard as the greatest argument in favour of covered yards. If cattle cannot be purchased, sheep will feed admirably under cover, provided due care is paid to their feet.

We have been converted to the merits of covered yards by actual experience some years since, as will be evident from our remarks on this subject in the first edition. We then held quite different views, and we remember being present at a meeting of the Farmers' Club many years ago, when a Leicestershire grazier stated that he would give from 1*l*. to 2*l*. a head more for graziers wintered in open yards than for animals that had been entirely under cover, because they would thrive faster. This was no doubt at the time a just indictment, arising from defective construction and want of proper ventilation. Indeed, the success of the system becomes very much a question of good ventilation. This must be supplied by air admitted under the eaves and above the animals, whilst the heated air should have an exit through the roof, either at the centre, when the covering material is of an impervious nature, such as corrugated iron, or from every part, which is much the best arrangement when the covering material is porous, as is the case with open slating or tiles. The colder air comes in freely at the eaves, and, being heavier, sinks and forces up the hot air from the animal's bodies. We may usefully summarise the principal facts in favour of covered yards, which we take from a paper read at the Farmers' Club in 1885. The subject was considered under five heads: (1) Increased Value of the Manure; (2) Saving of Litter; (3) Economy of Food; (4) Details of Construction; (5) Cost and Return.

1. *Manures.* — Recent investigations prove that the fœcal material should be preserved in a fresh unfermented condition; that the best manure is that which consists of a mixture of the solid and liquid excrements, with just so much litter as will cause the liquid portion to be absorbed by the cellular structure of the absorbing material, thus offering the most resistance to the fermenting action of the oxygen in the soil; so that the ingredients are gradually rendered soluble and available as plant food, and that manure should be applied either to growing crops, or immediately before the sowing of the crop for which it is intended. These facts being admitted, it is evident that the former practice of keeping manures till a year old, carting into heaps and frequent turnings to promote fermentation, separating the liquid from the solid excrement, must be condemned as unscientific and wasteful. Loss from excess of moisture, causing drainings from the manure in open yards, is proportional to the manuring value of the food, the more liberal our treatment the greater the loss. If manure is properly made under cover, the whole of the soluble matters, which are the most valuable, are preserved. The manure under the weight of the animals is so compressed that it is not affected by the air, and only when moved is there any perceptible heat or smell. It is, of course, richer, because the proportion of straw to fœcal matter is so much smaller, and it is uniform in quality for the food used, because the proportion of litter is uniform. Many have failed of success from using an excess of litter, by which the manure is injured; inasmuch as air gets access, fermentation occurs, and the result is fire-fanged manure, with great loss of manuring elements and probable injury to the animals, who rest on a hot bed instead of a perfectly cool bottom. Well made manure in covered yards is considered to be weight for weight worth double that in open yards, supposing the same class of animals are kept on similar food; but, owing to the smaller quantity of litter employed, the weight per beast is reduced fully one-third. Writers have estimated the quantity of litter required both for open courts and covered yards far too high. Mr. P. D. Tuckett, at the Surveyor's Institution in 1883, " supposed that a beast fed all winter under cover would convert from 3 to 4 tons of straw into 16 cubic yards of manure. Mr. Moscrop in his Essay,

Vol. 1., N.S., R.A.S.E., 1865, considers that a beast in an open yard would take 40lb. of straw as litter daily—which, for a period of eight months, would be about 4 tons—and half that quantity under cover. As has been explained before, the proportion will depend greatly upon rainfall, but both estimates are much in excess of reality. Taking one beast with another, we believe that 20lb. a day in open yards, and half that under cover will suffice, and that the production of manure cannot exceed 1 ton and 1½ tons per month respectively. Dealing with mixed stock, the average weight of excreta and straw under cover would probably average 75lb. a day, which gives close upon 1 ton a month. If these estimates are correct, and assuming a period of eight months' feeding, we have:

	£	s.	d.
Eight tons of covered yard manure, at 7s.	2	16	0
Twelve tons of open yard manure, at 3s. 6d.	2	2	0
Total gain in manure per head	0	14	0
Add saving in carting, turning, and heaping	0	4	0
Saving in manure per head	0	18	0

As bearing out these figures, better crops of potatoes have resulted from ten cartloads of covered yard manure than from fifteen cartloads of equal weight of open yard manure from animals receiving similar food.

2. *Saving of litter.*—According to the figures given above, this, during a period of eight months, will average about a ton per beast. How much this saves in money value will depend upon the market we have; but, estimating only consuming price, it cannot be less than 1l. a head.

3. *Economy of food.*—Upon this important question we have no exact experiments to rely upon. Mr Moscrop states in his essay, "That he proved in an experimental trial, that under cover, animals, each of which had a separate box, gained as much weight with something under one-eighth less food as others fed with the same description of food, but kept in the common form of court and shed, where the open part bore to the shedding the proportion of four to one. The gain was nearly 1s. per head per week, which was entirely attributable to the superior warmth, comfort, and repose enjoyed by the cattle

under cover." Treating of a breeding farm, and animals of all ages from a year upwards, we consider that the saving of food for a given result will certainly average 6d. a head per week, which for eight months comes to 16s. Therefore the saving by covered yards in manure, litter, and food comes to

	£	s.	d.
Increased value of manure	0	18	0
Saving of straw	1	0	0
Saving of food	0	16	0
Total per head	£2	14	0

Enough has been advanced to prove that the use of properly-constructed covered yards is one of the greatest and most important of modern economies, which should engage the careful attention of all who are interested in agriculture.

4. *Details of Construction.*—In a short sketch like this we are unable to do more than very briefly touch upon certain general considerations. We have a considerable choice as to method of construction and materials. Galvanised iron has been largely used, and as the price of iron is so moderate it is quite possible that as regards first cost, such are not dearer than wooden roofs and open slating; but there are three objections which militate against iron. Ventilation is less perfect; the galvanising coating is liable to injury from gaseous emanations, both from the animals and the manure when being removed. Painting occasionally will diminish this risk, but adds to expense, and, thirdly, iron roofs are hot in summer and cold in winter. Ordinary framed roofs, with principals, purlins, and rafters, may be used for small spans, and are necessary when a heavy covering like pantiles or Bridgewater tiles are used, but there is a limit soon reached as to the spans of such roofs. For large spans up to from 50ft. to 60ft., a series of ribs combining the bow string girders with rafters at a pitch suitable for slates placed 9ft. apart, supporting purlins carrying two rafters 3ft. apart, on which strong laths, $1\frac{1}{4}$in. by $1\frac{1}{2}$in., are nailed at intervals of about 8in. to carry the slates, forms a roof which, very light in appearance, possesses great strength, owing to the distribution of strain over every part which may be judged of from the fact that for such a span the maximum scantlings of

any portion except the purlins need not exceed 4in. by 1¼in. Open slating is used as cheaper, lighter, and allowing more perfect ventilation, which is further secured by placing the plates on pillars such a height above adjoining buildings as will allow of an open space of from 7in. to 9in. A roof of this kind will cost from 5s. to 7s. a yard of ground covered. A still cheaper, but less durable, roof consists of light principals 14ft. apart, carrying purlins 4in. by 3½in., about 4ft. 6in. to 5ft. apart, supporting rough ⅜in. boards, which are laid from the ridge to the eaves, with one-sixth of an inch spaces between. About ¼in. from the edge of the board is a groove ¼in. deep. The boards are nailed to the purlins, but actual contact is prevented by the presence of three clout nails under each board, and thus any moisture which collects on the under side of the boards can run down the face of the board without injuring the purlin. It is found that by leaving the spaces between the boards, less rain penetrates than if in contact. The globules of water run down the open space without breaking. Only very fine mizzle or a small driving snow penetrates, and that only to a limited and not injurious extent. The cost of such a roof should not exceed about 3s. a yard of ground covered. Though not so durable as the first described, such roofs are well adapted for construction when the surrounding buildings are old, or by a tenant on a lease or who has a protective agreement from his landlord.

5. *Cost and Return.*—Taking a farm with mixed stock, 120 square feet is an ample allowance of ground space for each animal. Assuming a cost of 7s. 6d. a yard, we have a first cost of 5l. per head, and, charging for interest and repayment of principal 6½ per cent., we have an annual charge of 6s. 6d. per animal per annum. If the saving is as we believe 2l. 14s., then it is clear that great economy results from the use of properly-constructed buildings; and this will be in proportion to the humidity of the climate and the value of straw. It should be remembered that covered yards effect a great economy in ground space; so much so that when new buildings are made we believe the cost for a given number of animals will not be greater than with open courts and sheds.

So much for the great advantage of covered yards. All,

however, cannot have this luxury, and hence we must consider how buildings without these modern adaptations can best be arranged.

The want of proper accommodation for cattle is very apparent in many of our principal dairy districts, cattle being treated much in the same way that they were a century and more ago. As a rule, the management is bad, or rather, to speak more accurately, there is a total want of any in many cases, the animals being left out in the grass fields all the year round. The want of proper accommodation is often the cause of this, and the landlord is more to blame than the tenant. Without good buildings, properly fitted up, both the farmer and his stock are more or less at the mercy of the elements. The buildings on most grazing farms are quite inadequate to the wants of the cattle they are intended to protect. We believe that by a little trouble and a small outlay they might be made much more comfortable than they are generally found. No one accustomed to travel about the country can have helped seeing the mess a lot of store cattle make in a field, especially near the shed and gates during the winter months. These sheds, which are so common in grass fields, and which are intended to protect a lot of beasts in winter from the inclemency of the weather, may be made much more useful at little cost. They now consist simply of a shed with manger and rack, a food house in one corner, and perhaps a small rickyard at the back; in few instances do we find a yard attached, consequently the animals cannot in the wettest weather be prevented from poaching the land, thereby rendering its surface more like a ploughed field than anything else. We would propose to have yards made to these sheds and a gate attached, so that the animals can be shut in; if the yard be walled in so much the better, from its being so much warmer than if merely railed round. By a little contrivance the shed may be made available for two or even four fields.

The yard need not be large, and must be made in proportion to the number of animals that the shed will accommodate, so as to give them all room without the stronger ones knocking their weaker brethren about too much. Unless they are tied up, this cannot be helped, the underlings getting less food than the others.

We strongly advocate the plan of tying animals up at night and when they are feeding. A considerable saving of space will be thus insured, for, strange as it may at first appear, a shed and yard will accommodate more animals if tied up than if they be left to range about. There are several other points to be urged very strongly in favour of having all sheds made so that the animals can be tied up, besides the economy of space. The equal distribution of the food is of great importance, all faring alike. When the animals are loose the principle of might being right holds good, and the weaker ones necessarily go to the wall. The effects of a winter spent among a lot of stronger beasts will be seen long after turning-out time; and it is doubtful if it be ever got over, however well the animal may be tended afterwards.

On most farms where sheds are built out in the fields for the accommodation of cattle, straw will be found to be very scarce. There are few farms where there is not some; but what there is ought on no account, if properly harvested, to be used as litter. And now comes the question, What are we to use instead? Sand has been recommended very strongly by Mr. Brereton, in his paper on "Stocking Land," in the "Royal Agricultural Society's Journal." There may, however, in many cases be very great difficulty experienced in getting sand, and, even if attainable, the price may be too great to make it of service. The same may be said of sawdust, which will answer the purpose equally well. We are not now speaking of littering the yard, but the shed where the animals are tied up at night. The amount of litter required will of course depend upon the floor of the shed and the fall from it. If the fall be imperfect and it lies wet, more litter will be required; but if a proper fall be attained the animals ought to require little or none. Sand and sawdust are merely useful in so far as they render the surface more readily cleaned, and not for their actual properties as litter. Irregularities in the surface of the floor will always hold moisture, and otherwise conduce greatly to the discomfort of the animals. Care should therefore be taken to have the floor as smooth as possible, and to provide sufficient fall to carry away all the liquid manure. This manure need not be wasted; but it is not our intention now to discuss how any waste may be avoided.

The more impervious the floor the better, and for this purpose we recommend asphalte, a good foundation of rough stones or some similar material being laid first, and then the asphalte on top of it. As some of our readers may be ignorant how asphalte may be cheaply made, we will attempt to explain its manufacture. Take either gravel stones or broken flints of about half an inch in diameter, and have the dirt carefully sifted out; then spread them on a wooden board with side ledges, and pour gas tar on a little at a time, turning over the stones or flints with a shovel until every stone is wet with tar. They cannot be turned too much. Care must be taken that no tar is left besides what the stones take up. The next process is to add enough sharp sand to make the mixture of the consistence of thick pudding, turning it over as before. The floor on which the mixture is to be laid should be carefully levelled, and then it should be laid on from three to six inches thick. A little sand must be sprinkled over it to prevent the roller sticking to it, and then roll with a heavy roller. This can be done during the summer months, when the shed is not required, and when there will be every chance of the surface getting quite firm and hard before the cattle begin to tread it. It must on no account be used for a month, and should be rolled as often as possible until perfectly hard.

Before dismissing the subject of tying up cattle in what would otherwise be open sheds, we would remark that no divisions will be requisite between the animals beyond the posts to which the chains are attached. If wooden divisions be put up they would interfere with the utility of the shed if it should be required as an open one at any time; whereas, if there are no divisions, it will be almost in the same state it was in before being altered.

And now we come to the mangers and racks. The latter we believe to be a most useless appendage, and one that we expect to see every year less common. The manger, if properly made, will answer the purpose of a rack, if long hay or straw be given, without distressing the animal in getting its food. Cows are not intended by nature to eat their food as a giraffe does, and therefore they cannot with ease to themselves pull hay out of a rack placed on a level with their horns when standing upright.

Every animal should have a separate manger, or rather division of manger, to itself. For this purpose, upright partitions must be placed in it opposite the posts to which the beasts are tied, and again in the middle between them. In order to prevent the chaff from being thrown out and wasted iron rods must be driven through from front to back; three of these rods will be requisite, in order to prevent any waste at all for each division of the manger. Unless this precaution is taken, the cattle will be found to toss their food out with their noses, and especially will this be the case if there be any pulped roots in it. In searching for the pulp among the chaff they waste the whole mass. The mangers should be at least from twelve to fifteen inches deep, and should be of uniform depth back and front, and will be best on the ground or nearly—just high enough to insure sweetness and freedom from waste.

The yards, if they are already in existence, will be generally found to lie very wet from having been hollowed out somewhat every time the dung was removed. To obviate this evil we would recommend their being filled up to a proper level with burnt soil ashes—the scouring of ditches, tussocks, and, indeed, all refuse, properly burnt, will answer the purpose. If the cattle are allowed the range of the yard during the day, it surely is of moment that they should have a dry instead of a wet bed, and if we do not tie up the animals at night we may be sure that, however dry and warm the shed may be, the yard will still be the resting place of the underlings, and ought to be made as comfortable as possible.

When a plentiful supply of ashes can be secured the yard may be cleaned out to the bottom every year, or, indeed, oftener, but where they are difficult to procure we can see no reason why they should not remain down until enough can be burnt to supply their place. These ashes make an invaluable manure for root crops, and, indeed for any crop, and are not expensive in manufacture. On mixed arable and grass farms a plentiful supply can always be secured, and on purely grazing ones, if care be taken, enough can be got for the purpose. The cost of burning will of course depend upon the ease with which the materials can be got together, and upon the nature of the soil—strong clay or calcareous soils burning the best, sandy soil being very difficult to burn. In some cases coal or wood will be required to help the fires. These ashes, if placed in a yard, should be laid so as to give a fall on all sides, and will require a small quantity of straw on them to prevent the cattle sinking during wet weather.

We are no advocates for costly premises; whilst advising that good durable materials be employed, we would have the thing that was necessary at the least cost, consistent with durability. Bricks and slates and foreign timber will supply our wants in most cases, but we may be able to wall with rough stone; or, in districts where coals are dear, we may be driven to use a substitute for bricks. If the soil is a retentive clay, for instance, we can use clay-clumps with advantage—the process is very simple. The clay is spread on the ground and thoroughly puddled by horse-treading; short straw, couch grass, or any fibrous vegetable matter being added in sufficient quantities to insure cohesiveness. When sufficiently tempered, the clay is made into lumps in an ordinary mould, and stacked up to dry, the stack being protected from rain by thatched hurdles. In this way it soon dries and becomes fit for use. The lumps of clay are fastened together with a mortar made of clay, and known as pug; and, provided the eaves are made to overhang, such buildings are sufficiently durable. In chalk counties we have another substitute for bricks: the lower beds are best for this purpose. Though quite soft when first quarried, exposure for a few months causes the surface to become hard, and, if protected from the action of moisture by well-spouted overhanging roofs, such walls are permanent, and when jointed with black mortar have a pleasing effect. Lastly, where gravel abounds, we may make use of concrete, which is probably the most durable material, next to granite, that can be found. Moreover, the process is so simple, and the knowledge of how to prepare the materials so easily acquired, that ordinary labourers can execute the work, under the supervision of a carpenter, and the cost is not more than half that of bricks. We commend Mr. Tall's lecture, delivered before the Society of Architects, as worthy of careful perusal by all who are about to erect farm premises.

It is a great mistake to pinch ourselves for room, especially in buildings intended for live stock. Several evils arise. The health of the animals suffers from want of sufficient air, and their comfort is interfered with from the confined position in which they lie. The feeding alleys and dunging passages are so narrow that necessary work is either imperfectly done, or takes up more time than would be the case if the space was larger. It should always be remembered that economy in operations that have to be frequently performed is more important than in those which are only occasional; indeed, this consideration should influence the arrangement of the buildings in reference to each other. Ventilation, however ably managed, cannot overcome the evils arising from insufficient area. The change of air, instead of being gradual, as it should be, is so rapid as to cause draught, which is more or less unhealthy.

In designing buildings for the accommodation of cattle, regard should be had to the method of feeding: if it is proposed—as assuredly is desirable—that pulped roots and chaff be used, then a good root-house, with chaff chamber over, must be placed in such a position, and with such access to the principal yards and feeding-sheds, as shall insure the least expenditure of labour in feeding. A building, 18ft. square, or 18ft. by 20ft., 9ft. high to floor, and 6ft. above, making the walls 15ft. in all, will be sufficient for a farm of from 200 acres to 300 acres. The root-house must have double doors, large enough to allow of a load of roots being backed in and shot up, and the position of the pulper and the entrances so arranged as to allow of the maximum quantity of roots being stored. A sufficiently large pitch door in the chop chamber must be provided, so that both hay and straw can be readily delivered from a cart. Of course it would be convenient if the roof-space of adjoining buildings allowed of the storage of straw, but such a provision entails great additional cost, and there are other objections to the plan, such as increased risk of fire, and the possibility of the fodder being injured by the moisture arising from the animals below. We prefer that the materials should be carted as required: a horse and cart, and one man to assist the feeder, is all that is requisite. The hay would be cut from the stack, and the straw come from the Dutch barn. Both pulper and chaff machine can be driven by a horse gear, and, if the operations are performed singly, one horse-power will suffice. Even where steam power exists, we should advise the horse machinery because the pulp must be made daily, and it would not answer to get up steam for such work. An old horse should be told off for this work, and with carting the roots, getting fodder, and working in the gear, the time will be pretty well occupied.

The particular arrangement of the buildings will depend upon circumstances, which it is impossible to define, except we have the case before us. The feeding-house should join the root-house on one side. Supposing we tie up our cattle, which is the cheapest way of housing them; the house should be 18ft. wide, walls about 7ft. 6in. or 8ft. from floor; feeding passage, communicating with root-house, 3ft. 6in.; manger, 2ft.; length of beast space from back line of manger, 6ft.; width of gutter, 12in.; passage way behind beasts, 5ft. 6in.; stalls for two animals, 7ft. in the clear; one or more doors, according to the length of the house, opening into the fold yard, for the removal of manure. For growing stock, in the absence of covered buildings, open yards with wide shelter sheds are desirable; the sheds not less than 15ft. inside measurement, and provided with a brick manger, so that the cattle can be fed under cover, and, if necessary, tied up at night or at feeding time. These yards should not be too large; ten or twelve beasts are sufficient for a yard. A yard 40ft. square, with a shed 40ft. by 15ft., would secure ample accommodation for ten or twelve two-year-olds. It can easily be arranged that the shelter sheds should form one side of the yard, easily reached from the root-house. Two of such yards, at least, will be required. And the cowhouse, if dairying is pursued, may be placed on the opposite side of the root-house, and form part of a second yard. Supposing that we adopt the system of box-feeding, which has much to recommend it, the main objection being the heavy outlay according to accommodation provided, the most economical arrangement is a double row of boxes, each box 9ft. square, with a feeding passage in the centre 4ft.; thus requiring a building 22ft. wide in the clear.

The animals should be separated by posts and rails, which may be made to remove. This will prove convenient when the manure is taken out. If practicable, water should be laid on. Animals eating a full supply of sliced roots will not drink; but if fed on pulp, and especially if the weather be hot, fluid is beneficial. The water troughs, like the mangers, should be movable, so as to be easily raised, as the manure accumulates, and should be fed from a pipe above. It will be found desirable to have a round rail to slide up and down above the manger, which can be raised by the beast when feeding, but which protects the manger at other times, and prevents what otherwise is unavoidable, viz., the frequent presence of the animal's fæces in the manger.

The height of the building is important in connection with proper ventilation; many an otherwise good feeding house has been spoilt from being too low. We think if the walls are 6ft. above ground, and the building open to the roof, ventilation may be satisfactorily arranged for by leaving spaces in the walls under the eaves, providing wooden shutters so as to regulate the admission of air according to temperature, and by raising every fourth ridge tile, making it overlap, thus allowing for the escape of heated air, whilst rain and snow are effectually kept out. This is so simple and inexpensive, that we are surprised to see the plan so seldom carried out. The roof, whether of slate or tiles, should be rain-proof. The joints between the tiles must be carefully pointed. And here we may diverge from our subject to remark that the addition of gas tar to ordinary mortar immensely increases its adhering properties. Half a gallon of common gas tar to a bushel of lime makes a cement which resists the strongest wind, and such a mixture is highly suitable for bedding ridge tiles, or pointing inside or out.

The buildings thus described would occupy the centre of the plan, proceeding from the root-house and opening upon each of the yards or courts, so that the manure could be removed, and the litter supplied; this could be effected by means of slide doors, one door answering for two boxes. Thus, then, a simple and economical arrangement of buildings is suggested; cattle boxes down the centre, on either side of the root-house open sheds, the third side of the yards being made up with stables, piggeries, and cow-house. It is impossible, without detailed drawings, to give further particulars, but it will be evident to our readers that we look upon farming as a manufactory of meat, and must arrange our buildings so as to contain the requisite machinery and raw material, so placed as to secure the maximum result with the greatest economy. The arrangements for the preparation of corn for market are quite of secondary importance; all the space formerly devoted to the storing of sheaf corn can be dispensed with. Dutch barns and portable machinery effect the object in a much simpler manner, and a granary over the cart lodge should be useful, not for storing the corn, but for dressing it for market.

And now as regards the comparative advantages of boxes or stalls, for feeding animals, we think it must be clear that the loose box, if properly ventilated, will prove the most comfortable to the animal, as offering less contrast to life out of doors than the stall. The young animal has room to move about, lie down, and rise up in any position, and freedom to rub—a point of considerable importance. When once settled, beasts thrive better in boxes than in stalls; and young animals especially, which grow their flesh as well as lay on fat, must have freedom to move about. The box system tends to a cheerful, contented state; the beast can see and even lick his neighbour, if inclined; there is no hindrance, though possibly *a bar*, to intercourse; and, thus being in company, the beast rests better than if quite apart. The minimum size of a box should be 9ft. by 9ft., the depth of the box from the surface not to exceed 2ft. Formerly it was customary to have the pits 3ft.; but this is a mistake, as it is inconvenient for getting the beasts in and the manure out; and, with care as to the supply of litter, we shall find that the depth named will hold the manufacture of three months. Nine feet square will be sufficiently large for ordinary beasts; but we should prefer 10ft. by 9ft., giving the extra foot in width on account of the space occupied by the manger. The great objection to the box is the cost. Two animals can be stalled in the same space. This, however, may be partially remedied by having a double row and gangway common to both. If young animals are stall-fed, they should be allowed to stretch their legs once a day, or, if this is not practicable, the skin should be regularly brushed over daily with a strong brush. The irritation encourages circulation and exhalation, and the healthy state of the skin has much to do with progress. What a sanitary lesson cattle teach us, as they stand for hours exposed to rain, though a warm and dry shed offers a tempting lair. Nature points out the importance of this *al fresco* bath; the rain causes reaction and increased circulation; the pores are kept open. When housed, nature's flesh brush is lost, and we must try and compensate. In the box, they can rub themselves in every direction, and so do well.

Instances may occur in which it is desirable to feed all the year round. Mr. Mechi and others adopted this plan as the only means of making sufficient manure; but it cannot be recommended with any idea of profit. Our animals live far more cheaply out, either in the grass fields or in yards, and generally come in about October. They should be gradually accustomed to the change; a yard and shed for two or three weeks will be an intermediate arrangement that must be beneficial. Great attention must be paid to ventilation; the cattle house cannot be too cool at first. The sudden change induces sweating, and we have known cattle brought at once from pasture into boxes lose more weight by sweating during the first two months than they gained by food. We think in the case of very strong coats the scissors might be used with advantage; certainly it is very desirable to check excessive sweating. If first placed in yards at nights, then kept altogether in yards, and so quieted down and accustomed to change of food and manipulation, a considerable step will be gained, and progress in the boxes will be rapid and without check. Indeed, we do not think a beast should be in the box longer than suffices to fill it with manure, and then, walking out fat on the level, he makes room for a successor. Therefore the preliminary attention in the yards is most important, bringing them on through the first stages; so the feeder secures a succession of animals, picking out as a start the forwardest, and working on through his lot.

Inasmuch as the principal object in cattle feeding is the manufacture of manure, it may not be considered out of place if we devote a few lines to its consideration, not only as to the actual making of manure, but also as to its preservation. Much misapprehension has and does prevail on the subject, illustrating the old proverb, that "a little knowledge is a dangerous thing."

Agricultural teachers of early days have something to answer for, since they dwelt upon the volatility of ammonia until the puzzled farmer believed that every smell proceeding from his muck heap meant serious loss of valuable elements. We remember when this was such a well-established point, that an ingenious theory was devised to reconcile the good effects of manure spread out and exposed for weeks on clover leys, in preparation for wheat. According to this theory, the exposure of manure, especially in a decomposed condition, to the action of the air, would result in a frightful loss of ammonia; but the under surface of the clover leaves were believed to possess a power of absorbing the ammonia, and so collecting it ere it could be lost in the air. Such ideas have long since been dissipated under the searching lights of modern science. Dr. Vœlcker studied the practice of successful farmers, and experimented to ascertain the scientific explanations, which he presented to the world in some admirable papers published in the "Journal of the Royal Agricultural Society." It is a consolation to know on so good an authority that, with ordinary care and management, manure may be manufactured and preserved without very serious loss, and this without the necessity for a heavy outlay in covering over yards or making boxes. Let us not be misunderstood. There are great advantages, as we have shown, in making manure under cover; and were the manure the only question, we should strongly advocate the erection of covered yards, especially in situations where the rainfall is considerable.

We must see how far manure can be economically made in open yards; and this will depend in great measure upon the construction of the yard. All surrounding buildings must be carefully spouted; the yard should not be large—40ft. by 50ft. is a good size, sufficient, with a proper shelter hovel, to accommodate ten beasts; the floor should be made water-tight, and should slightly incline towards the centre; the gradient should be very small; the yard as nearly level as possible; the walls and shed sufficiently high to allow of the manure accumulating without risk of the stock knocking off the tiles of the shed, or jumping the walls. Thus arranged, the absorption of liquid is a question of litter. There will of course be times when, owing to rainfall, a constant supply will be necessary, and must be used, otherwise the beasts will suffer or the liquor will escape, carrying with it a certain though small proportion only of valuable matter. The ingredients of fresh manure are neither very volatile nor soluble, consequently the high-coloured liquid which oozes from the fold-yard often looks more potent than it is; still, it should be retained, and when proper care is exercised, first in the form of the yards, and next in the distribution of litter, this can be done. When voided, excreta are not in a state of putrescence, and if properly incorporated with litter and sufficiently compressed, little or no fermentation takes place in the open yard. This is shown by the action of the air upon such manure when removed from the fold and placed in a heap. The fermentative process commences at once, and a few weeks effect a great change in the bulk and appearance of the manure. The manure is in a comparatively fresh state until exposed to the atmosphere. We are led to dwell on this subject in consequence of some incorrect views, as we think, promulgated in a lecture delivered to the Bakewell Farmers' Club by Mr. A. M'Dougall, who quotes Sprengel as to the supposed loss manure sustains in its manufacture. If urine is left exposed to atmospheric influence, no doubt putrefaction occurs, and volatile ammonia results; but if the urine is either collected in a tank, or, which is better, allowed to run into the yard and become absorbed by the litter, there is no loss whatever, for reasons that will be explained.

Whilst, therefore, strongly recommending the use of disinfectants, such as the compound prepared by M'Dougall, and which comprizes sulphurous and carbolic acids, to sprinkle the floors of cow-houses, pig-yards, &c., we cannot agree in the necessity for its use, either in the shelter sheds or open folds; although possibly such compound might be valuable to dust into the manure heap when being turned. We have reason to believe that M'Dougall's Patent Disinfecting Powder is valuable in a sanitary and economical aspect, and no stock-keeper should neglect its use. All we maintain is, that, foldyard manure not being being in a putrescent state, such an agent is not required. Whilst on this subject, we may be pardoned for explaining the action of the disinfectant, which is a compound of two salts, viz., sulphate of magnesia and lime, and carbonate of lime, to which are added bone flour—*i.e.*, phosphate of lime—used to reduce the strength and add a valuable fertiliser. The gaseous products of decomposition are sulphuretted and phosphuretted hydrogen, either free or in conjunction with ammonia, though, as we shall see by-and-by, the latter is, when the decomposition occurs in connection with vegetable matter, provided for in another way. Sulphurous acid causes decomposition, being in itself also destroyed, the resultant products being sulphur, phosphorus, and water. Sulphurous acid further acts as an anti-putrescent by its affinity for oxygen, absorbing that gas and preventing other bodies combining with it. Carbolic acid is the most powerful anti-putrescent known, and at once arrests fermentation in a remarkable manner. Its regular and frequent exhibition, not only on the floors and walls of the cowhouses, but by impregnating the air by wet cloths, &c., did more to keep off cattle plague contagion than any other preventive measure, and there is strong evidence to show it was effective. The magnesia is introduced in order to form a double compound of phosphorus and ammonia, highly valuable as a fertiliser.

The great point necessary for the due preservation of foldyard manure is its compression; hence the advantage of small yards, into which the sweepings of the stable, cow byres, &c., are carefully spread. Now, for strong land, manure cannot be in too fresh a state, and therefore it may be removed at once from the yard and spread upon the land; but on the light land, and in order that the manure may acquire a uniform condition, it is necessary that it first undergo a fermentative process in the heap; and during this stage loss may be incurred. A certain proportion of runnings, more or less, will break away from the heap; hence it is well to make the heap upon some good absorbent. The best we know of is red clay ashes, broken to powder; six inches of such a bottom is very valuable. The heap must be carefully built, the sides kept as plump as possible, and the top after a few days may be lightly covered with soil. Some moisture is necessary to start the process, but our soil prevents the washing effect of heavy rains. In a few days fermentation commences, and proceeds rapidly. Sometimes, in order to secure great uniformity, the heap is turned, and during the process M'Dougall's powder might be freely scattered, as it would probably be useful in decomposing the compounds alluded to, and possibly saving a small quantity of ammonia—very little, however, as the bulk that has been set free in the decomposing process is already united with the products of vegetable decay, forming highly soluble but non-volatile compounds. These are liable to be washed away; hence the necessity for placing the manure on an absorbent, and the utility of the soil covering. It will be seen from what has been advanced that the danger is not from volatilisation, but from solution; therefore, the sooner decomposed manure is spread upon the surface, the better for its safety; once there, the drenching shower may work in elements which the grateful soil will give account of. The wind and sun can only remove moisture, and not injure.

In some situations, where the rainfall is considerable, and where litter is scarce, it will be found impossible to absorb the moisture as it falls, and in such cases a drain from the centre of the yard is required to convey the liquid into a tank, from whence it should be carted or pumped on to the land. Such liquid in reality contains very little manurial matter, because the ingredients of fresh manure are, as has been stated, not readily soluble. Still there must be a portion of the urine, and it is important that, whether more or less, such should not be wasted. We have no great faith in tanks or liquid-manure carts—the first are seldom water-tight, and the use of the latter is frequently neglected; and therefore we prefer, where it is possible, the process of absorption by litter. The efficaciousness of the latter could be greatly increased if reduced into short lengths, as the absorbing surface is thus greatly increased. In the case of box-made manure, the value of litter when in short lengths, is very much enhanced, and the expense of reduction will be repaid in the saving of straw and the more perfect character of the manure, which can often be cut out like cheese, and is, when thus made, exceedingly rich. With properly constructed yards, and due attention as to the distribution of litter, good manure may be made, and all the moisture absorbed—manure that, according to the proportion of fæcal matter it contains, will be found as valuable as that which is made under cover.

CHAPTER V.

DAIRY MANAGEMENT, THE MILK TRADE, ETC.

OUR INTRODUCTORY CHAPTERS would be incomplete except we touched shortly upon Dairy Management. First, as to the method and cost of feeding. This will depend upon the quality of the soil, the proportion of arable and grass land, and the quantity and description of stock kept. On grass farms the cow lives entirely on grass and hay. If the land is rich, two and a half acres will be sufficient—viz., one acre for grazing, one and a half acres mown, and the lattermath fed. Many authorites consider that the hay from one acre should supply winter's food for five months; but it is not enough for a full-grown Shorthorn cow, which, when not supplied with anything else, will consume two hundred weight a week, if allowed to do so. On poor land, four acres will scarcely suffice for summer and winter food; and the medium soils, which range between these extremes, will feed a cow on about three acres, always supposing that the cow grazes in the ordinary way. Probably the most economical method of consuming our food is to keep the cows tied up in a well-ventilated shed, and bring the food to them, mowing the grass as soon as it is sufficiently grown, and, in the case of arable land, depending upon successive crops of Italian ryegrass, vetches, trifolium, lucerne, cabbage, sainfoin, clover, comfrey, &c. In this way we avoid that waste which is more or less inevitable when cattle seek their own food. Moreover, the cow is protected from flies, and lies cool and quiet; whilst in the fields they are made half wild by flies, which, in a woody country especially, are very trying; and this continual irritation powerfully affects the secretion of milk. The objection to this plan on grass land is that the quality of grass frequently cut soon degenerates, unless it is constantly manured. The plan, as we have seen it carried out in Switzerland, is highly successful. The meadows are never grazed, but mown three and often four times in a season. The liquid manure cart, however, follows the scythe with great regularity. A similar plan is now followed in England where forage crops are grown for the cattle. Upon the farm of Mr. Hunt, of Ingatestone, in Essex, where about 150 cows are kept (Shorthorns, Jerseys, Guernseys, and Ayrshires) soiling is carried out in an admirable manner. Tares, sainfoin, lucerne, Italian rye grass, cabbage and sorghum saccharatum, are grown for summer consumption, and, upon our visit to the farm on June 21, the cows were getting lucerne, the second cutting being ready before the first was finished. The previous food was rye grass, which was also ready to succeed the lucerne a second time. The first cut of lucerne produced sixteen piled cartloads per acre, the second twelve loads, and the third cut eight loads. One and a quarter loads fed twelve cows per day, and, therefore, an acre kept them for about twenty-nine days. An admirable liquid manure system assists Mr. Hunt to obtain the whole of the liquid excrement of the farm, and this is regularly distributed on the forage crops intended for second cuttings. It may be assumed that cows will consume from 100lb.—Ayrshires, Jerseys, &c.—to 140lb. (Shorthorns) daily, and therefore, a fairly good estimate can be made of the capacity of land to carry dairy stock when its quality is known. Under all circumstances milking cows should receive cake or corn, when being soiled, as well for their own welfare, as for the benefit of the lands which will more readily respond to manure thus enriched. Our own experience largely bears out that of Mr. Hunt; and, it may be assumed that, where a farm is carefully cropped for successions of food between May and October, an acre and a half of good land will provide the yearly requirements of an average cow. While, however, the expenses of hedging and ditching, and the trouble of driving the cows to and fro are diminished, there is much greater outlay in cultivation, and in the incessant cutting and carting during summer. Under Mr. Hunt's system the meadows, together with clover leys, are reserved for hay for winter consumption in conjunction with silage and a few roots. The old plan of feeding cows entirely upon a grass farm, by grazing in summer and upon hay in winter, although simple and labour-saving, is much too costly as compared with the plan described above. Another plan which has been tried with the Jerseys consists in tethering the cows, shifting them at stated intervals, sometimes as often as three or four times a day. By this system the grass is consumed very closely, and with economy. Heavier beasts, like the Shorthorns or Herefords, will not do on this plan, as they could not get enough, and, moreover, would not eat clean. Tethering is most useful upon stubble which will not bear cutting, and which is generally folded where sheep are kept. It is, however, only necessary in utilising aftergrowth upon lands which are not fenced off.

In all cases where we have command of sewage on sandy soils, Italian rye-grass should be grown and cut from time to time for cows which are tied up. Very heavy crops may be produced, and a great weight of cattle fed. It must not, however, be forgotten that, as the milk produced by sewage grass is not so sound as that made from ordinary grass, great care will have to be taken in its manipulation as well as in the management of the cream. The object of most dairymen is to have the calf born early in the spring, so that the cows may be ready to yield their milk when required. In such cases the cow should be allowed to dry, when she comes in about November, and during the winter she lives on straw, with a few roots and a modicum of cake, and is thus maintained inexpensively. When winter milk is required, the cow must be calved early in October. In view of the low summer prices and the higher winter prices of milk, winter dairying deserves more attention than it has received. Modern feeding has shown that cows can be induced to milk almost as well in winter as in summer, taking the average of the six months, and while a high average flow is maintained by means of sound and careful feeding, it must not be forgotten that the cow, at the end of her milking period, comes upon the spring grass, which never fails to give a stimulus to her supply. This the spring calver misses altogether. Nor is there any compensating advantage derived from the calf, for that dropped in October becomes strong and fit for turning out in the following May, whereas the spring calf, under careful management, is not sent to grass until the same period. The difference in the results between winter and summer dairying rather depends upon the system of feeding adopted, than upon the difference in yield. A stock owner cannot afford to underfeed even a dry cow; hence, it is questionable, whether her ration cannot be so improved at a slight cost as to bring in a more profitable return. We proceed to describe what we have found to be economical feeding, commencing with that period, whenever it may be, that the cow is dried off in order to prepare for the next calf. With a few exceptions, cows should not be milked closer than within about two months of calving, as they then have time to get up condition, and the calf is always the stronger for this rest on the part of the mother. Occasionally we find cows that are such abundant milkers that they will yield freely to the last, and, indeed, require milking to keep them quiet. This peculiarity is, however, as rare as that of a cow we once knew of in Northamptonshire, which required milking three times a day—a plan deserving more investigation. So far, experiments have shown that three milkings give a better return than two During the interval between drying off and the birth of the calf, the food should neither be rich nor too invigorating. If in winter, a run in a well-sheltered yard, or else a turn out on pasture during the day, and a supply of cut straw, with a few roots, night and morning; in summer, a short, rather poor pasture, will be most suitable. It is the custom of some feeders to omit all concentrated food from the ration between drying and calving. As the cow declines in milk this food may be diminished, but whatever she receives at drying may be given until calving, unless she is a gross or fleshy beast. A milking cow is proverbially the reverse of fleshy, hence, the extra ration will enable her to make flesh with advantage; but, to feed a high conditioned cow at such a period, and induce her to make more flesh, would only endanger her life at parturition. As this period approaches she will require particular care; for, if mending too fast, the risk of uterine inflammation is considerable, and from such attacks cows seldom recover. Some recommend taking a little blood as a precautionary measure, but we are averse to this as a barbarous expedient, and altogether wrong in principle, and greatly prefer to give a mild aperient in some warm gruel. For many years we adopted this plan, and gave, about three or four hours after calving, a dose consisting of from ¾lb. to 1lb. of salts, 1½oz. to 2oz. sulphur, and a ¼oz. of powdered ginger. In the case of very heavy milkers, and where the secretion commences some time before the cow is due to calve, it is a good plan to milk the cow once a day, commencing about ten days or a fortnight before.

Supposing our cow well over calving, and the calf removed after a week or ten days, the cow will now require good feeding, and on arable or mixed farms we should recommend a variety of material, such as hay and straw chaff, with a small quantity of pulped roots, just sufficient to moisten the mass some 12lb. or 15lb. daily would do well. Swedish turnips are generally objected to, as influencing the flavour of milk, but if the heads are cut off close to the root this objection will seldom be found valid, in any case the quantity mentioned would hardly have any effect; at any rate it will be worth trying when roots are plentiful. We want in addition some cheap concentrated food or mixture of foods. Palm-nut meal or cocoa-nut cake, especially rich in butter-making ingredients, have proved valuable. A mixture of the two would not be amiss. Cocoa-nut cake we have used with advantage in conjunction with cotton cake, the yield of milk and butter being maintained equally as well as upon any other kind of food. The butter was particularly delicious in flavour. Brewers' grains, when cheap, are principally valuable as removing the dry character of straw chaff. Bran is another fine milk food, especially when used in conjunction with maize, rice meal, or any foods rich in fat formers. It is, moreover, an extremely cheap food at 4*l*. 10*s*. a ton. Maize meal again is most valuable, and should be purchased when maize is cheap (it has frequently been lower than 4*l*. 10*s*. a ton). It is better used in a ration rich in nitrogen, such as cotton cake, bean or pea meal, although compared with cotton seed meal, Professor Armsby, an American authority of the highest eminence, has shown, by a lengthened experiment, that it can hold its own. Rice meal is another valuable food when it can be obtained pure. It is, however, being so rich in starch, better adapted for a mixed ration, and for a change.

Ground oats—whole oats are not digested—are an admirable diet, and from their composition well suited for milk production, either on a butter or a cheese farm. Beans and peas cannot be estimated too highly; both are believed by cowkeepers to "force" milk; they are well adapted for use where milk is sold or cheese made, and in conjunction with potatoes, roots, grass, and soiling. Malt coombs are nitrogenous, closely resembling peas and beans in this respect, and they have an especial value in a ration on account of their agreeable flavour and sweetness. Grains are valuable as a substitute for roots or any other succulent food, but while stimulating in the production of milk, they are deleterious when given in large quantities, or for a lengthened period, and they are moreover an improper food for cheese or butter-making dairies. We have distinctly found that foods which have undergone a thorough fermentation have produced cheese which will not properly mature, and this is the experience of French cheesemakers who, in cases known to us, absolutely prohibit grains and distillery refuse from consumption by their stock. Butter made from cattle fed upon grains will not keep so well as that made from ordinary foods. The following will be found a useful ration for medium-sized cows, for a milk selling or cheesemaking dairy:

8lb. Oat Straw...	1¼d.
8lb. Hay, at 60s.	2¼d.
48lb. Mangels, at cost (10s. ton)	3d.
1lb. Malt Dust, at 100s...	¼d.
2lb. Cotton Cake, at 130s.).....................................	1¼d.
3lb. Maize Meal	1¼d.
	10¼d.

The ration is based upon the scientific formula of five parts of carbohydrates or fat and heat giving foods to one part (not less than 2¼lb.) of albuminoids or flesh formers. For cows of large size it would simply be necessary to add 4lb. each to the hay and straw ration, when they would be satisfied both in bulk and in necessary feeding value. Various changes could be made by substituting bean or pea meal for the cake, and malt dust or palm-nut meal and ground oats for the cake and maize meal. Silage of good quality could be used instead of mangels, taking care to add a little more straw chaff to equalise the properties of the ration, and in this case cocoa-nut cake could replace the cotton cake, more especially in a butter dairy, for which both are valuable. Bran could be substituted for the malt dust, and rice meal for maize meal. If the potato is used it will not be necessary to give more than half the weight usually allowed when mangels are fed, its feeding value being more than double as great. Considerable benefit will be found by slightly fermenting the food. A day's ration is, in this process, commonly mixed on a concrete floor, the pulped roots, chaff, broken cake, and meal being well mingled together, and allowed to stand twenty-four hours before using. It is then more easily digested and highly relished by the cows. Some feeders, however, prefer to cook the whole, and this plan is often adopted in Scotland where we have seen it produce good results. A large quantity of water is added and the ration is given hot, and half liquid in winter, being wheeled directly to the mangers from the coppers in galvanised iron food barrows. The late Mr. William Bowly, after careful trial, adopted the following plan: The quantities are given for fifteen cows; a furnace containing 70 gallons of water, the water hot to the boiling point; then meal at the rate of 10lb. per cow, to be well stirred in and boiled gently for an hour. Half of this to be poured over chaff (three bushels per cow), placed in a long trough for the morning's meal; the remainder being used in the same way in the evening. The chaff and soup are thoroughly mixed, and left for about half an hour to cool before being used. One great point is to have the mixture as fresh as possible; all food that has been cooked is apt to turn sour if kept beyond twenty-four hours. No doubt the perfume is fragrant, especially when the soup is first mixed, but the question is how far the nutritive properties of the food are increased. It would be very serviceable if Sir J. Lawes, that prince of experimenters, would test the point with two lean animals, supplying the same amount of food, and weighing and analysing the excreta. If the digestive process is really so much assisted as some suppose, then a large proportion of nutriment would be extracted, and the dung would be poorer; the progress of each animal being ascertained by frequent weighing. Another plan is to place the chaff in a bin, and pour the soup over it in layers. The heat is thus kept in longer, and the fermentation is possibly greater. Lastly, waste steam is frequently passed through the chaff, roots, and meal, and the whole partially cooked. This is particularly desirable in case the fodder is inferior; mouldy hay is much sweetened and made more palatable. Food so treated must be presently used, or it turns sour and is not so readily eaten. The success of the cooking system depends mainly upon good management and constant supervision. Men are apt to become careless, and omit to thoroughly clean out mangers or coppers, or allow the food to become sour, and thus upset the appetite.

The chief point in this question of cooking is to consider whether or not the cows give an increased supply of milk, or whether they will maintain their usual supply upon a less costly ration. If so, is the increased return sufficient to cover the extra cost of labour and fuel, and to leave a larger profit than is available when cooking is not resorted to. It is also well worth considering whether the cows themselves are in any way constitutionally affected.

The cost of a ration for cows of and above medium size is generally from 1s. to 1s. 4d. per day in winter. These sums provide a liberal diet, and one which is a valuable assistant in the manufacture of good manure, and consequently in producing larger crops. Where a number of cows are kept the cost of attendance should not exceed the commonly accepted sum of 1s. a week. The tendency in feeding should always be to diminish the use of hay and roots, even though more artificials have to be purchased, as this encourages a corresponding diminution in the cost of labour and an increase in the productiveness of the land. As the cow loses her milk the quantity of artificial food may be reduced, and if she is grazed, during the summer, half of the year, she will probably not cost more for pasturage than 3s. 6d. a week or 4s. 6d. This sum, together with the artificial food she has received and the ration for the six winter months at 1s. a day, would bring up her total cost for a year to about 16l. 5s. This amount does not include straw for litter, but we are of opinion that the use of straw for such a purpose should be greatly reduced or entirely given up. Short hard earth standings and deep gutters behind the heels of the cow, with a plentiful use of dry earth or peat litter when it can be obtained direct from the land, will largely help to prevent any difficulty with regard to cleanliness, and will, moreover, save the whole of the liquid as well as the solid manure.

The next point, as to what may be considered a fair produce of milk and butter, is difficult to answer, depending so entirely upon the nature of the food and the peculiar disposition of the animal. In all such questions we must endeavour to get at something like an average result; and we have no hesitation in stating that well-managed cows should yield from 550 to 650 gallons of milk annually. Yorkshire shorthorns have been known to produce 1100, and Ayrshires as much as 850 gallons; but these are exceptional cases. A number of well-bred shorthorns belonging to Lord Warwick were not long since reported to have undergone a test of twenty summer weeks and to have averaged between nineteen and twenty quarts per head per day. It is not now a very difficult matter to find dairy shorthorns which will reach twenty quarts daily soon after calving. Shorthorns, however, have the reputation of yielding milk which is not rich in butter, 1lb. to twelve quarts being the commonly accepted ratio. The cream separator, however, has altered these figures. During the winter of 1885-6 Col. Curtis Hayward made a number of experiments. His herd consists of about fifty cows, two-thirds dairy shorthorns and one-third Jerseys, and during the worst months of the year, by means of the separator, he obtained 1lb. of butter from $18\frac{1}{4}$lb. to 22lb. of milk (a gallon is about $10\frac{1}{4}$lb.), the quantity varying each week. The writer made a similar experiment in April with a number of newly calved shorthorns, yielding from sixteen quarts (heifers) to nineteen and a half quarts (cows). Notwithstanding the heavy milking of the animals, they made 1lb. of butter to $18\frac{1}{4}$lb. of milk. The Ayrshires and the Kerries do not equal this, nor is it exceeded by the Red Poll or any cow but the Channel Islander.

The question arises, What would be a fair profit under good management? The value of a cow's produce may be put roundly at a sum between 17*l.* and 20*l.* We are not now speaking of the milk trade in the neighbourhood of large towns, because, without any addition from the cow with the iron tail, new milk often makes 8*d.* to 9*d.* a gallon wholesale, which leaves an average of 22*l.* to 23*l.* a cow. In such cases the cost of keep and attendance is greater than we have calculated, while additional outlay is necessary in the purchase of artificial manures to replace the phosphates removed from the land in the milk. But, assuming that the milk will yield, whether for butter or cheese, about 6*d.* a gallon, we arrive, with the calf, at an average of about 17*l.* to 20*l.* The cost of keep and attendance being deducted, we have, barring accidents, a fair net return, which varies with the cost of feeding and the breed of the cattle in addition to the value of manure, which is of considerable value, although difficult to calculate.

It may be asked whether, with cheese and butter at their present price, it will pay to give dairy cows corn or cake when out at grass. This important question has to a large extent been solved by farmers in the more famous dairying districts where bone and other manures are being more generally used.

Arable land is yearly improving in condition under better and more liberal management; but grass land, and more especially that devoted to the feeding of dairy and store stock without the addition of artificial food, is in many counties either at a standstill or is deteriorating, and every year becoming less productive. Artificial feeding will gradually remedy this state of things; and it cannot be too generally adopted, for it is as valuable for the cattle as for the land itself.

There are exceptional instances of pastures so naturally rich, having such an inexhaustible store of materials, that foreign manures are not necessary, and indeed the use of extra food would in such cases generally result in the accumulation of fat instead of milk. But upon all the medium and poorer soils cows will pay well for high feeding. Cows vary in their capacity, both for milking and laying on flesh. Some—and they are always the most useful to the dairyman as long as they are in milk—cannot be made fat. Like Pharaoh's lean kine, they swallow all before them; but, unlike them, they give a good account of their food. Such cows *must* pay for a reasonable amount of good food. Both the land and the milk pail must be enriched when it is certain that the animal is not making flesh. When cows are brought night and morning to the homestead and tied up for milking—a plan which, in all cases of central buildings, commands our warmest approval—there can be no difficulty in supplying each animal with the quantity of food proportioned to her capacity as a *milker*. In this way all have their proper allowance, which would not be the case if fed in troughs or cribs in the field. Moreover, there is a saving of labour and of loss to the food from exposure to weather.

In the next place we have to consider the arrangements of the dairy and details of management.

A regular and easily regulated temperature is of greatest importance in the milk room; hence the dairy should face north, and be sheltered from the south. It must also be well ventilated. This can be secured by introducing air passages in the walls near the floor, carrying the walls up, introducing under the ceiling a row of ventilating bricks, and having every space occupied by a movable casement covered with perforated zinc. We thus keep the temperature equable in summer, while in winter a hot-water apparatus, with circulating pipes, secures the requisite warmth. This plan is seldom adopted in England, although it is the most commendable of any in existence. A dry atmosphere is desirable; hence, sinking the dairy beneath the surface, thereby insuring dampness, is not recommended, though often practised. We should prefer taking the ground level, building the walls with a damp course, and laying the floor in concrete. A good drain round the building is desirable, but there should be no drain within. An open gutter to carry off water to the outside is all that is necessary. If the bricks are machine-made, and sufficiently even to make a good face, we should make the walls at least 14in. wide, with a hollow space of $4\frac{1}{2}$in. inside (the bottom ventilators may pass diagonally through this space), and cover with the finest cement, finished off with a steel float. Paint, although it makes a good finish, should never be seen inside a dairy. It should always be remembered that milk is subject to fermentation, and that it turns sour rapidly; hence, both in the dairy fittings and utensils, we should have surfaces into which moisture cannot soak; hence the objection to bricks, soft stone, and even wood, although this may be used with greater safety. Where outlay is no object, the walls are often covered with glazed tiles, which look remarkably well, are easily cleaned, and answer every purpose. The glazing must be well done, and every window should be made to open with ease to any distance, and furnished with a duplicate in wire of the finest mesh. The shelves, or dressers, on which the milk bowls stand, may be either of slate or wood. We much prefer the former; but the latter may be adopted on the score of economy in any dairy which is thoroughly dry and well aerated. Under these conditions the scrubbing brush will do all that is required to prevent spilt milk being absorbed. Slate looks very neat, and is easily cleaned; but the slates must be in one piece, and not jointed, or only in such a manner that the joints are moisture-proof.

We have said that the walls should be high, and the roof covered with red tiles, if straw cannot be used; the latter insures the most even temperature, and is the coolest in summer. Black slates should be avoided, or, if they must be used, straw should be packed underneath; but this is objectionable, as harbouring rats and mice. We have a considerable choice of cream-raising vessels; it depends, however, upon the size of the dairy and the arrangement of the milk room as to which system is selected. Shallow vessels are not suitable in a damp milk room, or one which is very hot in summer or cold in winter; they are only adapted to an equable temperature. On the other hand, deep vessels are more adapted to meet changes of temperature and damp dairies, doing their work well, provided always that water of at least 45° can be secured at all seasons. It is true that cream can be raised in shallow vessels which have jackets for warm or cold water, but these partake of the features of the shallow and deep systems without the advantages of either, and are not to be recommended. A large dairy of milk can be placed in a very small space in deep cans, and the cream all raised in twelve hours, even when shallow pans require thirty-six to forty-eight hours. In large cheese or butter dairies the cream separator is preferable to either systems, and where this is used the dairy should be constructed upon a slightly different plan. Glazed earthenware pans are cheap, and answer well, provided the glaze is even and good. Unglazed pans or vessels of wood are most objectionable, for reasons stated; glass or white porcelain are clean, but expensive and liable to break. Tinned iron pans are now more generally used, and they are cheap, strong, and satisfactory in use. They should not exceed 2½in. depth, as a loss of cream ensues if the milk stands in a deeper vessel. Another advantage in using shallow metal vessels is that the temperature of the milk is more rapidly reduced, especially if the vessels stand upon slate or stone shelves. It is very important that this reduction of temperature, say from 90°, at which it comes from the cow, to about 60°, should be as rapid as possible. It is, however, this feature that makes the deep-setting system so valuable. A rapid descent from 90° to 45° causes the cream to rise as fast as possible. The fatty elements, having a less specific gravity than the other portions of the milk, rise the more quickly, when the fall in temperature is greater and more rapid, because the fat feels the change of temperature first. Cream raised on this cold system is always thin and larger in volume than that raised in any other manner. Shallow pans, however, maintain a good consistency in cream only when the temperature is equable and temperate, and when this is the case the cream rises tolerably quickly and before the skimmed milk is sour. By the deep system, as by the separator, the skim milk is always sweet and sound, and this gives it a special value.

The invention of the cream separator marked an era in dairying. By means of this machine milk brought in from the cowshed at six o'clock in the morning can be deprived of its cream, and sold sweet for breakfast, while the cream itself can be churned and the butter made up for the same meal. The skimmed, or as it is now called, the separated milk, is peculiarly sweet and light from the aeration it has undergone, and for this reason, as well as on account of its freshness, it ought to prove an article of general and ready sale. In practice the cream, however, is not churned for at least twenty-four hours after separating, as it is too new. Sweet, fresh cream, neither makes so much nor such good butter, as cream which has been ripened. In the ordinary way, cream skimmed from the pans is put into a cream pot, and there it remains, every subsequent skimming being thoroughly mixed with it by stirring until it is ripe, *i.e.*, until it has commenced to sour. Souring, however, must not go too far. Ripe cream produces butter with a fuller and better flavour, keeping longer, and showing a much better per-centage to the milk than sweet cream. For this reason, where a separator is used, a small quantity of sour buttermilk should always be added to the cream in such quantity as will produce the proper degree of ripeness or sourness in twenty-four hours, the most convenient time in a large dairy. Although we have seen no less than eight different cream separators at work on the Continent—all the inventions of foreigners—there are only two which are in general use: Petersen's, made in Copenhagen, and the Swedish, or Laval, both of which are sold in London, and extensively used in this country. The system of the Danish machine is that the milk caused to revolve in a horizontal drum, at a speed of 3000 revolutions a minute, is divided by centrifugal force into two portions. The lighter portion, the cream, comes to the surface, *i.e.*, the face of the vertical liquid body. A skimming tube is fitted to plough this face, with the result that the whole of the cream passes through it to the outside of the machine. The skim milk is behind this cream column, and, to reach it, a flange encircles the drum near the top, space being left behind to enable the milk to escape upwards, between the flange and the periphery. This it is compelled to do by the pressure of the constantly entering new milk. Above this flange is another ploughing tube, which takes off the milk in the same way as the cream, with this difference, that it can be moved to take as much or as little as required. This regulates the thickness of the cream skimmed. When the skim milk tube is skimming as much as is entering the machine—no cream is being skimmed—it is, however, collecting, and comes off when permitted unusually thick. The Laval machine works by similar force, but, although neither milk nor cream are actually skimmed, both find their way out of the machine after division by the pressure of the new milk which is constantly entering. This machine is made in only one size, and revolves at a speed of 6000 to 7000 revolutions a minute, skimming 30 to 45 gallons an hour. The Danish machine is made in three sizes, revolving at from 2000 to 3500 a minute, and skimming from 20 to 120 gallons. It has two other advantages, the small machine can be worked by a pony with great ease, and the milk can be elevated to a considerable height (into a floor above for example) as it leaves the drum. The speed with which milk is skimmed varies with the temperature of the milk; hence, the great advantage of separating soon after milking, and when the milk remains at about 75° Fahr.

Of the remaining system of obtaining the cream from milk, the most popular is the cold deep setting system, known as the "Swartz." This plan has many advantages where water of a temperature not exceeding 45° can be obtained. A milk room is not necessary, inasmuch as a large dairy of milk can be placed in cans standing in a vat or tank of water about 6ft. by 2ft., by 2ft. in depth. The cans are 16in. to 20in. in depth, by 6in. to 7in. in width, the ends being rounded or oval in form. The great secret is in subjecting the milk, when brought in hot from the cows, to the sudden change of the cold water, by this means the temperature rapidly falls, and the cream rises within twelve hours, in time for the same cans to be washed and used for the evening's milk. The system has many advantages. It permits of the cream and skim milk being obtained perfectly sweet at all seasons; being covered, dust and insects are kept out. The plant required is very small and lasting. Very little washing is needed; the cream is obtained quickly, and more butter is obtained in the year than from milk set in shallow pans.

The Cooley system somewhat resembles the above. The cans, which are round, and generally about 20in. in depth by $8\frac{1}{2}$in. in diameter, being set in a refrigerator filled with water at 50°. They are, however, entirely submerged, and, consequently, the milk is free from all atmospheric influence. In both cases it is usual to keep up a continual flow of water, and this is very necessary in summer, especially with the Cooley. The cream is not skimmed as in other systems, but the milk is drawn from under it by means of an ingenious tap at the bottom of the can. The disadvantage of the Cooley creamer is that it does not receive the beneficial influence of the air, and that it is, consequently, more difficult to ripen the cream properly after skimming. Like the Swartz, it occupies little space, is simple and practical, but it is useless to adopt either, unless either water, always sufficiently cold, or ice, can be obtained.

Another very important method of obtaining butter is by

churning the milk. This plan has constantly been condemned, yet it has many admirers who declare it to be the best and most profitable system. Care must, however, be taken that the milk is properly ripened, it should be almost a sour curd, and that it is churned at not less than 66° Fahr. We have made numbers of experiments to test the relative value of churning milk against churning cream raised by different systems, and although sweet milk produced less butter than any, yet properly ripened milk equalled every system but the separator. Quantity of butter may, however, be obtained at too great a sacrifice, and this is undoubtedly the case in this instance. The butter is more difficult to make up than cream butter, it is less marketable, being deficient in flavour and containing more water. It is necessary to churn daily, and in most instances to employ power; and last of all, the butter milk being absolutely sour, it is quite unsaleable, except in a few districts where it is a *specialité* among the poorer classes.

In the three western counties—Devon, Somerset, and Cornwall—cream is raised by a plan common only to this part of England. A shallow milk pan, similar in shape, but some 2in. deeper than the ordinary setting pan, is allowed to stand in the dairy for twelve hours, often twenty-four hours in winter; it is then removed to a stove, either on an iron plate or set in boiling water, and the milk scalded. Care is taken to prevent the milk boiling, although it may reach any point short of this. A temperature of 175° to 190° is, however, quite safe. The pans are then returned to the dairy and skimmed in twelve hours. The cream is worked into butter by hand by the majority of farmers, although some still use a churn. The plan is troublesome and has nothing to recommend it; the butter has a peculiar flavour, which is admired by those who are accustomed to it, but strongly condemned by experts and the trade in general. Cream raised in this way is very delicious, and is not produced so largely as it might be.

In making butter, provided the cream is properly ripened and churned at a proper temperature—58° to 61°, the former in summer and the latter in winter—it matters very little what churn is used, so long as it is of a good make. We should select one with a large mouth, light joints, and which can be easily cleaned. It is, however, useless to attempt to do good work in winter by churning in an apartment at a temperature of 40° or in summer at 80°. The cream will rapidly rise or fall as the case may be, and endless trouble will be the result. Churning should be regular, and neither fast nor slow, and care should be taken to stop when the butter " breaks," i.e., when it comes, as it does at first, into small grains. This is the time to draw off the butter milk and to wash the butter, taking care to move the churn so gently that, by successive washings of cold water, each grain is brought into contact with the water, effectually removing the milk and making it crisp. A few turns of the churn will then suffice to convert the whole into a lump ready for the butter-worker, upon which the remaining water is pressed out as far as is possible, leaving it ready for making up. Salting may be performed in three ways, the salt being added according to taste. Some prefer to salt the cream, a plan which is an excellent one, costing very little trouble; others wash the butter in the churn with salt water, also a good plan; and others again prefer to salt the butter itself on the worker. The last is the least successful plan of the three. In either case the salt should be perfectly dry and ground in a mill until it is nearly as fine as flour. In preparing cream for churning, it should never be heated suddenly, it may be placed over-night in a warm kitchen, and so gradually raised to the temperature required. For cooling cream there is no better plan than hanging it down a well, or partially submerging the vessels holding it in a cistern of water below ground, a plan common in Holland. In the heat of summer butter should be made in the very early morning, and after working, left in a cold place to harden. It should never be coloured or streaked, the former being artificial and the latter the result of bad workmanship. Coarse salt worked into butter which has not been deprived of the water is retained in patches, thus giving the objectionable appearance.

In a brief sketch like this it would be impossible to enter into a lengthened description of cheese-making. The late Dr. Vœlcker, who prosecuted careful researches at the request of the Royal Agricultural Society, published the result of his investigations in the twenty-second and twenty-third volumes of the Society's Journals. Anyone who is about to commence cheese-making will do well to study these papers and to serve a short apprenticeship in one of the best dairies in the country. Dr. Vœlcker was convinced that the food has undoubtedly an influence on the quality of cheese; but the method of manufacture is still more important in determining its quality and character. The chief points deserving study are the complete separation of the whey from the curd, the heating of the curd, and the development of acidity. Various makers adopt different methods of attempting to obtain the ends in view; some lose a part of the fat, which is carried off in the whey, while others, by a too early or too pronounced acidity, produce a cheese which does not properly mature. The first idea of a cheese factory was derived from America, where the system is general, and the great and rapid improvement that has been made during the last few years in the quality of American cheese is undoubtedly due to more scientific treatment and the development of the sweet curd system. The experiment in this country was tried first in Derbyshire, where factories at Longford and Brailsford have been in operation since 1870; these have since been supplemented by several others, including two built by the Duke of Westminster, and one, the most important of all, at Sudbury, the property of Lord Vernon, who takes a deep interest in the subject. Notwithstanding prejudice and trade opposition, these undertakings continue, although the result during the past year or two does not warrant their imitation. The saving of drudgery to the farmer's family in escaping from the details of cheese making, may endear the system to farmers' wives, whilst the economy of labour should give better results in the long run than could be looked for from private enterprise; but, so far, factory cheese in England cannot compete with that made by private individuals, who devote greater study and intelligence to the work.

Milk consists of casein, albumen, butter or fat globules, milk sugar, and mineral matters. In the preparation of cheese, the curd, a term applied to the solid portions of the milk, is separated by means either of lactic acid, which forms when the milk becomes sour, or by the addition of rennet. In skim-milk cheese-making the curd is almost pure casein, in whole milk

it is casein combined with the fat, and a portion of the mineral matter; whilst in the resulting fluid, known as whey, remains the milk sugar, the albumen, small portions of the fat and casein, and most of the mineral matter. The division of the curd from the whey is brought about by the use of rennet, a liquid prepared from the fourth stomach of the sucking calf, which is especially arranged by nature for digesting curd. The value of rennet is not in its mere capacity to curdle milk, for this phenomenon will result from the use of almost any powerful acid. It is in the fact that it is a digestive agent converting the raw curd into a mellow and deliciously flavoured cheese. The importance of the purity of rennet cannot therefore be too highly estimated. The Cheddar process, which is the safest and most largely followed, is thus described by Dr. Vœlcker:—
"Immediately after morning milking, the evening and morning milk is put into a Cockey's tin tub, having a jacketed bottom for the admission of steam or cold water. The temperature of the whole is slowly raised to 80°, by admitting steam into the jacketed bottom. The rennet is now introduced, the tub covered with a cloth, and left for an hour. If annatto for colouring is used, it must be added before the rennet. Good rennet should properly coagulate milk at 80° in from three-quarters of an hour to an hour. If the milk fails to be coagulated within the hour, the curd produced will be tender, and not easily separated from the whey without loss of butter; whereas, on the other hand, if the curd is separated in twenty to twenty-five minutes, the cheese is usually sour or hard. Great care should be exercised, in preparing the rennet so as to ensure uniform strength. At the end of the hour the curd should be partially broken, and allowed to subside for half an hour, after which the temperature is gradually raised to 108° Fahrenheit, the curd and whey meanwhile being gently stirred with a wire breaker, so that the heat is uniformly distributed and the curd minutely broken. The heat is maintained at 108° for an hour, during which time the stirring is continued. The curd, now broken into pieces the size of a pea, is left for half an hour to settle; at the end of this time the whey is drawn off by opening a spigot near the bottom of the tub. As the curd should be quite tough, no pressure is at first

requisite to make the bulk of the whey run off in a perfectly clear state. The curd now collected in one mass is rapidly cooled, cut across into large slices, turned over once or twice, and left to drain for half an hour. As soon as it is tolerably dry, it is placed under the press, and most of the remaining whey is removed by pressure. After this the cheese is broken, first, coarsely, by hand, and then by the curd mill, which divides it into small fragments. A little salt should now be added and thoroughly mixed with the curd. The next operation is vatting. The cheese vat, carefully filled with the broken and salted curd, is covered with a cloth; the curd is reversed in the cloth, put back into the vat, and placed in the press. The cheese cloth should be frequently removed, and the cheeses are ready to leave the press on the sixth morning.

This is the process according to the Cheddar system. If carried out as described, and after treatment carefully attended to, viz., the turning and wiping of the cheeses and the maintenance of the cheese-room at a temperature of about 62° F., the result will be a well-made edible cheese, varying as to quality according to the richness of the milk and the care which has been taken in its manufacture.

The cheese tub or vat now generally preferred and used throughout Cheshire is rectangular and running on wheels, and sometimes on a tramway. It is on the American principle, and is lined for heating or cooling with either steam or water. Before using rennet—and all makes differ in strength—an experiment should be made to ascertain the quantity required to bring the curd from a given volume of milk in a given time, a finely graduated measure being used for the purpose. It must always be remembered that the quantity of rennet used is in an inverse ratio to the time the milk is set for curd; thus, to bring curd in half an hour, double as much rennet would be required as though the curd were brought in an hour.

The principal British cheeses of Great Britain are the Cheddar, the Cheshire, the Derby, Leicester, Gloucester, Dunlop, and the Stilton. Of foreign cheeses, those chiefly appreciated in England are the Gruyère, the Dutch, Gorgonzola, and the Roquefort made from sheep's milk. This great variety—and all differ, in some cases in a remarkable degree—is the result of variations in the temperature to which the curd is brought, of the manner in which the whey is removed, and the curd manipulated and ripened. In addition to the above, among the numerous kinds of hard cheese there are a host of smaller varieties known as "soft" cheeses, on account of the delicate texture of their flesh, the whey having been extracted by gravitation instead of pressure. These are chiefly the produce of France, but their manufacture has recently been introduced into this country. The choicest varieties are the Brie, the Camembert, Coulommiers, and the Neufchâtel.

In dealing with milk as upon the farm where it is produced for sale in large towns, it should always be passed over a cooler directly after milking and straining, the change of temperature enabling it to travel and to keep better; and it is not needless to say that a cowkeeper should keep a journal showing the yield of each cow, details of his feeding, and the weekly or monthly results based upon it. At the present moment milk must reach a standard which is supposed to indicate its purity; but this is of very little value to the public, who can be sold $11\frac{1}{2}$ per cent. of solids with impunity, although all the best cows give from 13 to 14 per cent. On the basis of seven years' trials at the London Dairy Show, the cattle which are known to give the smallest quantity of solids per cent. (casein, fat, and sugar) exceeded $12\frac{1}{2}$, while the races of cattle giving richer milk gave $13\frac{1}{2}$. The public should therefore demand of retailers milk of a higher quality and pay a good price for it, or decline to pay a full price for an inferior article. There is much ignorance in connection with the constituents of milk, and when those interested have educated the consumer they will be in a better position to charge him higher for what they produce.

CHAPTER VI.

SHORTHORNS.

By JOHN THORNTON,
EDITOR OF "THE SHORTHORN CIRCULAR."

THE SHORTHORN BREED OF CATTLE may now be fairly called cosmopolitan. Its "habitat" is everywhere. From one small spot in Britain, its native home, it spread through this country till it is found from John o'Groat's to Land's End; in Ireland it prevails everywhere; to most parts of the globe it has emigrated; and an importation of animals by the Government of Japan shows that even the exclusive East is ready to accept the breed as an imprint for the native races.

The "art and mystery" of breeding has worked marvels upon our native breeds of cattle; and the modelling powers of man have been so exercised upon Shorthorns that the gaunt, ungainly form, which seems once to have characterised the race has been fashioned into a parallelogram of symmetry and beauty. There seems little doubt that, from time immemorial, the breed existed, as a local type, along the rich grazing valleys of the Tees; in the counties of Durham and Yorkshire. Noblemen and squires, with a thoroughly English love of good stock, kept up the herds, on their estates, with as much pride as their own pedigrees. Numerous are the local records of the excellences and feeding properties of these cattle; and of their capability of attaining enormous weight when at full maturity. Mr. Chas. Colling, of Ketton, County Durham—(a follower of the great Bakewell, and a man of great judgment and sagacity)—was the first to fix national attention upon the merits of the breed.

Residing within the district, he collected the best specimens together; and, by careful selection and in-and-in breeding to the blood of one cow, Favourite or Lady Maynard, reared a herd of fine cattle, which were shown to arrive at maturity at a much earlier age than had previously been seen anywhere. His brother (Mr. Robert Colling, of Barmpton) was also an eminent breeder; and bred upon the same lines. Indeed, both brothers, at first, used the same bulls. The former bred an ox of wonderful dimensions, whose live weight was 34cwt.; whilst the latter exhibited a white heifer which obtained hardly less celebrity. The bull Favourite (252) was the sire of both. They travelled throughout England, and were shown in London, as well as at the leading country towns. Their appearance, followed shortly afterwards in 1810 by news of the sale of Mr. Colling's herd at an average of 151*l*. 8*s*. for forty-seven head, brought the breed into general notoriety; and, from the beginning of this century, Shorthorns continued to spread rapidly, until they may now be found in every county in great Britain and Ireland.

The breed is distinguished by its symmetrical proportions, and by its great bulk on a comparatively small frame; the offal being very light, and the limbs small and fine. The head is expressive; being rather broad across the forehead, tapering gracefully below the eyes to an open nostril and fine flesh-coloured muzzle. The eyes are bright, prominent, and of a particularly placid, sweet expression; the whole countenance being remarkably gentle. The horns (whence comes the name) are, by comparison with earlier breeds, unusually short. They spring well from the head, with a graceful downward curl, and are of a creamy white or yellowish colour; the ears being fine, erect, and hairy. The neck should be moderately thick (muscular in the male), and set straight and well into the shoulders. These, when viewed in front, are wide, showing thickness through the heart; the breast coming well forward, and the fore legs standing short and wide apart. The back, among the higher bred animals, is remarkably broad and flat; the ribs, barrel-like, spring well out of it, and with little space between them and the hip bones, which should be soft and well covered. The hind-quarters are long and well filled in, the tail being set square upon them; the thighs meet low down, forming the full and deep twist; the flank should be deep, so as partially to cover the udder, which should be not too large, but placed forward, the teats being well formed and square-set, and of a medium size; the hind legs should be very short and stand wide and quite straight to the ground. The general appearance should show even outlines. The whole body is covered with long, soft hair, there frequently being a fine undercoat; and this hair is of the most pleasing variety of colour, from a soft, creamy white to a full deep red. Occasionally the animal is red and white; the white being found principally on the forehead, underneath the belly, and a few spots on the hind quarters and legs; in another group the body is nearly white, with the neck and head partially covered with roan; whilst in a third type the entire body is most beautifully variegated; of a rich deep purple or plum-coloured hue. On touching the beef points, the skin is found to be soft and mellow, as if lying on a soft cushion. In animals thin in condition a kind of inner skin is felt, which is the "quality" or "handling" indicative of the great fattening propensities for which the breed is so famous.

The Ketton herd was of this character; the knuckles, or shoulder points, however, were rather strong and somewhat more upright than in some modern herds. The red colour of some of the early cattle had a yellow tinge, and this hue may still occasionally be seen. Some of the early breeders sought to remedy those defects which were thought to exist in the Ketton stock; although the public approval, as shown by the 151*l.* average, and the 1000 guineas for the six-year-old bull Comet (155), testifies to their general merit. Mr. C. Mason, of Chilton, Durham, improved the shoulders of the cattle in his herd, although, perhaps, at some sacrifice of the hind-quarters; and Sir Chas. Knightly, in more recent years, was very persistent in his endeavours to improve the formation of the fore-quarters of the animals from which he bred. During the last quarter of a century, fashion has run high, and there has been a constant adherence on the part of most breeders to some particular strain of blood; until different types have been produced within the same breed. In some strains, style and elegance have been mainly aimed at; the beautiful head is carried erect, the horns incline upwards, whilst the body has become elongated, and the shoulders have somewhat retained their uprightness; the whole animal bearing a most stylish attractive look. Animals of this type generally possess good milking properties, an abundance of soft rich-coloured hair, and a thin touch. By other breeders greater massiveness, with sloping shoulders and a greater disposition to heavy flesh, have been studied and attained. The adjudications in our recent showyards have been made to animals of the greatest substance; i.e., to those whose form has preserved most closely the type of the earlier Shorthorns, without the roughness of their shoulders.

The system pursued by the most eminent breeders has of late years been to couple animals of the same strain of blood. Several breeders readily sold, before the recent depression, bull calves for 500 guineas each; and the enormous sum of 4500l. has been paid for a bull; which, by a long and successful career at the stud, was shown to have realised 7000l. from fees paid to his owner by other breeders who sent cows to him. Another old established herd finds customers to hire the entire number of bulls produced, at varying rentals for the year, from 100 to 300 guineas each. The most effective method of producing saleable animals still seems that of somewhat close breeding, or, as it is termed, in-and-in breeding, from a few families all near akin to each other. So long as thick flesh, size, and constitution can be maintained, there does not seem any valid objection to this method. A common plan has been to use bulls of one strain year after year, upon a herd originally of different blood. It cannot be said, however, that this practice has been altogether successful. Attention to pedigree is, to this extent, found to be as efficacious as attention to form without pedigree, that it has produced animals which fetched the largest prices; still, the warning must be given that breeding for fashion, and the stimulating effects of high prices, have had a strong tendency unduly to deter selection, and to prevent vigorous weeding out of inferior specimens; so that although restrictions in alliances may have had the effect of perpetuating what is termed "pure" blood, and of avoiding the deleterious effects of rash and injudicious crossing, they certainly have been relied on too far. Indeed, it may be accepted as proved that as soon as individual merit decreases in a herd, the close-breeding method cannot any longer be relied on to uphold selling values in the most fashionable tribes.

The Herd Book has been the mainstay of this carefulness on the part of the breeder. Brought out in 1822, by Mr. George Coates in his old age, it was continued by his son. After his death it was taken up by the late Mr. Henry Strafford, by whose persevering labours the series was conducted up to its twentieth volume. This was published in 1873 with a record of 32,898 numbered bulls, and a proportionate number of cows. In 1874 it transpired that Mr Strafford, from advancing years and decaying health, was desirous of retiring from a position which involved great personal labour. A committee of breeders was formed to consider the position; and a meeting was held July 1, 1874, in Willis's Rooms, under the presidency of the Duke of Devonshire, to consider what steps it was desirable to take. As one consequence, a Shorthorn Society was formed; which, among other duties, undertakes that of maintaining the Herd Book, and issuing a yearly volume in continuation of the series. This has regularly been done; and the latest (Vol. XXXI.) appeared in October, 1885; bringing up the register of produce to the end of the December previous, and containing pedigrees of bulls up to No. 52,382.

The value attached to the best specimens, for a brief time, became almost fabulous. The late Mr. Whitaker, whose sound judgment and modest opinions entitle his observations to the most sincere respect, says that in 1829, 2000 guineas would not purchase ten of the most select animals in the country. More than twice that price breeders have been known to steadily and calmly decline for one animal. Following upon the great expansion of trade—and apparently of wealth, which the prosperous years of 1851-61 witnessed—farming became even a more fashionable pursuit than it had ever been. Throughout the country, estates, new or inherited, were stocked with pure-bred cattle. The retired merchant, or professional man, amused his leisure hours with model farming, and found a pleasing relaxation in breeding Shorthorns. The demand, for a time, in consequence, exceeded the supply; prices rose, and breeders of the more fashionable tribes endeavoured to meet the demand by offering, at intervals, portions of their herds. Some of these made enormous averages. Pure specimens of bulls and heifers were seen to make, at public auctions, above 4000 guineas each. All previous results, however, were eclipsed by the memorable sale of Mr. Campbell's Duchess and Oxford cattle, in 1873, at New York Mills, in America. Here eleven females of the Duchess tribe averaged 4522*l*. 14*s*. 2*d*., the highest price being 40,600 dollars paid for 8th Duchess of Geneva, which was bought for Mr. R. Pavin Davis, of Gloucestershire. It, however, proved that the purchase was in excess of that gentleman's instructions, and the cow was resold to an American breeder, in whose possession she soon afterwards died. The Earl of Bective, the Earl of Lathom, and Mr. Holford were the other English buyers. Three Duchess bulls averaged 1638*l*. 15*s*.; Second Duke of Oneida, a remarkably fine animal, making 12,000 dollars. With one exception the Oxfords were purchased by Americans. Six females averaged 1087*l*. 10*s*., and the bull calves 396*l*. 16*s*. 8*d*.

The excitement to which the American sale had lent "wings" culminated in 1875. In that year two auctions of Shorthorns were held, which (without equalling in individual cases the prices at New York Mills Sale) exceeded it in sustained demand throughout. The first was on the 25th August, when a selection from the Earl of Dunmore's herd was sold in Scotland: thirty-nine animals realised 26,223*l*. 15*s*., or an average of 672*l*. 8*s*.; the second sale, when the entire and large herd bred by Mr. William Torr, at Aylesby, in Lincolnshire, was dispersed upon his death, a week later. Here eighty-four animals of all ages fetched the enormous total of 42,919*l*. 16*s*., or an average of 510*l*. 19*s*. These figures must be looked at as highest water mark; not only for this breed of cattle, but for cattle of any variety. Nothing to compare with these sales occurred before, or have been witnessed since. The entire returns from Shorthorns sold, by public sale, in the year 1875 showed that 2355 animals had made a total of 220,321*l*. 13*s*., or 93*l*. 11*s*. 1*d*. per head.

The preponderance of the breed at the meetings of the Royal Agricultural Society of England was, for a time, remarkable. The result of seven years, ending in 1862, was 702 Shorthorns against 211 Herefords and 357 Devons; and, for the ten years succeeding, i.e., up to 1871, the numbers exhibited have been 1476 Shorthorns, 574 Herefords, 472 Devons. At the leading markets and fairs (except, perhaps, in the south-west) Shorthorns everywhere comprise the majority of the cattle shown; and it is estimated that there are more Shorthorns bred, fed, and grazed in England than all other breeds put together.

The great milking properties of this breed have made the cattle equally serviceable to dairymen and graziers; indeed, a recent company in London was started under the name of the Royal Shorthorn Dairy Company. Years ago it was customary for droves of the ordinary unimproved Shorthorns to be driven on foot from the north to the south of England. Farmers would meet the droves on the road and buy the best animals; and, in this way, many capital stocks were early established in the midland and southern counties. Of late years complaints have been frequently made that Shorthorns are not good milkers. This may possibly have been correct in some instances, but it has arisen, not from inherent inability, but from the pernicious effects of forcing young animals into a condition of premature fatness. There are, however, always to be found animals of all strains of Shorthorn blood, which are capable, not only of making their own calves fat, but of giving several quarts of milk daily in addition; and of sustaining such high condition, all the while, as enables them to compete successfully in our leading show-yards. And, at a recent conference on dairy matters, held by the British Dairy Association, a paper was read by an experienced owner of a dairy for town use, who bore witness that nothing had made a better return, to a milk-seller, than his pedigree Shorthorns.

The greatest record of the Shorthorn, however, still is its marvellous efficacy in crossing and improving other breeds. In Scotland many of the native black herds have been crossed generation after generation; and the produce is accepted as the very best beef which comes to London. In Wales, the "coloury beast" (as the Shorthorn is called) has gradually worked upon the Castlemartins and runts until it outnumbers them; and "pure-bred" herds are to be found in the south, as well as the north of the Principality. The marvellous improvement in the

Irish cattle, by the use of Shorthorn sires, has become proverbial; and prices there for good yearling bulls have ranged higher, until quite recently, than even in England or Scotland. Even the Isle of Man boasts its pure herds and a 400-guinea heifer; whilst the Orkneys and Shetlands are not destitute of pedigreed bulls.

But it is to the New World that the greatest exportations have been made. America imported pure Shorthorns upwards of fifty years ago, and every year numbers of cattle leave our shores for Canada and the States. Spirit and enterprise have been rewarded, and the offspring of animals imported a generation back have of late years found their way back to our own herds. The vast area and rolling plains of the Western States are affording fine fields for grazing and breeding; and what has for years been done in Australia—where numberless bulls and also heifers have been sent—is now being practised in the Far West. Canada, with its fields five months white with snow, finds the purest pay the best; and one energetic Canadian in 1870 spent 20,000*l.* in importing pure animals, which have been highly remunerative. The enormous plains in South America have recently been supplied with a large number of bulls. Although the first importation was made as far back as 1836, no number was sent out until about 1880, but during 1884-5-6 immense numbers, and often at very high prices, have been exported.

New Zealand has also its breeders and importers; and, coming nearer home, we find France took, thirty years ago, some of our best cattle, and one of her first acts after the Franco-Prussian War was the importation of four pure-bred Shorthorn bulls. Belgium, for nearly half a century, has annually imported a large number of young bulls. In Germany several pure herds are to be found; and Shorthorn bulls have been used among the native breeds of Russia and Bessarabia, and even in Egypt. It is in the prepotent powers of this impressive race that its greatest value lies. Its adaptability to all climates and soils, its marvellous faculty of growing and fattening at the same time, its maturity at an age when other cattle are considered but half grown, its faculty of raising its own offspring with a bountiful supply of milk, ensure for it a great and permanent superiority. The Shorthorn has been called the "Universal Intruder." Wherever Britons colonise, the Shorthorn makes its home; and in many a distant land, where the English tongue is comparatively unknown, its influence is extending; and the Shorthorn undoubtedly is the chief means of transmitting, to other countries and other nations, that great national institution "the Roast Beef of Old England."

CHAPTER VII.

THE HEREFORD BREED OF CATTLE.

By T. DUCKHAM.

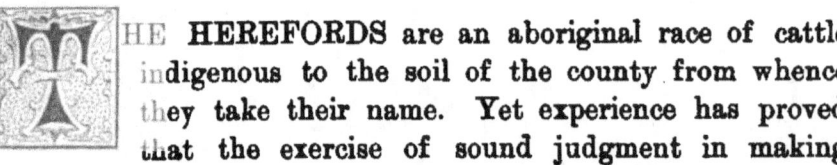HE HEREFORDS are an aboriginal race of cattle indigenous to the soil of the county from whence they take their name. Yet experience has proved that the exercise of sound judgment in making selections for breeding purposes is alone requisite to ensure the success of those who breed them in almost every known climate. They are of the middle-horn tribe, and have for ages past been highly esteemed for their fine quality of flesh, which, by the ntermixture of fat and lean, presents that marbled appearance so much prized by the epicure, and commands a top price in the market. The rapidity with which they lay on fat is certainly unsurpassed, if equalled. Experimental trials have been made with them and selected specimens of other pure breeds, which have ended in the uniform result that they yield the best return to the grazier for the food consumed. The value of the cattle of the district has been noticed by different writers for many centuries past. Speed, in his history of the county, says: "The county's climate is most healthful and temperate, and soyle so fertile for corn and cattle that no place in England yieldeth more or better conditioned."

The principal herds are in the hands of the tenant farmers of Hereford and adjoining counties, and have been handed down generation after generation from father to son in all their purity. The steers are looked upon as the rent-payers of the district, and perhaps no finer sight of cattle can be seen in the kingdom than that of the Hereford October fair, where several thousands line the streets of the ancient city, and by their distinctive marks and uniform appearance lay claim to each other as kindred of the same family. They have long been sought after by the graziers of the Midland and Eastern Counties.

Amongst those of bygone days the name of Westcar, of Creslow, Bucks (one of the most active founders of the Smithfield Club), stands pre-eminent. His forty years' attendance at the Hereford fairs was commenced in 1799, and his twenty first prizes in succession at the Smithfield Show was no mean achievement. The late indefatigable hon. sec. to the Smithfield Club, Mr. B. T. Brandreth Gibbs, in his tabulated statement of the prizes awarded, has shown that during the first fifty-two annual meetings of the club, when all breeds met in competition with each other, 185 prizes were awarded to Hereford steers or oxen, whilst only 190 fell to the lot of all the other breeds or cross-breeds put together.

The production of steers to meet the demands of the graziers being the chief aim of the breeders, and the well-known influence of the male animal for breeding purposes, has resulted in comparatively little attention being paid to the cow. It has been thought sufficient for her to possess the qualifications which long experience has proved to be necessary to ensure success with her progeny; her milking properties have been much neglected, and the calf usually allowed to run with its dam. Owing to the almost uniform adoption of this system, she has obtained the character of being a bad milker; but in other districts, where the milking properties are cultivated, it is not so, and, as her aptitude to fatten surpasses that of most other breeds, and she consumes less food in proportion to the quantity of meat made, she is gaining favour in many of the dairy farms of Dorset, Gloucester, Somerset, Cornwall, &c.

Duncombe, in his "Farming of Herefordshire," published in 1813, says: "Large size, an athletic form, and unusual neatness characterise the true sort; the prevailing colour is a reddish brown, with white face." But in 1845, when Mr. Eyton issued his first volume of the "Herd Book," he found it requisite to divide them into four classes—viz., mottle-faced, light grey, dark grey, and red with white face. As each of the three first-named classes are now nearly extinct, I will briefly notice the characteristics of the latter.

The face, throat, chest, lower part of the body and legs, together with the crest or mane, and the tip of the tail, a beautifully clear white. The horns yellow or white waxy appearance, frequently darker at the end; those of the bull should spring out in a nearly straight line from a broad, flat forehead; whilst those of the cow have a waive or slightly upward tendency. The countenance, pleasant, cheerful, and open, presents a placid appearance, denoting good temper and that quietude so essential to the successful grazing of all ruminating animals; yet the eye is full and lively. The head is small in comparison to the substance of the body; muzzle white and moderately fine; cheek thin; chest deep and full; shoulder blades thin, flat, and sloping towards the chine, and well covered on the outside with mellow flesh; kernel well up from the shoulder point to the throat, and so beautifully do the blades blend into the body that in a first-class well-fed animal it is difficult to tell where they are set on; the chine and loin broad; legs straight and small, the rump forming a straight line with the back; thighs full of flesh to the hocks; a well-sprung rib and deep flank. The whole carcase well and evenly covered with rich mellow flesh, distinguishable by its yielding with a pleasant elasticity to the touch, and a hide thick yet mellow, well covered with glossy hair, having a tendency to curl.

The "Herd Book" was commenced by Mr. T. C. Eyton, of Eyton Hall, Salop, in 1846, and the two volumes published by him contain the pedigrees of 901 bulls, but no cows. A revised edition of those volumes and seven others were published by the writer, who conducted the work from the year 1857 to 1878. At the commencement of his labours, he added that very essential part to all herd books, the pedigrees of cows with their produce, and illustrated the work with faithful likenesses of first-class breeding animals. The qualification for their being so placed was their having won a first prize at a show of the Royal Agricultural Society of England.

After the publication of the ninth volume he found it necessary to dispose of the copyright, and succeeded in forming the Hereford Herd Book Society. It was incorporated on March 5,

1878, when the copyright was purchased. The work is now an annual publication, and contains the pedigrees of 10,915 bulls, with a very greatly increased number of cows and heifers. As the cows with their produce are annually entered, a statement of their gross number would be delusive, seeing that there would be many duplicates. Some idea may be formed of the importance attached to it when it is considered that Volume 16, the last published, contains the pedigrees of 6300 cows and heifers.

About eighty years have passed since Mr. Grove first introduced the Herefords into Dorsetshire, and many valuable herds are now established in that county. The late Earl of St. Germains gave them a place on the west banks of the Tamar some sixty years or so ago, and several good herds now graze the pastures of Cornwall. They are fast supplanting the native breeds of the counties of Glamorgan, Brecon, Radnor, and Montgomery. The late Earl of Lisburne was foremost in placing them on the mountain sides in Cardiganshire, and the readiness with which they became acclimatised was such that his lordship's tenantry and others gladly availed themselves of the use of the bulls at Crosswood Park with the little black cows of the district. The result was most satisfactory, and the influence of the pure-bred sire is such that it is to be seen for several generations. The cross-breds are equally hardy as the natives; they feed more kindly, attain greater weight, and are more prized by the butchers. The late Mr. Lumsden, Auchry House, Aberdeenshire, first introduced Herefords into Scotland. Amongst his early purchases were some cows in the Hereford Fair and a 100-guinea bull, Matchless (415), at Mr. Hewer's sale in 1839. Mr. Lumsden made many trials with pure-bred Herefords, Shorthorns, and Aberdeens, and their various crosses, but always maintained the result was in favour of the "red with white faces." In 1870 the late Mr. Copland Mill, of Ardlethen, Aberdeen, wrote: "In regard to my experience of crossing with a pure-bred Hereford bull, I have now done so with from forty to fifty cows for the past seven years, and have no cause to regret it; for I find that my cattle are improved in weight, that they come sooner to maturity, and that their constitution is very much improved. So much is this the case, that a good many of my neighbours who were prejudiced against the Herefords are now breeding from them." In 1869 the Earl of Southesk established a choice herd in Forfarshire. He wrote of them that "the whole herd present a kindly appearance. A demand has been created for bulls at good prices, and those which have been sold have given great satisfaction. The generally acknowledged hardiness of constitution and fine character is well maintained and transmitted to their offspring." In consequence of the failing health of his lordship, the herd was dispersed in 1873. The late Prince Consort—that noble patron of all that was good and virtuous, whose desire for this country's greatness induced him, amongst his other pursuits, to become a pioneer in agricultural advancement—laid the foundation of the royal herd of Herefords at the Flemish Farm, Windsor, in 1855, and the marked success which has for several years past attended the exhibits from that herd at the various national shows is the best proof of the correctness of the judgment displayed in the selections.

Mr. R. W. Reynell, of Killynon, Westmeath, Ireland, whose herd was established a century or so ago, says that he has fed them with other pure breeds, and contends that the Herefords are the fastest feeders he knows, particularly on grass. The late Mr. P. J. Kearney was equally satisfied with the doings of his herd at Clonmel; and so was the late Mr. Gilliland at Londonderry, where it has been my pleasure to see them retaining their characteristics in all their excellence.

After the death of Mr. P. J. Kearney the herd was disposed of. There are now several important herds established in Ireland. Perhaps, for numbers and excellence, the herds of Mr. Fetherstonhaugh, of Rockview, Killucan, West Meath, and Capt. Purdon, Lisnabin, West Meath, are the most important.

Thus I have shown the satisfaction which the breeders of Herefords experience in their use from the extreme west and south of England to the north of Scotland, and from the south to the north of Ireland.

They were introduced in Jamaica by Mr. Malcolm in 1845, and in Trollope's "Travels in Jamaica" he says, "At Knockalva I looked at Hereford cattle, which I have rarely, if ever, seen beaten at any agricultural shows in England." The late Mr. J. Edwards, the manager, wrote of them: "The cross with the

Hereford bull and native cow is so direct that the bull carries all before him, and many of our half-bred cattle you would scarcely suspect as being any other than pure breds. Here we require a breed of cattle to be good workers, hardy, and of great aptitude to fatten, and I fear no contradiction when I say that no breed displays those qualifications in so eminent a degree as the Hereford."

Mr. F. W. Stone, of Moreton Lodge, Guelph, Canada West, has for many years been an extensive and successful breeder of Shorthorns; but, in addition to those, in 1861 he resolved to establish a herd of Herefords, and he writes that he believes the Hereford preferable to other breeds as grazers. Many valuable Herefords bred by him have travelled far away into the United States, and in various exhibitions, when competing with other breeds, have carried all before them. In 1876 the writer visited his herd, and was highly gratified to see how well the Herefords had acclimatised. At that time Mr. Stone had a very intelligent farm bailiff, a Scotchman, who had the charge of the two herds, Shorthorn and Herefords. Upon being questioned as to the suitability of the two breeds for that climate, he replied: "During my early life, when in Scotland, I knew nothing of any other breed than the Shorthorn, and my prejudice was naturally in their favour; but if I had to farm here upon my own account, I should keep the Herefords."

They were first introduced in the United States in 1817, and occasional importations followed. In 1876, when the writer visited the United States, the principal herds were in the hands of Mr. T. L. Miller, Beecher, Illinois; the Hon. J. Merryman, Hayfields, Cockeysville, Maryland; Mr. T. Clarke, Elyria, Ohio; Mr. W. W. Crapo, Flint, Michigan; Mr. H. C. Burleigh, Fairfield, Maine; and Mr. W. W. Aldrich, Elyria, Ohio. Mr. Sandford Howard, Secretary to the Michigan Board of Agriculture, in his report in 1868, gave a letter from the late Governor Crapo, in which he says: "The Herefords have done extremely well. They have no more than ordinary fair keeping, yet they are in prime condition. I have little doubt that the Herefords will yet be the stock for Michigan. They are docile and hardy, besides being very easy keepers, and I have no doubt will stand a long, severe winter, and come out ahead of the Shorthorns in the spring on two-thirds the cost of keeping. I intend, however, to give the Herefords, Shorthorns, and Devons a fair trial, both as full bloods and grades." Mr. W. W. Crapo has written me: "My father, the late Governor Crapo, of Michigan, and myself, have been for several years engaged in the breeding of Hereford, Shorthorn, and Devon cattle, having in view the testing of their relative merits. The result thus far is decidedly favourable to the Herefords."

It has been in the United States of America and Canada that the Herefords have found the greatest favour during the present century. In 1876 the International Exhibition of the Centennial at Philadelphia brought them prominently to the front in a manner they had never before attained in America. Some few years previously Mr. T. L. Miller, of Beecher, Illinois, turned his attention to the breed, and sent his then partner, Mr. Powell (a native of Herefordshire), to select some choice specimens, and amongst them he purchased the prize cow, Dolly Varden, from the herd of Mr. J. Morris, of Madley, Hereford, with her first calf at foot, a bull aptly named Success (5031). Mr. Miller also purchased several excellent cows from various breeders in America, and very quickly established a first-class herd of no mean proportions. His keen perception convinced him that the Hereford bull would prove an invaluable cross with the grade cows on the vast plains of Texas and adjoining States. They were first introduced there in 1867 by Messrs. Church and Prowers with most satisfactory results. In 1874 Mr. Miller sent as a speculation ten young bulls for sale there, they met a spirited demand, and such was the keen competition at the sale, that the following year he sent seventy-five with a similar result. Previous to the International Exhibition Mr. Miller had taken great pains in preparing a choice selection of animals for it, but he declared that he would not send them unless a judge from England was obtained to act upon the jury, and so strongly did he urge his views upon the executive, that a telegram was sent to the British Commission, which resulted in the writer's accepting an invitation to attend in that capacity. The following is a copy of the report of the jury:

"The Hereford cattle classes comprised many animals of high merit. The exhibits were from a very extended area, viz., Maryland on the south, Maine and Canada on the north, and Illinois on the west. Nearly the whole of the specimens were bred in those several districts; and it is most satisfactory for us to be enabled to report that we found they retained that fine character, form, and quality, for which they have for ages been so distinguished in the United Kingdom of Great Britain, thus proving that they readily became acclimatised; and we feel assured that they will soon occupy a foremost rank amongst the herds of the United States.

"We cannot close our remarks upon the Herefords without a brief notice of two pairs of gigantic working oxen from grade cows by a Hereford bull. They presented all the characteristic marks of pure breds. Their united weight was 10,199lb."

Feeling anxious to make the personal acquaintance of as many Hereford breeders, and see their herds, as he could in the limited time at his disposal, the writer travelled some 4000 miles, and visited nearly every herd at that time of any importance in America, including those from whence he had seen selected specimens. The object in view was to see how the Herefords acclimatised, and make the personal acquaintance of their breeders, whom he had for some years known by correspondence only. Several of the breeders had just returned from their state and county fairs (agricultural shows) flushed with the very marked success which had attended their various exhibits. At those gatherings champion prizes are offered for the best herd of any breed (a herd usually consists of a bull and four cows or heifers), and a very spirited competition is set up, as may be inferred from the fact that at the Michigan State Fair Mr. Clarke's success with his herd of Herefords, from Elyria, was declared against nine competitors, viz., seven Shorthorns, one Devon, and one Ayrshire herd. The triumphs of the Herefords that year for those important prizes were almost unbroken, and therefore it could be no matter for surprise that their breeders were somewhat jubilant at the very marked success which had attended their various exhibits. There were not any money prizes offered at the International Exhibition, merely certificates of merit, accompanied by bronze medals. This and other circumstances combined to render the number of exhibits there less important than at one time was anticipated, breeders generally preferring the more substantial awards of money offered at their state and county fairs. But the excellence of the specimens exhibited at the International Exhibition rivetted the attention of many of the distant stock owners who assembled there, and the correctness of the opinion expressed by the judges at the close of their report, that "they will soon occupy a foremost rank among the herds of the United States," has been very quickly verified. Such has been the demand for Herefords during the past eight years that many hundreds have annually been imported into the United States and Canada. The altogether unprecedented demand for them for America so popularised the breed at home that previously unheard-of prices have been realised for them. The value of the pure-bred Hereford bull upon the grade cows of the Far West having been fully established, a large trade has been done by the breeders in the Eastern States, whose extensive herds of imported Herefords have quite supplanted the more inferior herds that formerly grazed upon their pastures. Purity of blood, as proved by reference to the Hereford Herd Book, having been made a *sine quâ non* by purchasers on this side, numerous breeders who had previously not only refrained from entering their animals, but even ridiculed the value attached to pedigree, suddenly became impressed with its value and the importance of registering their herds; for this purpose their memories were severely taxed, memorandums, long since discarded, were anxiously searched after, that they might be enabled to so trace the pedigrees as to qualify them for entry in succeeding volumes of the Herd Book. Whilst that state of things prevailed on this side the breeders on the other side, with Mr. Miller for their guide, resolved to establish an "American Hereford Record," and under that title the first volume of their Hereford Herd Book was published in 1880. Three other volumes have since been issued.* The work is

---

* It is only common justice to Mr. Duckham's good work to state that the first Vol. of the American Herd Book contains his portrait, with a very complimentary notice of his labours in promoting the interests of the Hereford breeders.—ED.

very carefully compiled, and very beautifully illustrated. After the publication of the second volume, Mr. Miller, to whose indefatigable exertions the Hereford interest owes so much, resolved to place the work in the hands of an association of breeders. The Hereford Breeders Association was formed and the copyright disposed of. Mr. Miller is now its president. The members of the association soon became alive to the great change which had taken place with the breeders on this side, and at once passed a rule prohibiting the entry of any English-bred animal in the "American Hereford Record," whose pedigree is not clearly traceable "in the 13th or any prior volume of the Hereford Herd Book (English)." Had their prohibitory regulations ceased there, little or no just cause of complaint could have been raised upon this side, but last year another rule was added, which imposes an entrance fee of 100 dollars upon all cattle imported after Nov. 13, 1885. That heavy fee, added to the expenses attached to a ninety days quarantine, has materially checked the importations from England; in fact, it forms a prohibition for all except such as possess extraordinary merit, and thus a monopoly is created in a country that has, beyond all others, benefited from the Free Trade policy of this country, and is considered to reflect upon those who have been instrumental in establishing it. The Christmas fat stock show, established at Chicago in 1879, has greatly added to the interest manifested in the different breeds and their value as crosses with the grade cattle. In America they carry their researches much farther than in England. There, in addition to the close inquiry after age and register of weight, the carcases of the prize animals are judged and further prizes awarded according to the quality of the meat, the weight of offal, and the proportion of flesh to fat, all of which is carefully recorded. The adoption of that system has tended to emphasise the value that was previously attached to the cross of the Hereford bull with the grade cow, the flesh of the cross-bred animal displaying in a marked degree that beautifully marbled character so greatly admired by epicures, and so thoroughly characteristic of the Hereford.

Mr. E. Maclean, Butley Manor, Auckland, New Zealand, purchased a lot of Hereford cows in Australia, and bulls from

Her Majesty's herd. He writes: "I now breed about 100 calves a year, mostly Herefords, as I am gradually weeding out all others. I have bred Devons and Shorthorns mostly, the latter for the past twenty years in this province, but I much prefer Herefords."

Several other herds are now established in New Zealand, Mr. Maclean's being by far the most extensive. In consequence of the total prohibition of importations from Great Britain to Australia, selections have been made of animals from Mr. Maclean's herd for breeding purposes to meet the requirements in the sister colony.

Many large and valuable herds are established in Australia, and numerous importations from several of the best herds in England have been made by Mr. F. S. Reynolds, Tocal Patterson, the Hon. J. White Martindale, Hunter's River; Mr. J. Nowlan, Eelah, West Maitland; Messrs. Barnes and Smith, Dyraaba, Richmond River; Messrs. Livingstone, Learmonth, Ercildoun, Victoria; Mr. Angas, Angaston, South Australia; Mr. Robertson, Lake Colac, Melbourne; Mr. A. Bloxsome, Ranger's Valley, New England; Messrs. Mort and Co., and Messrs. Dangar and Co., Sidney; Mr. J. Price, Hindmarsh Island, South Australia; Mr. G. Loder, Singleton; Messrs. W. and F. Fanning, of Wooroowalgan, Richmond River, Mr. R. Wyndham, Leconfield, Branxton, &c.

The *Sydney Morning Herald*, Aug. 12, 1870, in its report of the Singleton Show, says: "A glance at the cattle pens could not fail to establish incontestably one fact of great importance to the cattle breeders of this district, viz., the marked superiority of the Herefords as contrasted with the Durhams (Shorthorns). They were not only most numerous, but in better order, showed better breeding, and were in every way superior to the Shorthorns. There is no mistaking a pure-bred Hereford." The same paper (Sept. 7, 1870), in an able article on the Agricultural Resources of New South Wales, says: "The debate concerning the merits of Shorthorns or Herefords is very strong. Both breeds have many advocates. It is generally admitted, however, that the Hereford travels better than the Shorthorn, and better endures periods of dearth and drought. A vast quantity of cattle of this colony having to travel 500 to 800 miles to the slaughterhouse, this quality is a consideration of the utmost consequence."

The *South Australian Register*, Oct. 11, 1870, in the report of the Adelaide meeting, says: "The first place in the programme is conceded to Mr. Price's imported Hereford bull. Never was priority of position more deserved." He was acknowledged to be the monarch of the yard. The *Maitland Mercury*, Aug. 17, 1871, in the report of the Singleton Show, says: "Such beasts as were exhibited are seldom seen in this or any other district; it was impossible to view them without thinking of the roast beef of old England;" and thus describes Mr. White's prize bull Prince of Wales: "With a coat of velvet, a mild, gentle eye, and quiet temper, that seems more like that of a lamb than that of a bull." And Mr. A. A. Dangar's colonial-bred heifer: "We think, in quality and condition, this animal is perfect." And recently, in commenting on the splendid white-faces at the sixteenth annual show of the National Agricultural Society of Victoria, calls attention to the fact that both Sir Roger and General Gordon, the first-prized, aged, and yearling bulls shown by Mr. J. H. Angus, were bred by Mr. Reynolds, at Tocal, who is spoken of as one of the most successful breeders of Herefords.

The *Australasian*, in the report of the Sydney Show, 1871, says: "The feature of the exhibition was undoubtedly the cattle, and, of the cattle, especially the Herefords. Without disparaging the exhibition of Shorthorns, it is but fair to state that they did not equal the Herefords; nor, if the exhibition of fat stock be any criterion of the success of the breeders, did they prove any essential adaptability to either climate or pasturage. Mr. White's pen of fat oxen were magnificent animals, and they appeared to assert their supremacy as being the primest of the prime; and the Hereford cattle will remain on the annals of the exhibition of 1871 as the main feature of excellence." A considerable number of Hereford bulls have recently been exported to South America, and their value as a cross upon the native breeds have been most satisfactorily tested.

In 1878, the importation of cattle from the United Kingdom of Great Britain to Australia was prohibited. For several years a ninety days' quarantine was imposed, but, in consequence of the continued prevalence of "foot and mouth disease" in Great Britain, the stock owners in Australia became so alarmed that they felt their only safeguard against the disease consisted in total prohibition. Its effect is seriously felt by our home and colonial breeders. Seeing that nearly twelve months have passed since a case of the disease has been known to exist in Great Britain, there can be now no valid reason why they should be longer subjected to the losses the prohibition imposes.

During the past decade the Herefords have been introduced into South Africa. The late Mr. Barry, of Cape Town, Cape of Good Hope, who was an ardent admirer of them, purchased a choice young bull from the writer in 1879. He is still doing good service, but, unfortunately, Mr. Barry was destined never to see the result. Mr. Sutherland, of Toise River, East London, also made some selections from the writer's herd in 1879, and in 1884 he wrote of them: "I have every confidence that they are the breed of cattle most suitable for this colony."

The Herefords are rapidly gaining favour in South America. Several herds are now established in the Argentine Republic, and a considerable number of bulls and heifers have been shipped to Buenos Ayres and Monte Video during the past few years, with most satisfactory results.

Wherever Hereford bulls have been used the influence of the purity of the blood and impress of character upon the grade cows have been uniform, and their value universally acknowledged.

CHAPTER VIII.

DEVON BREED OF CATTLE.

By LIEUT.-COLONEL J. T. DAVY,
LATE EDITOR OF THE "DEVON HERD BOOK."

THE BREEDS OF CATTLE reared on farms are very numerous, and often approximate one to another by a series of the nicest and almost imperceptible gradations. Where a breed has found a congenial soil and climate, it seems to flourish almost in spite of neglect. The early history of cattle speaks of three kinds, viz., the long-horned, found in the midland counties and in Ireland; the short-horned, in the eastern and northern counties; and the middle-horned, in the western part of England, in Sussex, and in Scotland.

The Devon breed belongs to the middle-horned variety, is evidently an aboriginal one, and there is little or no doubt that Devon, Hereford, and Sussex cattle, and probably also those of Wales and Scotland, were originally descended from the same stock. They have all the characteristics of the same breed, changed by soil, climate, time, and by being subject to man's will and control. These influences change the capabilities and characteristics of most breeds of animals coming under the denomination of stock. The late Mr. Youatt, in his valuable work on cattle, speaking of the skulls found in different parts of England, says: "There is a fine specimen in the British Museum; the peculiarity of the horns will be observed resembling smaller ones dug up in the mines of Cornwall, and preserved in some degree in the wild cattle in Chillingham Park, and not quite lost in our native breeds of Devon and Sussex, and those of the Welsh mountains and the Highlands."

The middle-horned varieties are fairly good milkers, but are remarkable for the *quality* rather than for the *quantity* of their milk, which yields a large proportion of cream and butter. As a general rule, the better the milking properties of cattle, the more are they disposed to internal accumulation of fat; and it should be understood that excessive accumulation of this kind are the farmer's loss and the butcher's gain.

It is not difficult to observe that the Devonshire and Sussex races are of the same extraction; so nearly do they resemble each other in colour and length of horn, that if anyone unacquainted with the distinctive features of those two breeds were shown two animals, one a Devon, the other a Sussex, he would find it difficult to detect any material difference between them, except that the Sussex beast might appear rather the larger or the "taller," from the greater length of leg. A more experienced observer would notice in him a less finely chiselled head, coarser eyelids combined with a less pleasing expression of the eye, and a crescent-shaped, upward horn, instead of the deer-like head and gracefully curved waxy-looking horn of the Devon. In their shape and size, as well as in the curve of the horn and the heavy eye, the Sussex cattle bear a strong resemblance to those formerly bred about Taunton, before the latter were so much mixed with Hundred Guinea (56) and other North Devon bulls.

Particular breeds and their varieties were formed long before the modern scientific system of breeding was established. We find that large breeds and bulky varieties of the same are co-extensive with a warm climate and rich herbage, and that smaller breeds and their varieties pervade those districts where the pasturage is more scanty and the climate colder. To wit, we find a larger variety of Devons with long straight hair bred in the fertile vale of Taunton Dene; while the Devon reared in North Devon is noted for his soft, rich, curly coat of hair, which he frequently loses when taken into other and richer districts. It is well known that the general appearance and hair of the bull Hundred Guinea (56), who was purchased by Messrs. Bult and Bond, near Taunton, of the breeder, at Molland in North Devon, at the foot of the Exmoor hills, altered considerably after two years' residence in his new home. The effect of soil, climate, and water on the colour and hair, and in developing changes in the form and physical structure, is well known and duly appreciated; so much so, that a great number of animals are sent to North Devon for summering, with the twofold object of grazing and changing their coats.

As an illustration of the influence of climate and food, we may refer our readers to the difference between the ordinary Devon, and the South Hams cattle, a splendid collection of which were exhibited in competition for the Devonshire Agricultural Association prizes, at the Plymouth meeting of the Bath and West of England Society, in June 1873. These animals are much larger and coarser in bone, and closely resemble the Sussex cattle, though hardly exhibiting so much quality; they are hardy, milk well, and are esteemed in their district, which comprises the country on the right of the railway from Plymouth to Totnes; they require more time to mature than the more compact North Devon, and would hunger where they would thrive. Amongst others we may name Mr. William Coakes, of Charleton Court, Kingsbridge, as a highly successful breeder.

Notwithstanding his curly hair, the skin of a Devon must be mellow and elastic. Experience shows that some animals fatten faster than others. On "handling" them, we find the skin and parts beneath soft and "mellow." This "mellowness" is a kind of softness or elasticity perceived upon pressing the skin with the fingers, and is a favourable sign of the aptitude of an animal to fatten. These parts are the cellular membranes, which in fat animals are full of fat, and the possession of this mellow feeling by store stock denotes that there are plenty of membranous cells ready for the reception of fat. None have been more thoroughly successful than the Devon breeders in attaining this desirable object; they consider an animal of little value if it cannot be fattened without very extraordinary food.

The general form of a Devon is very graceful, and exhibits a refined organisation of animal qualities unsurpassed by any breed. The expression of the face is gentle and intelligent; the head small, with a broad, indented forehead, tapering considerably towards the nostrils; the nose of a creamy white; the eye bright and prominent, encircled by an orange-coloured ring; the jaws clean, and free from flesh; the ears thin.

The horns of the female are long and spreading, gracefully turned upwards, and tapering off towards the ends. The general aspect of the head should in many points resemble that of the deer. The horns of the bull are thicker set and more highly curved, in some instances standing out nearly square, with only a slight inclination upwards.

Red is the true Devon colour, which varies from a dark to a lighter, or almost to a chesnut shade. In summer the skin is mottled with beautiful spots of a slightly darker shade than the ground colour of the skin.

The outline of a fat Devon very nearly approaches a parallelogram. The frame is level from the tops of the shoulders to the tail; the belly is longitudinally straight, and well filled out at the flanks. The breast is wide, coming out prominently between the fore legs, and extending downwards almost to the knee joint. The neck is long and thin, increasing towards the shoulder, which is tapered off to meet it. The ribs project at right angles to the back, with wide flat loins, and long rumps well filled out, thus enabling them to be loaded with more beef in the most valuable parts than almost any other breed.

As converters of vegetable into animal food, breed against breed, they return as much per acre, or for weight of food consumed, as any. Animals possess no magical power of producing beef, except from the food which they consume; it is therefore contended that, if the herbage of any given number of acres were to be consumed by Devons, they would produce in the aggregate as much beef as any other breed, a greater number being required to consume it; at the same time there would be a greater weight of the most valuable beef, and less of the coarse joints and offal. This is the reason why Devons and Scotch cattle sell first in the morning, and command the best prices, in London and other markets. Mr. Wainwright, a Devon breeder in the State of New York, says: "Their beef is of a fine quality, and brings a high price in the markets. They withstand extremes of temperature. On a poor pasture, from their peculiar build, they are enabled to travel rapidly over the ground without fatigue, and get sufficient nourishment where a heavy Shorthorn or Hereford would starve. The very best of this breed are the best in the world." Mr. Steinmetz, of Pennsylvania, writes me, as Editor of the "Devon Herd Book": "I find North Devon cattle the most profitable breed in America; I can raise more valuable beef on them with the same amount of food than any other breed."

In reply to some inquiries made in 1883, a well-known Devon breeder writes: "My herd consists now of 130 Devons. We always calculate a cow capable of making from 3½cwt. to 5cwt. of best cheese per season. I don't profess to say that the Devons are a milking breed, but I *do say that the milk is rich*, and that you can make as much per acre from the Devon as from any other mixed breed or Shorthorn, and they are more hardy. I have just tested our day's milking of forty of my Devons, which gave forty-seven gallons of milk, which made 61¼lb. of whole milk cheese. My neighbours forty cross-bred cows gave sixty-one gallons of milk, and that produced only 56½lb. of whole milk cheese. The keep was similar. I make butter for the London market up to May, I then begin making whole milk cheese. I think many farmers are beginning to see that they can make more per acre from the Devon at less cost and care, as in many cases they are going in for Devons as the most suitable and paying. From our altered point of view, the high price of meat shows them to be the most saleable animals, and the quickest feeders.

What is meant by a gold medal beast at our shows? That animal which most nearly approaches to the form and quality of North Devon; it is the length, depth, and width, not the height of a beast, which constitutes size. The cry has been for the animal that will be the first ready for the butcher, and the Devons have answered it. They bear the change of soil and climate well, *thrive* where many breeds would *starve*, and rapidly outstrip most others when they have plenty of good pastures. That they are a good rent-paying breed, especially in cold, hilly districts, is clearly proved by the fact that the majority of the oldest and most successful breeders are tenant farmers, whose ancestors have kept them for the last 150 or 200 years, in most cases on the same farms, in North Devon and West Somerset, frequently at an elevation of 600ft. or 800ft. above the sea level.

A few years ago the late Prince Consort established a herd of Devons, and they are also patronised by several noblemen and gentlemen in various parts of the United Kingdom. They have also been conveyed to new homes in the United States of America, where there are a great many herds of the purest descent; and to Australia, Natal, Mexico, Jamaica, Canada, and France, in all of which places they are answering a good purpose. "The Devon Herd Book" was first compiled by the writer and published in 1851, the 8th volume, bringing up the number of bulls to 1671, and females to 5739, was published in 1881.

Towards the end of 1883 Col. Davy sold the copyright of the Devon Herd Book to Messrs. Hawkes and Risdon, shortly after which "The Devon Cattle Breeders' Society" was formed, who now publish "The Devon Herd Book." Mr. John Risdon, of Williton, near Taunton, is the secretary to the society. Volume IX. has been published by the society.

Every breed and its varieties possesses peculiar merit, each answering a better purpose than the other, according to the soil, situation, and other circumstances in which it may be placed. To succeed, we must study to keep animals which are suited to our soil, pasturage, and climate. Those animals which will thrive in cold, bleak, hilly districts cannot fail to flourish in more favoured situation; and North Devons are never seen to greater perfection than among their native hills, the last haunt in England of the wild red deer, and where

<blockquote>At morn the blackcock trims his jetty wing.</blockquote>

[It is hardly doing full justice to the Devon to conclude a notice, however short, without reference to their superior qualities for draught purposes. It is true that at the present time working cattle are the exception. The value of meat and the advantage from getting our animals early into market having operated against the practice, but no one can say how soon circumstances may arise which will render it desirable to recur to bullock labour. The extreme price of horse flesh lately may well cause us to reconsider the question, and if, as is possible, the supply of cattle either from increased breeding or foreign sources should exceed the demand, a part of our surplus stock may advantageously be employed in this way. Without disparaging other breeds, we are bound to state that the North Devon one is quite unrivalled as a worker, and this is due to his activity and strength. We have not a ponderous over-weighted animal, good at a dead pull no doubt, but hardly able to crawl under its own weight, but we have a class of cattle that, with proper training, are capable of walking as fast and getting through as much work as heavy draft horses. The late Mr Plumb, of Ashton Keynes, near Cirencester, whose land was of a light sandy nature, worked Devon oxen in pairs, and his teams could hold their own at the ploughing matches, the work being done quite as expeditiously as where horses were used. These bullocks were yoked with collars and guided by reins. When a dead pull was required, or when a load of sheaves got fast in the fold yard, a bullock was much more efficacious than the strongest horse. Mr Plumb knew the value of good food, and whilst he worked hard he fed well. These bullocks on an average consume a bushel of corn a week, with plenty of bulky food, his practice was to buy in each year three-year-olds, work for two years, and send off fat when turned five years old. Another point that may be urged in favour of the breed is that, though careful breeding and selection has only been practised of late years, they have always been noted for symmetry and quality.—ED.]

CHAPTER IX.

THE LONGHORNS.

By GILBERT MURRAY, Elvaston, Derby.

THE present position of the Longhorn breed of cattle illustrates the old saying that "every dog has its day." Confined now to a few amateur farmers in the midland counties, it is difficult to realise that a hundred years ago they were the most valuable breed in this country; yet such is the fact. Yorkshire has the credit of giving rise to the Longhorn and their supplanters, the Shorthorn. The latter, however, originated in the eastern division, whilst the district of Craven (the original home of the Longhorns) is in the West Riding, bordering on Lancashire, from whence they spread out into the latter county and the south-eastern portions of Westmoreland. Like the Durham cattle, they enjoyed a considerable local reputation, those bred in the fertile vale of Craven being considered the quickest feeders, as they were the handsomer beasts; but it required the genius of Bakewell to draw them from their comparative obscurity, and give them a reputation which at that time seemed unassailable. Culley states that before Bakewell's time "The kind of cattle most esteemed were the large, long-bodied, big-boned, coarse, flat-sided kind, and often lyery or black-fleshed." This, however, is rather a sweeping denunciation, and, though applicable enough to the general run of Longhorns as they appeared in the various counties to which they had gradually spread, either tolerably pure or incorporated with the prevailing breeds of the district, must not be taken as a fair account of the Craven cattle, many of which were noticeable for rotundity, length of carcase, mellowness of skin, and quality of their milk. The improvement of the breed dates from 1720. At that time a Sir Thomas Gresley had a choice selection at Drakelow House, near Burton. A blacksmith and farrier of the name of Welby, who resided at Linton, in Derbyshire, on the borders of Leicestershire, purchased some valuable animals from Drakelow House, and took much pride in improving the stock. After carrying on for a few years with manifest success, a disease broke out and carried off the greater part of the herd. Mr. Webster, of Canley, near Coventry, is the next name of note in Longhorn history. How comparatively unknown now are those who worked for their sort as successfully as Collins, Mason, Bates, &c., did for the Shorthorns. The latter have become familiar in our mouths as household words, as helping to illustrate the rise and progress of a breed of world-wide celebrity, whereas the pioneers of the Longhorns are known only to the student—*sic transit gloria mundi*. Mr. Webster also worked upon Sir Thomas Gresley's stock, using bulls from Lancashire and Westmoreland. He bred a celebrated bull named Bloxedge, which produced some remarkable stock. It is unfortunate for posterity that Bakewell was not large-minded enough to leave us a record of his work. It would be of great interest as well as advantage to know how he set to work to develop the improved Leicesters, as his Longhorns were soon christened and known for many years. We, reasoning by analogy, can only surmise that he went for quality rather than size, and, as in his sheep, strove after correct outlines, fine bone and offal, with great aptitude for feeding. This last quality he paid most attention to, and naturally sacrificed to it other points, more especially hardiness and yield of milk. Bakewell found that feeding properties were to a great extent hereditary and could be perpetuated by close breeding; he, therefore, preferred improving his stock by selecting animals of the same kind rather than run the risk of crossing. He commenced his cattle breeding with two heifers from Canley, using on them a Westmoreland bull; and as far as is known he never went further, at first breeding very closely, but as the herd increased he was able to unite more distant affinities. In a few years his stock became known for rotundity of outline and aptitude to feed. They were much prized for feeding, but did not fill the pail like the old sort. Twopenny, out of one of the Canley heifers, by the Westmoreland bull, was very celebrated. His offspring, a bull named D., was even more remarkable; he was very closely bred, being by a son of Twopenny out of a daughter and sister of the same bull. Following in the steps of Mr. Bakewell, came Mr. Fowler, who farmed in Oxfordshire. His cows were of Canley breed; whilst his bull, Shakespeare, considered the best he ever bred, was by D. out of a daughter of Twopenny. Mr. Marshall, in his "Economy of the Midland Counties," gives a good description of this bull, which, save in his horns, did not resemble the Longhorns so much as the Durham of that day: "His head, chap, and neck remarkably fine and clean; his chest extraordinarily deep; his brisket down to the knees; his chine thin, and rising above the shoulder points, leaving a hollow on each side behind them; his loin narrow at the chine, but remarkably wide at the hips, which protruded in a singular manner; his quarters long in reality, but appearing short, occasioned by a singular formation of the rump. At first sight it appears as if the tail, which stands forward, had been severed from the vertebræ by the chop of a cleaver, one of the vertebræ extracted and the tail forced up to make good the joint; an appearance which, on examining, is occasioned by some remarkable wreaths of fat formed round the setting on of the tail—a circumstance which in a picture would be deemed a deformity, but as a point is in the highest estimation. The round bones snug, but the thighs rather full, and remarkably let down. The legs short, and their bone fine. The carcase throughout (the chine excepted) large, roomy, deep, and well spread." The value of this bull was fully appreciated by Mr. Fowler, who, except letting him for two seasons to a Mr. Prinsep at 80 guineas a season, retained him for his own use. In the year 1791 Mr. Fowler had a public sale, at which fifty head of cattle produced 4289*l*. 4*s*. 6*d*., being an average of over 80*l*. a head, prices that would be regarded as satisfactory for a choice selection of shorthorns at the present time, and which show the high estimation in which the improved Leicesters were held, and the advance made by such men as Bakewell, Prinsep, and Fowler. Mr. Marshall thus describes the character of the improved Longhorns: "Fore-end long and light (this we may observe is a fault apparent both in the few herds remaining in this country and in the Irish imports, unmistakably of Longhorn origin, of which more anon); neck thin, head fine but long and tapering, eye large, bright, and prominent. The horns vary with the sex; those of bulls comparatively short, from 15in. to 2ft. The oxen extremely large, from 2½ft. to 3½ft. Cows nearly as long, but fine and more tapering. Most of the horns hang downward by the side of the cheeks, and then if well turned, as in many of the cows, shoot forward at the point; the shoulders fine, thin, and well placed—this was particularly noticeable in the Dishley herd—girth small, as compared with Shorthorn and Middlehorn breeds; the chine remarkably full when fat, but hollow when low in condition; loin broad, and hips wide and protuberant, the quarter long and level, fleshy thighs, with small clean, but comparatively long legs; carcase round, and ribs well sprung, flesh of good quality, hide of medium thickness, and colour various—the brindle, the finchback, and the pye most common. As grazier's stock, they undoubtedly rank high: the bone and offal small, and the forend light, while the chine, the loin, the rump, and the ribs are heavily loaded, and with flesh of the finest quality. In point of early maturity they have also materially gained; in general they have gained a year in preparation for the butcher." Such was the character of the improved Longhorn as established by these leading breeders. With such excellence, how is it they so soon disappeared from this prominent position? Possibly those who followed were not able to maintain the character and quality of the stock; but more probably the increasing popularity of the Durham or Shorthorns caused them to be shelved; and it is a noteworthy fact that, at the present day, as far as we know, Leicestershire does not possess more than one herd of the old sort, whereas in their original homes, viz., in Westmoreland, Lancashire, and Yorkshire, the Shorthorn is dominant. In Ireland the Longhorn influence was undoubted; whether this was due to importation of English cattle, or *vice versâ*, we are unable to say, as records are wanting. Anyhow, they were at one period the prevailing breed in a large portion of Munster; and we can trace their influence in the cattle from that district to this day. Shorthorns, however, are rapidly becoming predominant through the length and breadth of the Emerald Isle, and to them is due in great measure the marvellous improvement of late years which is rapidly placing the Irish on a footing with the home-bred.

A century ago the Longhorn was the most important and fashionable breed of cattle inhabiting the counties of Derby and Stafford, and there still lingers in the district wondrous tales of the quantity of milk yielded by some favourite cow, or the more marvellous weights which the oxen and heifers attained when grazed on the rich alluvial pastures of the Trent, the Dove, or the Derwent. Viewed by their narrators through the mists of a long series of bygone years, their merits became magnified to a degree sufficient to awaken in our minds a feeling bordering on incredulity. It will give the reader some idea of the state of agriculture in this district when we state that down to about the period to which we refer the cattle were all wintered in the fields, and within five minutes' walk of where we write stands the first cowshed ever erected in this parish, or within a radius of several miles.

The farmers attending the markets of Derby and Loughboro' would no doubt freely discuss all the floating rumours concerning the doings of their neighbour of Dishley. In pre-railway days enthusiastic breeders made long pilgrimages by stage coach, or road waggon, and frequently by the more independent mode of travelling on horseback booted and spurred, with whip in hand, and equipped with all the travelling paraphernalia of wallet and saddle bags, men from all parts of the kingdom flocked to Dishley to inspect the stock and gather information from so popular and authentic a source. The *vivâ voce* reports of these visitors wafted the name and fame of Bakewell to the most remote corners of Great Britain; hence his cattle, like his sheep, became famous. What a change the revolving years of a century have brought about. Having officiated at the local show, where the present tenant of Dishley Farm was a competitor and successful prize winner, our drive homewards, in the cool of an early September evening, led us past the roofless church and tangled and dilapidated church-yard where the forefathers of the hamlet sleep. A lot of well-bred, fat, two-year-old Shorthorn steers were congregated round the cattle tubs in the home pasture, whilst a large herd of breedy-looking, large-framed dairy cows were slowly wending their way to the stalls where the milkers were waiting their approach; a solitary pigeon was keeping watch on the topmost pinnacle of the ancient church; this appeared the only probable lineal descendant of a century ago. The Longhorn has now been superseded by the universal intruder, the Shorthorn, and the Leicester has been supplanted by the ubiquitous black face. The shrill whistle of a railway engine in the distance awoke us from our reverie.

Twenty years ago the late Mr. Bakewell, a descendant of the celebrated breeder of that ilk, kept a large herd of Longhorns on his farm at Lockington on the south banks of the Trent, and just within the confines of the county of Leicester, whilst down to a more recent period the late Sir John Harpur Crewe, Bart., kept a large and fine herd at Calke Abbey, his seat in Derbyshire. This herd was sold in the autumn of 1885 through the death of the owner. The late T. W. Cox, of Spondon, near Derby, had a small but select herd which had been carefully cultivated for generations by his ancestors. Mr. R. H. Chapman, of Upton, Nuneaton, was long known as a skilful breeder. Mr. Richard Hall, of Thulston Grove, Derby, was for years a large breeder and successful exhibitor. Mr. Prinsep, of Croxhall, was another well-known name in the annals of the breed. With the exception of Mr. R. Hall, who still retains a nucleus, all the other herds have been dispersed through the death of the owners or other causes. Long before root cultivation had become general in Derbyshire, and prior to the use of artificial foods or stall feeding was generally practised, we have it on authentic authority that Christmas oxen which had never known the shelter of a roof-tree, and whose only food from weaning time consisted exclusively of hay and grass, would frequently reach the great weight of 360lbs. per quarter; this would be at the mature age of five years. Such results could not but stamp their merits and enhance their reputation amongst breeders.

The Darwinian axiom is as clearly exemplified in the case of the Longhorn as it is in many other of the ancient races of domesticated animals: in times of keen competition the survival of the fittest becomes a dire necessity. The Longhorn, though surrounded by a halo of ancient prestige, is nevertheless a slow grower, hence finds little favour in the eye of the rent-paying farmer. The breed is now chiefly in the hands of rich men;

they form an ornamental and highly-interesting feature in the parks of noblemen and large landed proprietors.

The largest herd is that of his Grace the Duke of Buckingham at Stowe, numbering at the present time nearly 100 head; they are directly descended from the Bakewell, Canley, Rollright, and some of the purest old Warwickshire families. The herd is of long standing, and has been bred with great care and judgment. Animals from this herd have frequently distinguished themselves in the show yard. They come to hand mellow to the touch. The skin, though thick, is covered with a profusion of rich, soft hair; the rib is well sprung, chine broad, shoulders well placed, barrel round and deep, the general appearance in unison, denoting a healthy and vigorous constitution. They are moderate milkers, and, as a rule, shy breeders. They are longer in arriving at maturity than the improved Shorthorn; consequently they give a less return for the quantity of food they consume. By a careful selection of breeding animals with a view to early maturity the Longhorn might yet regain much of their ancient popularity. We have seen the effect produced by a cross with the Longhorn bull on the cross-bred dairy cows of the country We would much like to see a cross between a good pure-bred Shorthorn bull and a Longhorn cow; such a cross could hardly fail to produce a superior animal, at least for the purposes of the graziers. At the shows of many of the leading Agricultural Societies, where there is no separate class for Longhorns, they are placed at great disadvantage by being compelled to compete with other established breeds; the system of comparing the merits of totally distinct races of animals, each of which is specially called to fulfil widely different purposes, is always a difficult and generally disappointing calculation.

Mr. Legh, of Lyme Hall, Cheshire; Mr. Shaw, of Fradley; Mr. J. German, Mr. Burbery, Mr. Godfrey, and Mr. Hall are all owners of small herds. They are no longer held in estimation for the purpose of crossing with other breeds, or for use in the dairy. Their hardy character well adapts them to high and exposed situations, yet they have failed to attract the attention of the cattle kings and great ranche owners of the Far West, in whose hands they might prove of some value.

CHAPTER X.

THE SUSSEX BREED OF CATTLE.

By A. HEASMAN, Angmering, Arundel, Sussex.

THE origin and date of the Sussex cattle may be a matter of uncertainty. Was William of Normandy attracted by the fine oxen grazing in the rich marshes of Pevensey, or did he import them? It is generally understood they date back to the time of the Conquest, and it is well known that Pevensey and the surrounding district has always been their principal home. There is no direct evidence as to whether the Sussex cattle were an original breed, or are descended from the Devons, which are supposed to represent the oldest breeds in England. The matter is not of much importance, but it is easy to believe that the heavy draught purposes to which they were subjected might cause alteration of type and appearance even greater than the originals of these breeds presented.

This useful class of stock were formerly bred principally for draught purposes, being converted into food for the public after they had cultivated the soil of the Weald of Sussex and Kent—some of the heaviest tilled land in the kingdom—and at times been required to start the heavy carriage of the county member from the same muddy district, when it was necessary for him to attend to his parliamentary duties, before the locomotive came into operation, or the Highway Act had been amended. Even in those early days the Sussex cattle were fully appreciated, and, always possessing the finest quality of flesh, were never neglected by the grazier.

When they had been worked for several years, and age at

last rendered it necessary that they should be drafted from the team, the farmers of the western part of the county would pay a visit to their brothers in the east; attend the fairs held at Battle, Lewes, or on the borders of Kent, in order to buy up the aged oxen; and, after grazing them a year, supply the markets with animals weighing from one hundred and eighty to two hundred stone.

Times have very much altered, and the Sussex beasts are no longer what they were, neither are they reared to the same extent, or for the same purpose. They have given place to horse and steam power, and take up their position as one of the useful and established breeds of the kingdom to meet the pressing and increasing demand for beef. Their colour was formerly both light and dark red—in some instances so dark that it almost amounted to black; but the intermediate or cherry colour is now the favourite, denoting good flesh and better quality for fattening. In attempting to trace the progress of improvement from the long-legged, strong-boned, coarse, but useful beast of burden which prevailed when the century was young, nay, even before it had dawned, the work of Mr. John Ellman, of Glynde, must not be passed over. It is matter of history that so far back as 1797 this spirited agriculturist sent a pair of his best oxen to Woburn Abbey, and in the previous year he, in concert with Lord Egremont, called a meeting at Lewes, to collect money for prizes to successful breeders. This action led to the formation of the Sussex Agricultural Society, and two years later we find Mr. Ellman's name associated with those of the Duke of Bedford and Mr. Astley as the founders of the Smithfield Club. For many years Mr. Ellman was the leading spirit of the county society, and his cattle were so successful that after a time he ceased to exhibit, so that others might be in the running. Sussex breeders are much indebted to Mr. Ellman for the good work he did. Another most successful breeder was a Mr. Selmes, of Knell Farm, Bickley, who, it is said, once challenged Lord Althorpe to show a certain number of Shorthorns against his Sussex cattle.

The breed has been too well appreciated by the tenant farmer to be allowed to die out, and great pains and attention have been taken latterly in endeavouring to alter the style and type by breeding from the smallest bone with the greatest amount of flesh; this seems to have been successful when we compare the present animals with what may be called the old-fashioned sort, one of which was fattened many years ago at Burton Park, near Petworth, and called the Burton ox. A portrait of this animal was dedicated to the gentlemen of the county of Sussex by Mr. Spilsbury, of Midhurst; its height was 16½ hands, and it measured 8ft. from the back of the horns to the tail; and from hip bone to hip bone, across the back, 2ft. 8in.; the depth of shoulder, 4ft. 7in.; girth behind the shoulder, 10ft.; and it weighed 287st. 4lb. Although this was considered a wonderful animal at the time, the meat was not in the right place; its bone was enormous, its back rib shallow instead of deep, with a spare thigh and small twist, and it was not to compare with the class of cattle now exhibited, saving in the matter of weight, which has always been a great feature in the breed. The Sussex cattle are second to none as regards early maturity and weight for age; this is proved by the weights of the animals shown at the Smithfield Club meetings. The Sussex are great favourites with the butcher and consumer. At three years old well-fed steers will weigh from twelve to fourteen score pounds per quarter. Their general features may be described as follows: Nose tolerably wide; muzzle of a golden colour, thin between the nostril and eye; eye rather prominent; the forehead rather wide; neck not too long; sides straight, and not coarse at the point of the shoulder; wide and open in the breast, which should project forward; girth deep; legs not too long; chine bone straight; ribs broad; loin full of flesh and wide; hip bones not too large, but well covered; rump flat and long; tail should drop perpendicular; thigh flat outside and full in; the coat soft and silky; with a mellow touch.

The Sussex cross well with any breed; by using the male animal, substance and firmness of flesh are imparted, and the colour of the offspring is generally red. They are of themselves a hardy breed, and have been found to surpass all others in the poorest pastures of their native county. The cows are not good milkers; those with the heaviest flesh are the worst, but produce sufficient to rear their calf. The most successful way of breeding is to calve them down in October and November, let them have their own calf through the winter, which can be weaned in the spring, and another calf put to the cow. If managed in this way, each cow will rear two calves, and the number of barreners be greatly diminished, which is one of the greatest evils when cows are allowed to drop their calves all the year round.

Great credit is due to Mr. Edward Cane, of Berwick Court, for the energy he has displayed in the improvement of the breed. Mr. Cane at one time was one of the largest breeders, and always ready to give a good price for the best cow brought to the hammer. He was, however, very unfortunate with his yearlings, and after losing a great many, was induced to sell off his herd. A cow purchased at this sale produced one of the best steer specimens of the breed, which was in 1867 exhibited at Smithfield, obtained the first prize in its class, and was one of the most formidable competitors for the cup, finally awarded to Mr. McCombie's celebrated Black Prince. In the report of the show published in *The Field* of Dec. 14, 1867, this animal is thus alluded to: "The older class of steers contains eight entries, and is decidedly good, Messrs. Heasman's entry being probably their *chef d'œuvre*; so good that it remained out a long time competing with Mr. Foljambe's steer as second-best male." And again: "The Duke of Sutherland's very smart, level, and well-fed young Shorthorn was much fancied for second place; but he gave way to one of the best Sussex steers yet shown." Mr. Cane was the first to introduce the Sussex cattle to the notice of the Smithfield Club, and from that time much improvement has taken place. The Smithfield Club have been very liberal in their support by offering good prizes, the result of which is that the classes are well filled, and the breed year by year becomes a more prominent feature in the Christmas gatherings.

The Sussex men do not feel inclined to spoil their best animals by over-feeding.

The Public Herd Book of Sussex Stock (without which no breed is perfect) has been established about thirteen years, and promises to be of great assistance. Each year it becomes of more importance, and as recorded pedigrees increase it will enable breeders to select a cross with a degree of certainty which has hitherto been a difficulty. It will also help the sales.

The public are thereby assured of the pedigree of the stock they purchase, instead of having to rely on the statement of the vendor. The book at the present time numbers seven hundred and thirty-two bulls, and three thousand six hundred and twenty-three cows. It is also a chronicle of all pedigree prize animals, for breeding purposes, and records the names of the breeders who have gained honours at the Royal Agricultural, Southern Counties, or Sussex, and Kent County Shows.

CHAPTER XI.

NORFOLK AND SUFFOLK RED POLLED CATTLE.

By THOMAS FULCHER, ELMHAM, NORFOLK.

IN searching for data to determine the origin of this breed, we have come across a book entitled "The Rural Economy of Norfolk," by Mr. Marshall, resident upwards of two years in Norfolk (from 1780 to 1782), acting, he informs us in his preface, as agent to the Gunton Estate. Under the heading "Cattle" he says: "The native cattle of Norfolk are a small, hardy, thriving race; fatting as freely and finishing as highly at three years old as cattle in general do at four or five; they are small-boned, short-legged, round-barrelled, well-loined, thin-thighed, clean-chapped; the head in general fine, and the horns clean, middle-sized, and bent upward; the favourite colour a blood-red, with a white or mottled face. The breed of Norfolk is the Herefordshire breed in miniature, except that the chine and the quarter of the Norfolk breed are more frequently deficient. If the London butchers are judges of beef, there are no better fleshed beasts sent to London market."

Suffolk cattle, according to the earliest records on the subject, were polled, and, originally, dun in colour; later on they are described as red, red and white, and brindled.

From a very early period large numbers of polled Galloway cattle were brought into the counties of Norfolk and Suffolk. There can be little doubt that these were crossed with one or other (probably both) the native races, and that thus the present breed of Norfolk and Suffolk red-polled cattle was called into existence.

The writer is by no means disposed to accept the theory propounded by the author of the article on "Scotch Polled Cattle" (page 98), that our Norfolk polls are simply red Galloways. True enough, there is a resemblance between the heads of the two sorts, each being furnished with a thick tuft of hair, covering the forehead. In the Norfolk beast this appendage will, however, be frequently composed of a mixture of red and white hair. More rarely, a large spot of pure white makes its appearance in the face. The deep, blood-red colour of the Norfolk polls is, moreover, many shades darker than we have seen in any specimens of the Galloway breed. These two peculiarities go far to support the conclusion we have arrived at—that the old native race had a due share in the concoction of the present breed. As to by whom this cross was first resorted to, we have no precise information. Marshall, indeed, mentions that long before his time polled Suffolk, Galloway, and even West Highland bulls were used for crossing with the Norfolk home breeds; but so highly did he appreciate the good qualities of the latter, that he only refers to crossing, in order to condemn the practice.

In the absence of documentary evidence, we have it on the authority of Mr. Money Griggs* of Gateley (now in his hundredth year, and for upwards of eighty years a tenant on the Elmham estate), that from his earliest recollection red polled cattle were kept in the neighbourhood of this place.

One of the oldest breeds in the county is that of Mr. George, of Eaton, near Norwich. The son and successor of its founder, in reply to inquiries from the writer of this article, says: "I think about 1800-10, my father commenced to collect what were at that time called the blood-red polled Suffolk cows; but why *Suffolk* polled I do not know, for to the best of my recollection they came chiefly from Norfolk farmers, from Mr. Reeve, of Wighton, Mr. England, of Binham; and I well remember he had a cow picked up for him by a Mr. Walne, of Foulsham (four miles from Elmham) not long after he commenced the

---

* On Jan. 29, 1872, this fine old yeoman died, after a few days' illness; until within a short time before his death he was in the daily habit of walking over the farm; long after ninety he would ride to meets of the Norfolk foxhounds in his neighbourhood.

breed, which cow was called Foulsham, and was one of the best my father ever possessed, costing 25 guineas—at that time thought a frightful price. She bred some very good blood-red calves; one, a bull, was much prized for some years. After this my father went on breeding in-and-in for many years, not being able to find bulls to his liking. The only bull I recollect his buying proved a regular brute, whom, with his offspring, he got rid of as soon as he could. After this came Mr. Birkbeck's and the Elmham blood. Mr. Etheridge, of Starston, had first and last several bull calves for himself, and the late Sir Edward Kerrison, and some went farther into Suffolk. As to Norfolks and Suffolks being the same breed, I can form no opinion, except that I know they have been a good deal mixed if only through my father's blood."

In Youatt's work on cattle, published some forty years ago, appears a portrait of a very handsome cow, bred by Mr. George. The Eaton herd still flourishes, and as might be expected from such careful breeding, exhibits great uniformity of character. Mr. Reeve and his herd have passed away, and the Binham herd no longer exists; but Mr. England writes: "My grandfather came to Binham in 1792. I have heard my father say the polls were much improved by Mr. Reeve and my grandfather, but whether they were red or not I cannot say; they were as alike as possible when I first remembered a herd of thirty cows here, and a beautiful red. I doubt if there are any better at the present time."

Other old-established Norfolk herds are those of Lord Sondes (Elmham), Money Griggs (Gatcley), the late Col. Mason (Necton), Mr. Henry Birkbeck (Stoke Holy Cross), the Messrs. Hudson (of Quarles and Blakeney), Mr. Nicholson (Gressenhall), Mr. Savory (of Rudham), and Mr. Lombe Taylor (of Starston). Of herds established within the last twenty-five years the most important are those of Sir Willoughby Jones, Bart., Mr. Colman, M.P., and Messrs. Brown (Thursford), and Hammond (of Bale), all of whom have been highly successful in the show-yard of late years. Mr. Tom Brown (of Marham), famous for his Cotswolds, is getting together a very pretty lot of cattle, beginning with heifers from Elmham, and an Eaton bull. Altogether, the red polls are now bred, with more or less care, on upwards of one hundred farms in the county.

The principal herds in Suffolk are in the hands of the Earl of Stradbroke, Col. Tomline, M.P., Sir Edward Kerrison, Bart., Mr. Arthur Crisp, Mr. M. Biddell, and Mr. R. E. Lofft, of Troston. Mr. Badham, who at one time possessed a choice herd, and may therefore be accepted as a competent authority, is of opinion that the red-polled cattle of the two counties are now precisely the same breed. The original Suffolks are still represented by the herd at Riddlesworth.

Amongst the good qualities that may be fairly claimed for the redpolls are hardness of constitution, enabling them to thrive on scanty pasturage, and to withstand the severe winters and piercingly cold springs usually experienced in the eastern counties; their milking properties are unquestionable, and they have not that tendency to go dry which belongs to the Alderney, Ayrshire, and most other breeds having a reputation as dairy cattle. It not unfrequently happens that a cow will continue to yield a good quantity of milk from one calving to another.

The unimproved "home-bred" of twenty-five years since was open to the objection of being flat sided, thin on the loin, light in the hind quarter—in short, a somewhat rugged-looking animal generally. What the breed of the present day is like is well described by Mr. J. K. Fowler, of Aylesbury, who, in the capacity of Judge at the last meeting of the Norfolk Agricultural Association, held at Dereham, June, 1871, was pleased to say of them: "He was struck with the remarkable usefulness and value of the cattle of this district; the cows had good useful udders, so that they were likely to be capital cows for the dairy; while the bullocks had capital chines and good backs, but they were somewhat deficient in the springing of the ribs and in the hind quarters. Amongst the lot they scarcely found an animal that was not fit for a show-yard. As a Shorthorn breeder, he wished he could put some of the good points he found upon the Norfolk polled cattle on the animals which he was breeding." It would be interesting to know the particular points Mr. Fowler wished to transfer to his Shorthorns—possibly the "good useful udders" would be one of them. Although there is believed to be not the remotest affinity between the two breeds, yet many of the female polls very nearly approach the Devon type in their sweet deer-like heads and general thoroughbred appearance.

The vast improvement which has taken place in the polled breed is probably owing, in no small degree, to the liberal prize list of the Norfolk Agricultural Society, supplemented, as it is, by the cups and gifts of private donors. So recently as ten years ago Lord Sondes was almost the only Norfolk exhibitor, Mr. Badham doing battle for the sister county. Now there is no lack of competition; of late the chief prizes have gone to tenant farmers.

The Royal Agricultural Society of England at the great Battersea show of 1862, accorded a separate class to the polls. Previously to that date, they had been relegated to that heterogeneous collection—the class for "other established breeds." In response to a memorial from the breeders, separate classes were granted at Oxford and Wolverhampton. On each occasion the breed was so creditably represented, that it may be hoped a similar concession will be made whenever the show is held within reasonable distance of the eastern district.

Cows and heifers for dairy purposes have been sold from the Elmham herd to buyers in the counties of Beds, Berks, Bucks, Chester, Hants, Northampton, and Sussex; whilst exportations of breeding stock have been made to Egypt, Germany (north and south), and Austria, where, strange to say, on an estate of Prince Leichtenstein, a breed of red polled cattle has been in existence from time immemorial. Also during the last two years a few cattle have been sent to the United States, and there are orders on hand at present from the same quarter.

This breed is now known as Red Polled, having from its remarkable revival and distribution to every part of the United Kingdom, fairly earned a name apart from locality. Forty years ago the breed was very nearly extinct, whilst at the present time pedigrees of 1110 bulls and 3842 cows are recorded in the Red Polled Herd Book, which was started in 1874. It is a matter for regret that a few breeders are so lazy and blind to their own interests, as not to take advantage of this most useful publication.

The registered herds are thus distributed:

In England	83
In Ireland	1
In Scotland	1
United States of America	28
Germany	1

For this wide diffusion many thanks are due to the Royal Agricultural Society of England, who of late years have given separate classes and substantial prizes. At the Shrewsbury Royal, 1884, the official reporter, Mr. William Housman (albeit a Shorthorn man), thus writes in the journal of Red Polled bulls: "Some very good and useful-looking bulls, level and thick-fleshed, of rich colour, with the style which only carefully-bred animals can possess, competed in the three classes. . . . But the bull of the really highest style and character, the beau ideal of a thorough-bred animal, is Mr. R. Harvey-Mason's Napoleon, whose merit (although particulars of which the butcher can take no cognisance, go to make a most attractive appearance) consists mainly in properties which the most practical man can appreciate. There is a frame of ample size, not overgrown, truly moulded, moulded to the most perfect proportion of each part to the whole; straight limbs set on in the right places, and the joints most beautifully turned; the hocks are especially straight and neat, and the short tapering of the ends of the massive thighs to the hocks, with flesh as far as flesh can go, and then no lumber, but a nice clean joint, make quite a pattern of refined form. The touch discovers a rich layer of lean flesh, spread everywhere evenly, and the hair is of the richest red, deep in colour, but not blackened. The head is gay and full of life; the neck (a most expressive feature in a bull, if the term may be allowed, as his way of using it tells his character) is sufficiently substantial without coarseness, and extra arched, with a bridling side-way half turn of the head on the approach of a stranger. Vigorous life, not nice, appears in this instant consciousness of notice."

The bull thus so eulogistically and faithfully described, was at the time about eighteen months old, but did not subsequently distinguish himself in the show yard, Mr. Harvey-Mason having wisely determined to keep him in active service at home, he has thus when shown been in somewhat low condition.

As might have been anticipated, the largest and best lot of Red Polled cattle ever got together in a showyard appeared at the Royal, at Norwich, July, 1886, where no less than 144 males and females competed for the handsome prizes given by the R.A.S.E. and local donors. The champion bull and cow illustrate this paper, and speak for themselves. [We take this opportunity of acknowledging our indebtedness to the proprietors of the *Live Stock Journal* for the use of their blocks for these pictures.] The bull Falstaff (303), at 10½ years, looked "as fresh as a four-year-old." This grand animal, "the bull of the century," so far, was bred by the Rev. A. G. Legge, vicar of this parish (Elmham, Norfolk), sire Rufus (188), bred by the late Lord Sondes, dam Fairy Bradfield (891), bought by Mr. Legge from a tenant on the Elmham estate, the late Fisher Bradfield, who for very many years occupied an adjoining farm to that held by "Old" Money Griggs for upwards of eighty years on the same estate. Falstaff was bought, when a calf, from Mr. Legge by the present writer, but is now at the head of the largest herd of Red Polls extant, that of Mr. Garrett Taylor, of Trowse House, Norwich, who is the happy possessor of some two hundred head, half of which are milch cows, getting to be "as like as two peas."

The first exportation of Red Polls to the U.S.A. occurred in the year 1873. As may be judged by the number of herds now in existence in the States, a considerable export trade has been going on since. Our American cousins see in this a breed combining milk and beef production—milk not thin and watery like that of one at least of the breeds they have gone in for, and beef of superlative excellence.

In closing these notes I would impress upon my brother breeders to keep the milking properties well in view. This may be done by selection, by early breeding from heifers, by allowing the herd plenty of fresh air and exercise—not by coddling and over-feeding—aim at cows giving their 40lb. to 50lb. of milk per diem five to six months after calving—there are plenty that do it, and their number may, with care, be added to.

CHAPTER XII.

GALLOWAY CATTLE.

By GILBERT MURRAY, Elvaston, Derby.

GALLOWAY proper comprises the counties of Wigton and Kirkcudbright, forming the south-western seaboard of Scotland. Geologically it rests principally on the Silurian rocks of the primary formation, attaining in some parts an elevation of nearly 2000 feet. The uplands are principally devoted to pastoral purposes, to which they are well adapted; the mild humidity of the climate and peculiar nature of the soil tend to produce a luxuriant growth of the coarser kinds of grasses and herbaceous plants, thinly interspersed with patches of heath. The average yearly rainfall is 35in.

The cattle, which derive their name from the district they inhabit, are possessed of distinctive and strongly marked characteristics. We learn from authentic sources that so early as the beginning of the sixteenth century they had attained considerable celebrity, which they have successfully maintained to the present day. The Galloway is generally classed with the mountain breeds, though we think he might more appropriately be placed amongst those of the plains. With ordinary keep, the ox at three years old will weigh from nine to ten scores per quarter. To persons but slightly acquainted with the breed they are very deceptive, as they weigh heavier in proportion to their size than any other classes of cattle. The hides weigh well, and when well fed the beasts produce internally a large quantity of loose fat. The beef is tender, and has the fat and lean well mixed together; hence they are held in high estimation both by the butcher and the consumer.

The distinguishing characteristic of the race is the absence of horns, both in the male and female; they are mostly of a black colour, though we occasionally meet with individual animals of unalloyed purity of a red or brown colour; others again, of equally pure lineage, have white faces, and are marked with white under the belly. The trunk is symmetrical and well formed, supported by short legs; the flesh is evenly distributed over the frame, even down to the knee and hock; the shoulders are thrown well back, which gives breadth to the chine, and depth and expansion to the chest; the sides are long, ribs well sprung; the hips round, and less prominent than in some of the improved races; the skin, though thick, is mellow to the touch, and covered with a profusion of long silky hair.

Down to the first decade of the present century, the Galloway was by far the most numerous and important breed of cattle inhabiting the wide pastoral range, stretching from the river Tweed on the east to the Irish Channel on the south-west. At the same date they were the principal breed cultivated in Carrick, the south-western division of Ayrshire. Long prior to the date of railways, and before artificial manures were known and appreciated (which gave an impetus to the cultivation of root crops), a large and important cattle trade was carried on in the south. Some idea may be formed of its magnitude at the period of which we write from the fact that, taking Dumfries as the great mart and principal centre from which the southern droves took their departure, from this one point alone there passed yearly from 25,000 to 30,000 head to be finished off on the rich pastures of Leicestershire or Northamptonshire, or the salt marshes of the eastern counties. Several weeks were occupied on the journey; the drovers, being well acquainted with the country they traversed. In order to save turnpikes and the feet of the animals, and at the same time snatch a morsel of food, preferring the devious windings of the cross-country lanes to the more direct lines of the turnpike. There were likewise a goodly number fattened off on the rich pastures of Milton, Howell, and Baldoon, in Wigtonshire, and on the sheltered and productive lowlands of Dumfriesshire.

One of the distinguishing features of the breed, the absence of horns, has by some been attributed to the physical conditions by which they were surrounded. The practical man who has closely studied the different races of our domesticated animals well knows that soil and climate produce wonderful effects in altering their appearance. This, and a careful selection of males and females of approved type through many generations, could not fail to stamp the race with a distinct character.

The polled breed of the eastern counties of England, which is now attaining such well-merited repute, raises a lingering suspicion in the mind, strengthened by the fact of long-continued and extensive importations of Galloway cattle into those districts, that at some distant period it derived its origin from that source. It is true that the colour is distinct, but any person wishing to improve or establish a race of a red colour would have experienced no difficulty in selecting pure-bred animals of that colour to start with. There is still a striking resemblance in some points between the two races, more particularly in the formation of the head; and it is clearly within the range of probability that the existing difference of appearance may have arisen from the effects of soil and climatic influences.

Down to within the last forty years, large numbers of cattle were bred in Ireland bearing a close resemblance to the native Galloway, with the exception of horns, which the greater portion of the Irish-bred animals possessed. As many as 10,000 of these were yearly landed at Portpatrick, a small seaport on the south-western promontory of Wigtonshire, at a period antecedent to that at which the harbour was made available for the admission of loaded vessels of from sixty to one hundred tons burden. The practice was then to let go the anchor as near to the shore as the depth of water would permit without endangering the safety of the ship, and then to sling the cattle into the sea and swim them ashore. On a rough tempestuous coast this often proved a hazarduous, and not unfrequently a disastrous undertaking; at all times it required the services of several well-manned small boats to guide them ashore, for if left entirely to themselves they were as likely to swim out to sea as they were to make for the land. The greater part of the cattle landed here were from one to two years old. As soon as they were sufficiently rested they were driven on to some market or fair. Nearly all fresh arrivals were pitched at

Stranraer, that being the nearest market to the port of disembarkation, and at that time the great mart for Irish cattle in Scotland, and was regularly attended by graziers from long distances. In those days, and down to a much later date, the father of the writer was constantly in the habit of riding on horseback out of Ayrshire, a distance of thirty miles, to attend this market, returning home the same night. The great object of the purchaser was to select those most nearly resembling the native Galloway. As already stated, great numbers were possessed of horns; some were very small and loose to the touch, appearing to be only attached to the skin, whilst many were of a fixed and more prominent character. The matter of horns was considered of minor importance compared with that of hair and colour, and the closeness in general *contour* with which they approached the approved type. In order to obviate the difficulty and remove the objectionable appearance, the Irish breeders and cattle dealers had recourse to the following stratagem: the animal was cast in the usual way, and when properly secured, a sharp knife was passed round the base of the horn, severing the skin and fleshy part; a fine-toothed handsaw was then inserted into the incision, and the horn and pith cleanly sawn off level with the skull. To prevent hæmorrhage, the protruding ends of the arteries were either secured by a needle and fine silk, or else they were seared with a hot iron. A circular patch of coarse linen, corresponding in size to the base of the horn, was dipped in moderately hot tar or pitch, and placed on the wound; three thicknesses of this were generally used, and stitched round the edges to the hair, to prevent them peeling off before the wound had healed up. This was undoubtedly a cruel operation, yet, when skilfully performed, we believe it was seldom attended with fatal results. It is but fair to state that the operation was only performed on those animals more nearly resembling the native Galloway. At that time many of the Irish-bred cattle had a line of white or ashy brown along the ridge of the back; nothing could be gained from removing the horns of such, as the merest tyro could at once recognise their lineage.

It has been asserted by some that the Galloway breed was greatly deteriorated from its having been crossed with the Irish.

This we believe to be a fallacy, as the Galloway breeders had always a strong aversion, I may say hatred, to the Irish; the prices obtained for the pure-bred Galloways would of themselves be sufficient to deter them from adopting such a course. These cattle, when from one to two years old, would be of the average value of 4*l.* to 9*l.* per head, and always realised from 1*l.* to 2*l.* per head above the Irish of corresponding size and age. Occasionally an Irish cow might be found in the hands of a small cottager too poor to purchase a Galloway, but, as a rule, they were never used by the farmers for breeding purposes. It was then the universal custom to spay all the heifers at the age of one to two years, with the exception that occasionally one or two of the most promising were reserved to take the place of those removed by death or accident, or which increasing years had rendered unfit for the duties of maternity. There was no fixed principle of drafting out the old cows; if good breeders, they were kept to the age of ten to twelve or more years. All the Irish heifers were also spayed, either before they left Ireland or shortly after their arrival. The distinguishing mark of a spayed heifer was a triangular piece cut from the top of the right ear. The advantages gained by spaying was that the animals rested better, and consequently fed faster, and steers and heifers could with safety be grazed together in the same field. The system of spaying might still be carried out with great advantage with many other breeds beside that of the Galloway.

Galloway cows have the reputation of being bad milkers, due in a great measure to mismanagement. The old breeders invariably allowed the calves to suck their dams, depending more on the amount realised by the stock reared than they did upon that obtained by the produce of the dairy. Farmers possessing tolerably large herds were content if they could only obtain sufficient milk to supply the wants of their family, and furnish them with butter and a small quantity of skim-milk cheese; we have frequently seen the maid milking on the one side of the cow, whilst the calf was sucking on the other. During the summer and autumn they were generally milked out of doors in the open yard. When the calf had reached the age of five or six months, it was usual to place a muzzle on the nose; this muzzle was simply a head stall, the nose piece of which was a broad band armed with sharp spikes, the object of which was to set the cow kicking whenever it attempted to suck, and therefore defeat its object; at milking times the muzzle was removed, and the milkmaid and calf started on equal terms. When the milk became exhausted before the appetite of the calf was appeased, it became rampageous, and not unfrequently charged and overthrew the milkmaid and her pail. The cows and their calves were never separated, and except in cases where the means of restraint were used which we have related, they had free access to their dams at all seasons. Every practical man well knows the deteriorating effects such treatment would produce on the most prolific milker.

When regularly milked in the usual way we find them little inferior to many of the other established breeds, and the milk is invariably rich in butter. It has been noticed in highly cultivated cattle that the calf's sucking the dam prevents the latter becoming again in calf; whereas in animals less improved, it appears to exercise but little effect, as the Galloways as a rule were regular breeders.

In former days, cattle of all ages were generally wintered in the fields. They received a foddering of rough hay or straw twice a day. During the winter months they made little progress; as the season of grass advanced they grew and improved rapidly. Within the last thirty years the entire system has undergone transformation as if by the wand of the magician; the improvement and reclamation of large tracts of waste lands, the extended growth of root crops and of the cultivated grasses, led to the introduction of the Ayrshire breed of dairy cattle and the feeding of cross-breed sheep in large numbers, which tended for a time to lessen the number of the indigenous race of polls.

The establishment of a herd book in which is faithfully recorded the ancestral genealogy of the race, has tended greatly to improve the breed and attract public attention to their superior merits. The cattle kings and great ranche owners of the Far West are recognising the value of the male for the purpose of crossing or grading the native races, their fine constitutions and hardy nature are well adapted to withstand extreme climatic changes. The late Duke of Buccleuch, Mr. Jardine, and others, have done much to encourage improved breeding. The Rev. John Gillespie, through whose indefatigable efforts the herd book was established, has done more to popularise the breed than any other individual.

A cross has been tried between the Galloway and other established breeds. That between the Galloway bull and Ayrshire cow is held in high esteem, the produce have the reputation of being kindly feeders, arrive early at maturity, and attain a medium weight. The prepotency of the male is most potent in the blood of ancient and unalloyed races, hence few, if any, are equal to the Galloway in this respect. Put a well descended Galloway bull on a herd of Ayrshire cows and every calf will come black as a crow without exception, on the other hand put the purest bred Shorthorn bull it is possible to obtain on a herd of Galloway cows, and calves of all imaginable colours will be dropped, some black, others black and white, generally two-thirds of the number will be of bluish-grey colour. For feeding purposes the Shorthorn cross is not surpasssed by any pure breed in the kingdom; when well kept they attain great weight at an early age; for lightness of offal and superior quality they cannot be surpassed.

During the spring of 1884 I sold thirty Galloway steers from the stalls at an average of 33*l.* each, many of these were under three years old, they realised ninepence per pound, the top figure in the market. As a further proof of their popularity in the London market all other breeds are comparatively neglected by the West End butchers as soon as the Scotch begin to make their appearance in the markets.

CHAPTER XIII.

THE ANGUS-ABERDEEN CATTLE.

SINCE the issue of the first edition of "The Cattle of Great Britain," published in 1875, the Angus-Aberdeen cattle have become so much more widely known, and have spread so considerably in various directions from their national pastures, that it is necessary to give a more complete history and description than was then done. Not only have several herds been established in England, from some of which animals have been shown with success, even at the meetings of the Highland and Agricultural Society; but their aptitude for laying on flesh, combined with good constitution and hardy character, render them favourites on many ranches in the Western States of America, and the fact of being hornless allows of closer packing in the trucks, an important consideration.

Most important testimony as to their value was given to the writer by an extensive breeder and importer, Judge Goodwin of Belvoir, Kansas, viz., that on the Goodwin Park Stock Farm, which is worked by himself and his brother, the relative and comparative merits of Shorthorns, Herefords, Angus-Aberdeens, Galloways, Jersey, and Holstein cattle had been tested on the natural pasture without any additional food, and the result was that the Shorthorns were first drafted, next the two dairy breeds, then Herefords, and lastly Galloways. The evidence in favour of the Angus-Aberdeen cattle was overwhelming, especially as regards adaptability to climatic conditions, hardiness of constitution, and ability to thrive on moderate diet; indeed, when once in condition it was found difficult to starve them.

We were further informed that this firm now possess one hundred and twenty-five head, including many choice animals; indeed they owned Judge the Jilt bull, which was the champion at the Paris International Show of 1878. Messrs. Goodwin breed partly for the ranches, but their principal demand is from the small farmers of Kansas, who buy their bulls to improve the common stock of the country. It would be easy to multiply evidence from American breeders and importers as to the increasing demand for these cattle in the States. It is sufficient to state that, previous to the great depression in the cattle trade, many of the best and highest priced animals were exported either to the States or to South America.

History is silent as to the origin of the breed. Their general resemblance to the Galloway cattle points to a common origin; the former, being natives of the South of Scotland, were earlier known in the London market, but there is no evidence that their lineage is more ancient; and although the two breeds have been, as far as any record goes, kept distinct, it is reasonable to suppose that with the red polled cattle of Norfolk and Suffolk they all came from one original stock. The question has no practical importance. The fact remains that whether sprung from a common source or not, the influence of climate and cultivation has resulted in well marked distinctions, and it is well nigh as unsuitable to class them together as to mix up Jerseys and Guernseys under the general title of Channel Island cattle.

Black polled cattle were first known in the counties of Aberdeen and Forfar as the Buchan and Angus "doddies;" they have now largely spread into Kincardine and Banff as well as in other directions alluded to. Though originally varying considerably from each other according to local influences, the combination of the best blood from different sources has resulted in establishing a uniformity of type and character, which may well challenge comparison with any other breed of cattle. For the following details of the history of the Angus-Aberdeen cattle we are indebted to the late Mr. McCombie's charming little book, "Cattle and Cattle Breeders," than whom it would be impossible to find a more qualified historian, or one who from long experience and rare judgment, attained greater success in making the cattle notorious. Very little is known previous to the beginning of the century, when a Mr. Hugh Watson, of Keillor, established a herd, from which all succeeding breeders derived benefit. He commenced exhibiting in 1810, and for a long period was well nigh invincible, his champion heifer at the Smithfield Show of 1829 was so perfect in outline that she was modelled, and her portrait appears in the volume "Cattle," published by the Society for the Diffusion of Useful Knowledge. An ox, which gained the champion prize at Belfast, was purchased by the Prince Consort, and died at the age of seventeen years at the Royal Farm, Windsor. But more remarkable for longevity was the cow, Old Grannie, the first cow in the Polled Herd Book, who died at thirty-five years of age. Mr. Watson had many celebrated animals, but his greatest success was Old Jock, a bull that was never beaten, and was sold in 1844, at the then large figure of 100 guineas. Mr. McCombie himself benefitted largely from the influence of the Keillor bred Angus. Mr. Walker bred many other celebrated animals of their day, including Strathmore, Windsor, Pat, and Second Jock, which last beat all the bulls in a sweepstake at Perth in 1852, when he was over thirteen years of age. The work of improvement was ably carried on by Mr. Bowie, of Mains of Kelly, to whom Mr. McCombie was indebted for Hanton, who, with Angus and Panmure, he describes as his herd's fortunes.

A great many prizes fell to Mr. Bowie's herd. And Lord Southesk, who at one time had a most valuable strain, derived great benefit from Mr. Bowie's Cup Bearer. Probably no one did more to make known the merits of Angus Aberdeen cattle than Mr. McCombie himself, whose champion steer at Smithfield, Black Prince, was such a remarkable animal both for scale and symmetry, that he travelled to Windsor for Her Majesty's inspection, and a baron afterwards graced the royal table. And it will be in the memory of our readers, that amongst the last triumphs of the Tillyford herd, was the special prize at the Paris International in 1878 for the best group of beef-producing animals. Amongst other extinct herds must be noticed that of Mr. William Fullerton, first of Ardvie, and afterwards of Ardestre. This celebrated breeder owned Black Meg (766) and the bull Panmure (51), from Brechin Castle; from both descended a host of notabilities; and Mr. Thos. F. Jamieson,

Mains of Waterton, Ellon, a great authority, states that, having devoted much attention to the subject, he found that all the best blood traced back to three fountain heads—viz. 1st, Mr. Fullerton's Black Meg; 2nd, the bull Panmure; and 3rd, the Keillor Jocks. Mr. Fullerton's description of Black Meg, which bred calves till close on twenty years, is as follows: "She was low on her legs, but of lengthy and heavy build on small bone. Her back was straight as a rush, and her tail so well set that you could never tire to stand behind her and look along her back. Then her hooks were so level, wide enough and not too wide. Then her ears and eyes full and sticking well out; then her beautiful jaw and muzzle, with fine good-natured expression of face, were such, and when taken as a whole, why one could stand and look at Meg and not weary for a whole hour as she chewed the cud." Mr. McCombie states that he was indebted to Mr. Fullerton for his best stock of the female line, and purchased at his sale the Queen, whose descendants in the female line achieved extraordinary success in the show yard. Space precludes more than the mention of Mr. Robert Scott's herd at Ballwyllo, Forfarshire, which was well-known in the prize ring. Of Lord Southesk's cattle at Kinnaird Castle much might be said, but it must suffice to note that he bred in 1857 Erica (843), by Cup Bearer (59). Sold in 1861 to Sir Geo. Macpherson Grant for 50 guineas, and the foundress of the most fashionable and highest priced family that has yet been produced. Terrible havoc was made by rinderpest in most of the leading herds in 1865, and nowhere were the losses more severe than at Kinnaird Castle, where the herd was practically destroyed. Coming down to present times, the place of honour must be awarded to the Ballindalloch herd, which, though of ancient date, was made notorious by the present Sir G. Macpherson Grant, whose first purchase was the Erica cow in 1861. Jilt (973) was bought soon afterwards from Tillyfour; then Sybil from Castle Fraser, and valuable members of the Pride tribe; and mainly from these foundations the herd has been built up, which has for many years occupied as foremost a place as the Warlaby cattle amongst Shorthorns. Great care has been exercised in the selection of bulls. Out of a host of celebrities we may name the Tillyfour bred Trojan (402), dam Charlotte, by Black Prince of Tillyfour, and largely imbued with Panmure blood, described in a private catalogue as having done more good to the herd than any other sire. The Jilt bull Juryman, and the several Ericas bred at Ballindalloch, as also Young Viscount from Montcoffer, all did good service and helped to stamp Ballindalloch stock with a character and style which has resulted in the realisation of high figures. Both Ericas and Jilts, which are probably the families of highest value in the herd, trace back to Keillor origin. A large number of herds were founded from twenty-five to thirty years since, amongst which may be named the Cortachy herd of Lord Airlie; the Drumin herd by the late Mr. James Skinner, father of the present owner; Mr. T. Ferguson, of Kinochtry; Mr. C. Grant, Mains of Advie, from whence came the Advie Roses, a celebrated family; whilst far older, indeed probably the premier existing, herd is Mr. A. Bowie's, Mains of Kelly. Space would fail even to mention the half of the herds which are to be found in the counties named. But those who desire to gain more information should read the History of Polled Aberdeen or Angus Cattle by Macdonald and Sinclair, published by Blackwood and Sons. A very recent herd, but one that has come rapidly to the front of late years, is that of Mr. Geo. Wilken, of Waterside of Forbes, who has been a large exporter of cattle to the States. We must also notice the late Mr. Walker's herd, at Montbletton, as one tracing back over fifty years, which is now carried on in reduced numbers by his niece, Miss Cruickshank. The Mayflowers, descended from Lady Craigo, were well-known in the showyard. That we have not named others must not be taken as any reflection on their merits; what we desire to indicate is that in Aberdeen and surrounding counties the cattle have been almost unanimously adopted by breeders, and the book we have referred to contains information on some one hundred and thirty distinct herds. We would rather trace their introduction and progress across the border, where they promise to be hereafter more popular than at present; and if this occurs it will be in no small degree owing to the exertions of Mr. Clement Stephenson, who has had unprecedented success at the fat shows, and more recently has taken several prizes at breeding exhibitions.

The Balliol College farm herd, at Long Benton, near Newcastle, is of comparatively recent origin, and its existence is due to the fact that a polled Angus bull on the common cows then on the farm impressed his characteristics on 80 per cent. of his produce. Mr. Stephenson, in his first catalogue, says: "Having for many years been engaged in a large veterinary practice, with special opportunities for forming an opinion on the merits of the different breeds of cattle, and having for the last eleven years been a farmer and feeder of stock, I believe this breed of cattle stands pre-eminently forward, both to the farmer and the butcher, as being hardy and healthy, good milkers both in quantity and quality, easily fed, good beef producers, coming early to maturity, and highly prized by butchers. In 1881 the Erica bull Englishman, then a calf, by Young Viscount out of Edith by Juryman, was bought from Sir G. M. Grant, and so splendid an animal did he turn out that when his services were no longer required by Mr. Stephenson he was sold for 300 guineas to Mr. Owen Wallis, of Bradley Hall, where we saw him in the spring of 1886. He is succeeded by Evander, another Erica, and also from Ballindalloch; he is by the Jilt bull Julius (1819) out of Evening, by Elcho, and promises well.

There are at Long Benton at present very few Ericas or Jilts. The principal families are the Prides, Lady Idas (Montbletton); Prides from Tillyfour; Abbess family from Easter Tulloch, &c. Mr. Stephenson has done a feat hitherto unparalleled in showyard annals, viz., winning the Elkington Challenge Cup three years in succession. His last triumph was with Luxury, a two-year-old heifer out of a cow called Lemon, which came from Mr. Walker's herd at Portlathen, which also took Champion in London, 1885, as best animal in the hall. We do not remember her exact age when shown, but her live weight at Islington was 1714lb., and her carcase weight without inside fat or appurtenances was 1318lb., being at the rate of $76\frac{3}{4}$ per cent., which speaks volumes as to thickness of flesh, quality, and lightness of offal. Mr. Stephenson's success in founding a herd encouraged his neighbour, Mr. O. C. Wallis, of Bradley Hall, Ryton-on-Tyne, who commenced operations in 1881, by purchasing all the Ericas he could lay hands on, supplementing with members of other well-known tribes. He has now a valuable and extensive herd, and the influence of Englishman and the Jilt bull Justinus by Challenger, born in 1884, should tend to increased uniformity. The animals are well managed, and we have the elements of a very valuable herd. Mr. W. B. Greenfield, of Beechwood, Dunstable, who has a valuable herd, began operations by purchases from the Glamis sale in 1880, to which other valuable animals have been added. Col. Godman, of Smeaton Manor, near Northallerton, has a good herd, which also dates from 1880, selections having been made by that good judge, Mr. Robert Bruce, comprising promising females of the Easter Tulloch May Queen, Prides, Kinochtry Princess, and Favourite, &c. Mr. Arthur Egginton, of South Ella, near Hull, and Major Dent, of Ainderby House, near Thirsk, have small herds of good cattle; and we believe Mr. Loder cultivates the herd as well as Shorthorns at Whittlebury. Thus it will be evident that these cattle will thrive out of their native districts; indeed, one of the strongest points in favour of the Angus-Aberdeen is their hardy character and adaptation to high situations and moderate pastures. At Long Benton, on strong land, cows that are not in profit are given a winter's run, with a shed at night, when they have some oat straw, and on such moderate diet condition is maintained. At Bradley Hall, where the climate was too severe for shorthorns, half-a-dozen females which principally from bad management were not breeding properly, were, during the winter of 1884 and 1885 turned out on a rough hill-side on the higher part of the land, which was very much exposed, and here they lived throughout the winter without any shelter whatever, or any other food than the rough grass they could pick up, except that for ten days, when the ground was covered with snow, a little hay was supplied; thus entirely supporting Judge Goodwin's experience as to hardy character. As to this and ability to thrive on high, exposed pastures, there is probably little to choose between these cattle and the Galloways, but the latter are probably more suitable for a very wet climate, and as a rule have more hair. The typical colour is black, and all black is preferred, but many animals of excellent lineage and valuable personal qualities have some little white under, chiefly on or about the udder. Occasionally we find a shade of brown, which is not rejected. Now and again a red calf is dropped, which favours the conclusion that originally colour was various; such sports are never bred from. The absence of horns is even more typical than colour, and hence the presence of "scurs," *i.e.*, small rounded pieces of horn, sort of rudimentary horns, is always most objectionable, and no animal having the least tendency in this direction should be bred from. In form the Angus-Aberdeen is more cylindrical than the Shorthorn; the carcase should be lengthy, deep, wide, and well-proportioned, but the points should be more rounded off—for example, whilst the quarters should be long, level, thick, and deep, they should be rounded off at the tail and not square, and this because approach to the Shorthorn type indicates the probability of a cross at some period. The following description is taken from Macdonald's and Sinclair's work: Head of the male not large, but well set on; muzzle, fine; nostrils, wide; only a moderate length from eyes to nostrils; eyes mild, large, and expressive; the poll high; ears of fair size, lively, and well covered with hair; throat clean, not jowly; neck pretty long, clean, and rising from the head to the shoulder top, surmounted by a moderate crest. The neck should pass neatly and evenly into the body, with full neck vein. The shoulder blades should lie well back, fitting neatly into the body; this structure tends to good fore ribs, a very important point; chest wide and deep; the bosom well forward between the fore legs; crops full and level, with no falling off behind them; ribs well sprung, barrel-like, and neatly joined to crops and loins; back level and broad; loins well covered, not so wide in the hook bones as the Shorthorn; quarters long, even, and rounded; the tail coming neatly out of the body, not higher at the root than the line of the back, should hang straight down close to the body; on both sides of the tail the quarters should turn away in a rounded manner, swelling out downwards, and passing into thick deep thighs; twist full, and hind legs wide apart; flank full and soft, and under lines even, bones fine, flat, and clean. All over the frame there should be a rich and even coating of flesh, no patchiness, indeed, in the best specimens of the breed, symmetry is perfect, and not exceeded by any other breed. The skin is moderately thick, but soft and pliable; hair short, but thick and soft. In the cow the head is finer and longer, the neck thinner and cleaner, shoulder top sharper, the bone altogether finer, the skin not quite so thick, the udder of fair size, and milk vessels large and well defined. The Angus-Aberdeen is not so celebrated for volume of milk, as for yielding a fair quality; but it is as a meat producing breed capable of early maturity that they are most remarkable. It is claimed for them that they will give as good an account of themselves for food consumed as any of the most cultivated breeds, and that the quality of flesh and the proportion of carcase to offal are superior to any other breed. The figures we have given as to Luxury are, we believe, unique in this respect. We think it will not be disputed that, according to actual bulk Angus-Aberdeen cattle give greater weight than any other breed. They are also valuable for crossing purposes. Some of the best feeding animals ever produced have resulted from the first cross between them and the Shorthorn, and, whichever way they are bred, the prevailing colour is black or grey. The most usual plan is to mate the polled cow with the Shorthorn bull, and this is probably the safest plan; but we have known good animals from the reversed parentage.

As regards management it is much the same as in other improved breeds. Those who keep their breeding stock in the most natural condition will assuredly get the best results, and it is especially important that animals with such a natural tendency to lay on flesh should not be too much forced in early life. With steers destined for the shambles it is different, the calf should be kept going on. Mr. Wm. Anderson, Wellhouse, Alford, thus describes his system of preparing bullocks for the London Christmas market: The calves get milk for at least six months; but after six weeks old, if fed by hand, the milk should be mixed twice a day with a small allowance of porridge made from bruised linseed or bruised oil cake, quantity to be increased as the calf grows older and stronger; a daily supply of cut turnips and straw should also be given. For the first fortnight after birth the calf gets a small quantity of milk four times a day, after that it gets milk three times a day, on to twelve weeks at least, and from thence till weaned twice a day.

Before this it should have learnt to eat linseed cake, getting one pound a day up to one year old, after which decorticated cotton cake may be substituted with good results. In the first winter turnips should be given twice a day, and plenty of oat straw. When on grass neither yearlings nor two-year-olds have any cake; the former remain out till October 1st, but the latter are taken up about the end of August or beginning of September, to be specially prepared for the London market. Plenty of turnips and straw with two pounds a day of decorticated cotton cake, and later on bruised oats in addition sufficed to produce prime animals weighing from eight to eight and a half hundredweight. Mr. Anderson's experience is that he can make 5s. a hundredweight more of the pure bred than of their crosses, because the bone and offal are less, and the quality of the beef first rate. To an English feeder the quantity of artificial food is small, but the roots and straw are of much superior quality. Moreover many might consider that whole or sliced roots was not the most economical method of feeding. One thing is quite certain, viz., that healthy good breeding stock can only be secured by very natural treatment, and, whilst the young bulls and steers for the grazier cannot, in reason, be done too well, dairy cattle and heifers from a year old must be kept on the natural produce, and not allowed to become fat, if we would maintain the hardy, healthy, and prolific character for which the breed has been famous. Like the Hereford, too little attention has been paid to dairy properties, and this is a mistake which breeders are beginning to recognise, and will, we trust, correct.

CHAPTER XIV.

THE AYRSHIRE BREED OF CATTLE.

By GILBERT MURRAY, ELVASTON CASTLE, DERBY.

WHATEVER improvements or alterations were made in the breed of live stock during the unsettled state of Scotland prior to the Revolution in 1688 arose either from accident or from natural causes, as the principles of breeding were then unknown. The powerful influences which soil and climate exercise upon every species of live stock, more particularly on those which are constantly exposed to the elements, are so great as to have fixed the breed of animals in every quarter of the globe. So completely is this the case that, though great improvements have been effected by the superior intelligence which nature has conferred upon man, yet through all the different varieties of live stock we trace the distinguishing peculiarities of their several districts which were originally stamped on them by Nature. The connection between the soil and climate and the cattle that are reared and fed in each situation is so intimate that they cannot be separated; at the same time we do not deny that they may be greatly modified by artificial means.

The breed under consideration is indigenous to the county from which it derives its name. The county of Ayr is divided into three separate districts; that of Carrick, the southernmost, embraces the whole of the county south of the river Doon; Kyle, the central division, occupies that part lying between the rivers Doon and Irvine; whilst Cunningham stretches north of the river Irvine to the confines of the county. This latter division claims to be the cradle of the improved Ayrshire dairy breed of cattle.

Prior to the year 1780 the inhabitants of this part of Scotland passed through a long period of religious feuds and dissensions, entailing on the people a great amount of suffering and privation. Possessing education and intelligence considerably in advance of the age, the farmers of those days were ever the warmest defenders of their homes and their religion; even at this distant date many a solitary moss-covered stone is still held in reverence, and marks the last resting-place of one who had fallen in support of the cause he had espoused.

Until the country emerged from this condition but little attention could be given to any branch of husbandry. The cattle in the best parts of Cunningham were then of small stature and badly fed; they were mostly black, with white spots on their faces, back, and other parts of their bodies. The cows had high-standing crooked horns, marked with very deep ringlets at their base—a true indication of their meagre fare. The improvement of the Ayrshire breed dates from about the year 1780, first by a cross with a stranger breed, combined with a better system of feeding. Aiton, who wrote a survey of the county, and who was himself a farmer in the district of Cunningham, could well recollect the appearance and condition of the cows in that district as far back as 1766. After great pains taken to inquire into the origin of the present celebrated breed, he was of opinion that they are descended from the native stock of the district, changed in their colour, and partly in their shape and qualities, by being crossed with the Teeswater or Dutch breeds. It is impossible to trace out all the crosses that were made between these strangers and the native cattle of Cunningham, and even to say explicitly who it was that first brought them into the district. In 1750 the Earl of Marchmont purchased from the Bishop of Durham several cows and a bull of the Teeswater or some other English breed, of a light brown colour, spotted with white. These his lordship kept for some time at his seat in Berwickshire. Bruce Campbell, who was then factor on his lordship's estates in Ayrshire, carried some of the breed into Kyle; from thence their progeny spread throughout the county. A bull from this stock was sold to Mr.

John Hamilton, of Sundrum, who raised a numerous herd from that strain. About the same date Mr John Dunlop cultivated the stranger breed at Dunlop House in the Cunningham district. However valuable the breed has now become, it is said the first offspring of the cross was far from being of the best shape. The race was chiefly propagated by coupling bulls of the stranger with cows of the native races, and, as the former were far superior in size to the latter, as might naturally be expected, the progeny had at first an ill-shaped mongrel appearance, with bones large and prominent. But these cattle soon toned down, accommodating themselves to the state of the pastures; and the improvements that began about that time to be made on the soil of the western counties rendered the pastures capable of supporting much heavier stocks.

The most desirable quality of dairy cows, of any breed, is that they should yield a large quantity of milk in proportion to the quantity of food they consume, and that when dry they should feed quickly. The pure-bred Ayrshire certainly excels all other in the former, and as to the latter she is no way inferior to many of the best established breeds inhabiting these islands. Of the quantity of milk which an average Ayrshire dairy cow yields it is difficult to speak with precision. There is not only a great diversity between some of those animals and others, but the quantity and quality of the food, the size, age, and habit of the animal, distance from or to the time of calving, all exercise a marked influence on the quantity of milk yielded at any given time. Whatever is said on the subject is open to contradiction by such as are disposed to cavil. Aiton, in his survey of the county, says that some of the dairy cows in Ayrshire yield for a time from five to six gallons of milk per day. Such returns are, however, rare; yet many, when in their best plight and well fed, will yield four gallons per day for three months, and during the season produce a total of 800 to 900 gallons per cow. Many will, however, not yield more than half that quantity, and probably 600 gallons per cow during the year may be taken as a fair average of the Ayrshire dairy stock. Some of the best farms show an average of 620 to 650 imperial gallons as the average of the year; whilst on others, where high feeding is practised, the average has run up to 680 and 700 gallons. On the majority of farms the average yield is 600 gallons. As a matter of personal experience, during the past summer I have had two young Ayrshires, who dropped their second calves the latter end of April, and gave the large yield of five gallons each per day for upwards of two months. I need scarcely say they had a liberal allowance of artificial food. Aiton goes on to say that, since the publication of his survey, the farmers have satisfied him that he has underrated the produce of their cattle, and that they have furnished him with satisfactory proofs of various cows having produced from six to seven gallons per day for several weeks. These, he remarks, are, no doubt, extraordinary returns.

Then, as now, the farmers were in the habit of letting their cows to dairymen at a fixed rent per head, the farmer furnishing the dairy plant and the necessary food for the stock, the dairyman performing the whole of the labour. At that time the rents were from 15*l*. to 17*l*. 10*s*. per cow per annum; the calculation then was that 30 gallons of milk produced 24lb. of marketable cheese, or 12¾lb. of milk to each pound of cured cheese. Descending to our own times, the following is the result of a milking competition held at Ayr on the 26th and 27th days of April, 1861, viz.:

Name of Owner.	Greatest Milking.		Average of Four Milkings.		Weight of Butter.	
	lb.	oz.	lb.	oz.	lb.	oz.
A. Wilson	28	12	24	3¼	2	2
J. Hendrie	26	0	24	5	2	14¼
W. Reid	25	7	20	8¾	2	9
W. Reid	30	15	27	5¼	3	6½
R. Wallace	28	14	28	8¼	1¼	9¼
R. Wallace	25	5	23	8¼	1	15

In this case, the greatest yield at a single milking was rather over three gallons, which produced at the rate of 15lb. of butter per week. We have here no record of the quantity of milk required to produce 1lb. of cheese. In the Derby cheese factory, where the milk of three hundred and sixty cows was manufactured into cheese during the year 1871, taking the average of the season, 11½lb. of milk produced 1lb. of cured cheese.

The most approved points of the Ayrshire cow are: Head small, but rather long, and narrow at the muzzle; eye small, but quick and lively; horns small, clear, and crooked, and placed wide apart at their base; neck long and slender, tapering towards the head, with no loose skin below; shoulders thin; fore-quarters light; hind-quarters deep and large; back straight, and broad behind; the joints rather loose and open; carcase deep; pelvis capacious, and wide over the hips, with round fleshy buttocks; tail long and thin; legs small and short, with well-bent joints; udder capacious, broad, and square, stretching well forward, but neither fleshy, low hung, nor loose; the milk veins large and prominent; teats short, and all pointing outwards, and at considerable distance from each other; skin thin and loose, and the hair soft and woolly; the head, horns, and all those parts of least value should be small, and the general figure compact and well proportioned.

The Ayrshire farmers prefer their dairy bulls to possess the feminine aspect in their heads, necks, and fore-quarters, with broad hook bones and hips, and full in the flanks. They likewise pay particular attention to the formation of the small teats of the bull, and also to the colour of the scrotum. If this were any other colour than white, though the animal might otherwise be possessed of great merit, he would immediately be rejected by the best breeders.

The farmers of Ayrshire have long devoted great attention to the improvement of their dairy cows. When cows are kept solely for the dairy, and are profitable in proportion to the quantity of milk they yield, self interest would stimulate the farmer to acquire the most correct knowledge of cultivating the desirable qualities in his stock. If one cow excelled in milking, they would look out for others in which the leading characteristics were fully developed; they would rear the calves of the best milkers, knowing that they would to some extent inherit the good qualities of their dams. It has been chiefly by these means, and not by changing the stock or crossing with bulls of other breeds, that the Ayrshire dairy stock of the present day has attained its unrivalled perfection.

The improved breed was first planted in Carrick by a Mr.

Fulton, in the year 1790; and in 1802 the first herd was established by Mr. Ralston in Wigtonshire, on the south side of Lockryan. They were introduced into Dumfriesshire towards the end of the last century, gaining a footing on the estate of Hope Johnstone, of Annandale. Dairy-farming has spread rapidly in the south-western counties; in these parts the Ayrshire breed is gradually taking precedence of all others. The Ayrshires bred and reared in Galloway are generally longer, thicker on the chine, rounder in the chest, and heavier in the fore-quarters, and less capacious behind, than the native-bred Ayrshire. They seem better fitted for the purpose of the grazier and the butcher than that of the dairymaid, thus furnishing another proof of the effects of soil and climate on the natural propensities of animals.

The superiority of the Ayrshire for dairy purposes is now generally admitted; they are to be found in every county from John o' Groats to the Land's End. The demand on Irish account is steadily increasing. Hitherto the Irish farmer, as a rule, has devoted more attention to breeding and feeding than he has to the products of the dairy. For many years large numbers of the best animals have been exported to America, where they are said to succeed remarkably well; and, as a proof of the value attached to them on the other side of the water, the Americans have established a herd book, in which the pedigree of all the best-bred animals is entered. An Ayrshire Herd Book is now an established fact, which will eventually tend to enhance the value of the breed.

At present the price of ordinary dairy cows ranges from 14*l.* to 21*l.*; there is always a keen competition for the best class of cows. Many dealers hold commissions to purchase all the most promising animals for exportation; show cows often realise from 50*l.* to 70*l.* each. Where so many excel, it would not only be invidious, but would occupy too much space to enumerate the breeders' names. We would strongly recommend those who wish to see the true representatives of the race to attend one of the agricultural shows which are yearly held in the county, and feel sure the most fastidious would be gratified. A visit to the Ayr Spring Show will well repay the trouble and expense by those interested in the breed.

There are many interesting features in dairy management peculiar to the south-western counties of Scotland. The cows are frequently let to men who either pay a fixed rent per cow, or deliver over to the farmer a stated weight of cheese! these men are provincially called "bowers." The farmer owns the cow, and furnishes a stated quantity of food, the bower and his family performing the whole of the manual labour of feeding and attending to the cows and making the cheese. On many of the Ayrshire dairy farms there is a very limited area of permanent pasture, many of the farms being under arable culture, and managed on a five or six course rotation. The cows are principally pastured on the one or two years' seed layers, which on good land keep a large quantity of stock. We have known twenty-four imperial acres of second year's seeds to pasture twenty-two Ayrshire dairy cows and a bull from the first of May to the end of September. The Scotch dairy farmers, as a rule, use hay very sparingly; on most farms oat straw is substituted, and of this they have an abundant supply. When the cows are let to a bower, the usual allowance is from five to six tons of roots per cow, in about equal proportions of swedes and common or Aberdeen turnips, and $2\frac{1}{2}$cwt. of bean meal to each animal. The rent per cow varies in accordance with the quality of the pastures and the merits of the herd. From 3cwt. to 4cwt. of cheese per cow when rendered in kind, and from 10l. to 12l. per cow when paid in cash are the average rates which are now obtained. On many of the high-lying farms, where the land is less suited to arable culture, with a large breadth of inferior land in permanent pasture, and at a low rent per acre—on this class of soil, it is now considered to be the most profitable system of management to combine dairy farming with stock rearing. Hence on many farms of this description the Ayrshire cows are crossed with a polled Galloway bull, and the whole of the produce reared on the farm, and either sold off as stores to the grazier, or made off fat to the butcher, at from two to two and a half years. The crosses prove kindly feeders, and attain from eight to nine scores per quarter, at the ages mentioned above. On some of the better qualities of land the Shorthorn bull has been used to cross the cows. With good keep the produce attain heavy weights at an early age; they are considered less hardy than the Galloway cross. There is no breed of dairy cattle in these islands that will produce an equal quantity of milk, butter, and cheese from a given quantity of food with the pure-bred Ayrshire.

CHAPTER XV.

WEST HIGHLAND CATTLE.

By JOHN ROBERTSON, BLAIR ATHOLE, N.B.

THE breeding and rearing of cattle have in recent times become subjects of special importance. The increase of the population, the wonderful prosperity in trade, and the consequent high rate of wages which prevailed for some time prior to 1874, combined to promote the consumption of animal food among the working classes to a very high degree, while the home supply did not at all increase in proportion; and, concurrently with these causes, the fall in the price of grain, consequent on foreign competition, forced the British farmer to abandon tillage to a great extent, and to take to the production of animal food, in the hope of coping successfully with the foreign producer. This hope was realised for a time, but for the last few years the importation of cattle and sheep—dead and alive—from America, Australia, and New Zealand has seriously handicapped the British farmer in this department of his calling also, and it therefore becomes a matter of vital importance to him to produce the best description of animal food at the least possible cost.

Notwithstanding the sad depression in trade all over the country for the last ten years, the habits of good living, beneficially contracted by the working classes during their previous great prosperity, still prevail, as far as means will allow, and, fortunately for the consumers, the supplies of cheap foreign meat bring animal food still within the easy reach of working men. But the quality of imported dead meat cannot compare with home beef or mutton, and the consequence is that, although importation places animal food of a kind within the reach of almost all classes, the best home-fed meat will always bring the highest price, and be used by those whose means enable them to command it. The means of internal transit by rail and steamer are now so developed that distance from the great centres of population and demand, is a matter of comparatively little moment, and the traffic in live and dead stock has become one of the most important and active in this country, especially between the north of Scotland and the great English towns.

The breeds of feeding cattle chiefly reared in Scotland are the Shorthorn, the Black Polled, the West Highland, and crosses.

In Scotland, as in other countries, Shorthorns for long held the field against all other breeds, but of late years the Black Polled breed of Angus and Aberdeenshires has been coming prominently to the front, especially in the north of Scotland, and as fashion appears to prevail in stock-breeding, as well as in lighter and less practical departments of human industry, it is to be hoped that the day of Highland cattle is coming too. As a step in this direction, there was formed, chiefly through the efforts of the Earl of Dunmore, who has himself a large fold of Highland Cattle in the Isle of Harris, the "Highland Cattle Society of Scotland," having for its object to maintain the purity of the breed of Highland cattle, and to establish a Herd Book as a record of pedigrees. The society was established at a meeting held in Edinburgh at the Centenary Show of the Highland and Agricultural Society in July, 1884, with the Duke of Athole as president, and Mr. Duncan Shaw, W.S., of Inverness, as secretary, and in 1885 was published under Mr. Shaw's care, the first volume of the Herd Book recording the pedigrees of 561 Highland bulls, the record being mainly retrospective; and a volume for cows is now in forward preparation.

Of the purely Scotch breeds of cattle the Highland, though not now so numerous as it once was, nor so highly prized as some of the southern breeds, deserves special attention, not only on account of its being the original breed of the west, and of a great part of the north of Scotland, but also on account of its importance in point of numbers, its suitableness to the climate and herbage of a great extent of country, its quality as food, and its value as a breed to cross with. It is very difficult to trace the origin of any distinct breed of cattle, because climate, soil, treatment, and other conditions combine through time to impress special physical characteristics; but there can be no doubt that the horned, shaggy, hardy, and comparatively small breed of cattle now best known as the West Highland, has for ages been the breed peculiar to the mountainous districts of Scotland, although it is now chiefly confined to the counties of Argyll, Inverness, and parts of Ross, Nairn, Perth and Dumbarton.

Until about a hundred years ago, the mountains of the north and west of Scotland were depastured with "black cattle," as the Highland breed is still frequently called from black having long been their prevailing colour; but from that period they have been gradually displaced by flocks of sheep. Previous to that time the stock of a Highland farmer consisted of cattle, horses, goats, and a few sheep, the goats and sheep being generally penned at night, and the remains of those "pens" are now frequently seen in the form of grassy mounds or the foundations of ruined walls on the green sites of the old "shealings" or summer quarters of the Highland tenant of olden times. About the end of May the women and children migrated with the cows, ewes, and goats to the "shealings," always located in the grassiest and most sheltered glen on the holding, and there they remained till harvest time, making butter and cheese in abundance, the men coming and going between the home in the strath—where the tillage, such as it was, had to be attended to—and the summer quarters in the mountains. The grazings were usually held in common by a number of tenants, and a "shealing" therefore formed a small hamlet of huts, chiefly built of turf on a stone foundation, with an earthen floor, and with beds of heather, and generally with such sanitary arrangements as would now call for an inspection from the "local authority" of the district, but yet, as Burns says,

> Buirdly chiels and clever hizzies
> Were bred in sic a way as this is.

This migration enabled them to get the benefit of the high and distant pastures and to save the grass in the valleys for hay and winter food for the stock. But the provision made for winter keep was unmethodical and precarious, and the frequent result was that a severe winter, or, still worse, a severe spring, cut off by sheer starvation a large proportion of the stock. The only compensating consideration was that at that time the stock did not represent much money, and that the rent was not difficult to pay, being partly in kind (*i.e.*, in stock) and partly in labour, as well as in money. It is not over a hundred and twenty years since a grazier in the district of Rannoch, in Perthshire, not reckoned at the time by any means a wealthy man, lost from starvation in one spring a hundred and twenty head of cattle. Now, however, the system of management—if system the old practice could be called—is entirely changed. In many parts of the Highlands, especially in the north and west, sheep farming on a large scale, chiefly conducted by men of capital from the south of Scotland, has entirely superseded the cattle-rearing practice of former times; and, indeed, over the whole Highlands sheep have now for at least sixty or seventy years very much displaced cattle on hill pastures. Whether this sweeping change has been physically beneficial appears now to be doubtful, for grazings which were fifty years ago notably healthy for sheep have now in many districts become very much the reverse, and many shrewd observers attribute this unhealthiness in a great measure to the exclusion of native cattle from the rougher and ranker pasturage of the lower grounds, especially of woodlands and marshy lands, which pasturage is very suitable in summer for Highland cattle, but very injurious to sheep in the end of the year when they descend to the lower grounds. This, among other influences, points to a return, in some degree, to a more mixed system of grazing than has for some time prevailed, and no breed of cattle is so well adapted to this mode of farming as the native Highland breed.

Over a great part of the mainland and islands of the counties of Argyll and Inverness, in the north-west of Perthshire, and in the highlands of Dumbartonshire, Highland cattle are extensively bred and reared on the lower lands, generally with marked improvement and success, and in many instances to great perfection; and there is every reason to believe that in the extensive districts named this breed of cattle is the most profitable to cultivate, because from its hardy character it will thrive both in summer and winter under circumstances in which the smoother coated and softer southern breeds would pine or perish. Highland cattle are easily fed, and the quality of the beef is admittedly superior, and consequently in great demand; and a score or two of splendid three-year-old Highland oxen are no unusual feature in a noble English park, which they adorn when in life, and at the owner's table they are no less appreciated when they appear as roast beef at Christmastide, for, like Goldsmith's venison,

> Finer or fatter
> Ne'er roamed in a forest
> Nor smoked on a platter.

The objection to Highland cattle as compared with the southern breeds is, that they do not mature so early, and that therefore there is not in them the same quick return of capital, and the objection is so far perfectly sound. A Lowland farmer therefore finds it more profitable to breed shorthorns, polled, or cross cattle; but the Highland farmer rears his Highland cattle at little or no expense beyond the value of the hill or rough natural pasture on which they are kept in summer, and of such meadow hay or straw as they get in winter, and he is thus enabled to sell them at a paying price at the age of six quarters, or two years, to the southern farmer or dealer, who buys them and carries them on to profit until they are prime fat for the butcher. They are not, however, so well adapted for court or yard keep as for the open field, or for tying up, and that is a disadvantage.

The characteristic appearance of this breed of cattle is well known. A well-bred animal of almost any species is a pleasing object to behold, but there are perhaps few animals familiarly known to us so graceful in form and movement, and so picturesque in colour and coat, as a thoroughly well-bred and well-conditioned Highland steer or heifer in its free condition. In form it possesses all the points so much and so justly prized in the shorthorn—the straight back, the short, straight legs set well apart, the broad chest, the breadth of loin, the well-sprung and deep rib—in short, the squareness and solidity of form which imply strength and weight whether in man or beast, while the nobly branching horns, widely sprung from a broad, flat, and well-haired forehead; the large, full, and steady eye; the short, broad, and well-bred muzzle; the large and well-opened nostril; the thick coat of strong (not coarse) hair of richest colour, in black, or red, or dun, or brindled, impart a picturesqueness which is still further enhanced by that grace and deliberation of movement so distinctive of all animals well reared in perfect freedom.

In former times a very extensive trade was carried on, chiefly by dealers or "drovers," who bought Highland cattle from farmers at home and at district fairs in the north and west, and sent them to England and to the southern Scottish counties, the great mart for this trade being the Falkirk trysts, held in August, September, and October, and at which the southern dealers and farmers met their brethren from the north. Within the memory of men not yet very old, before railways or even fast coaches were in existence in Scotland, and before the distinctive peculiarities of districts and races were so much effaced as they are now too rapidly becoming, the scene presented by a Falkirk tryst to an observer of men and manners was a very animated and striking one. There, year by year, met crowds of men remarkably different in form and face, and dress and language, any one of whom might be selected as a type of the shrewd and active man of his own district or race, to transact business on a large scale in the open field, amid the strange din of lowing herds, barking dogs, flourishing of sticks (and sometimes more), frantic torrents of purest Gaelic, broken English, vigorous Lowland Scotch, and honest Saxon English, and not unfrequently ankle-deep in mud—the whole scene not unlike a mimic and bloodless Flodden. "Falkirk trysts" are still held, but the improved means of communication, and the establishment of auction sales of stock at many centres, have almost superseded fairs, and the "trysts" are rapidly falling off. Whether the new state of matters is better than the old for the farmer is very doubtful, for certain it is that stock now, in the course of its short life, changes hands many times, and every change means expense and profit or loss to somebody. A story is told of a remarkably shrewd Highland grazier of the last century, who, from his wealth and position in his own district, was styled "the Baron," and who had a fine fold of "black cattle." The Baron told a friend that he was going to Falkirk tryst with his cattle. "What!" said his friend, "*you* go to Falkirk to sell cattle, without a word of English in your head?" "Never mind," said the Baron, in his native tongue; "I have no English, it is true; but my cattle will speak for themselves and me too."

The English trade in Highland cattle has, from various causes, such as diminished supply and preference for other breeds, for a time very much declined; but great numbers of the better class of cattle are still bought for English pastures, where they thrive and fatten to a great weight—bullocks to 14 or 15 stones per quarter.

The largest folds of Highland cattle are in the Long Island —in Harris, Uist, and Barra—and in the Isle of Skye; but in all the islands of the west coast of Scotland this is the breed almost exclusively reared, and in no other part of the country are its leading characteristics more fully developed. The nature of the pasturage, the moist climate, and the comparatively mild winters consequent on vicinity to the sea produce hair and horn such as the inland pastures of the mainland cannot, except under very favourable conditions, rival; but, on the other hand, the inland pastures produce much heavier cattle than the island or seaboard pastures ever do. Perhaps one of the finest large herds ever seen in the west was that which belonged to Messrs. Donald and Archibald Stewart, Perthshire men, in the island of Harris, fifty or sixty years ago. By judicious selection from the best folds of the day in the counties of Inverness, Argyll, and Perth, they brought the breed to notable perfection, and showed what could be done by careful breeding; and their three-year-old bullocks, of large size, with hair like goats and horns like buffaloes, used to attract great attention in the districts through which they were annually driven to Falkirk tryst. The herd of Mr. John Stewart, of Ensay, now worthily maintains, at Highland Society shows and elsewhere, the reputation of the old Laskintyre fold in Harris, and of what was the well-known Duntulm herd in Skye. Mr. Alexander Macdonald, of Edenwood, Fife, has at Balranald, in Uist, one of the largest folds of pure Highlanders in Scotland, numbering over a hundred breeding cows with their followers. This fold is of long standing and of high repute, and the present owner has introduced new blood from the best inland herds, which will improve the stock in size. The Earl of Dunmore has within the last twelve or fifteen years established a large and select fold of Highland cattle at Rodil in Harris, and the sales of drafts from this fold at Inverness within the last few years worthily attracted the attention of breeders. His lordship's well-known energy and attention to pure and judicious breeding will make this very soon one of the leading herds in Scotland.

The late Dr. MacGillivray, in Barra, had a large and well-managed herd of Highlanders, which is still kept on. In the islands of Mull, Jura, and Islay there are fine herds of Highland cattle, Islay especially being noted for its rich and fattening pastures; and in Morven there are also some good herds.

At Ardfinaig and at Kilfinichen, in Mull, there were excellent herds, which have been dispersed, but they were chiefly bought by local breeders, who have at the same time been importing cattle from the best mainland herds. On the mainland of Argyll the famed herd of Poltalloch maintains a foremost place among the herds of Scotland. This stock has been carefully managed and improved by judicious selections from the best inland herds, the last noted purchase having been the champion prize bull of the Centenary Show of the Highland Society, Calum Riabhach, bred by the Duke of Athole. Indeed, throughout the higher grounds of Argyll, the class of cattle which is seen on almost every farm is superior, and there is no better place for getting good Highland heifers than the June fair at Dumbarton, which is still a good market for Argyllshire Highlanders. Doune fair, in November, used to be the best market for Highland stots or bullocks; but this fair, like others, is falling off. On the mainland of Inverness-shire Highland cattle are not so much bred as they are in Argyll, probably because the climate and pasture are not so well adapted to cattle as to sheep, and although the pastures on many farms would rear fine cattle, they are generally reserved for winter feeding for sheep.

In the valley of the Spey, however, from Cluny Castle to Grantown, Highland cattle of a very good kind are largely reared; but of late the black-polled have been freely introduced, and they should well suit the climate and pastures of lower Badenoch and of Strathspey. At Castle Grant the Earl of Seafield, a good many years ago, established a herd of Highland cattle, which is still maintained in high efficiency under the rule of the Countess of Seafield. In the county of Perth Highland cattle have very much decreased in numbers within the last forty or fifty years; but there is now a reactionary tendency, and they are again receiving deserved attention in the upper glens of Athole, Breadalbane, and Balquhidder. Before the period just named there were numerous remarkably fine herds in the districts of Glenlyon, Rannoch, Breadalbane, Callander, and Balquhidder, and the names of Messrs Donald, John and Charles Stewart, Glenlyon; McLarens, Braes of Rannoch; M'Laren, Callander; M'Donald, Monachyle, and others, were familiar over the Highlands as famed breeders; but at that time the means of communication were limited, and animals of rare excellence were as a rule known only to a local few, and therefore they were but seldom removed to distant folds for change of breed, as is now so easily and so frequently done. Notwithstanding, a bull was bought from a Rannoch herd for 120*l.* nearly fifty years ago by Mr. Macneill, of Colonsay, and the result was visible in an excellent herd sold a few years ago after the death of the late Lord Colonsay. From the best of these Perthshire herds was selected with great care and judgment the nucleus of the famous stock owned by the second Marquis of Breadalbane, which, at the time of his death in 1862, was probably the finest in Scotland. For many years Lord Breadalbane took a personal interest in his Highland cattle, and in both their selection and management he had the assistance of his friend and neighbour, the late Mr. Stewart Menzies, of Chesthill, than whom there was not perhaps in Scotland a sounder judge of Highland cattle. The Breadalbane stock was carefully drafted every year, and the annual sales in October afforded for many years an excellent opportunity to farmers and other breeders of improving their stocks by purchases of pure blood. When the Breadalbane herd was sold in 1863 the principal purchaser was the late Duke of Athole, who then established at Blair Athole a fold which has since well maintained the character of the Taymouth cattle. Mr. Malcolm, of Poltalloch, was also a purchaser at the Breadalbane sale of 1863, and the Athole and the Poltalloch herds, and that of the late Mr. Robert Peter, of Aberfeldy, were very representative of the Breadalbane stock. The sale of this stock attracted much attention at the time, and the prices realised were very unusual for Highland cattle. The late Duke of Athole bought the celebrated red bull Donull Ruadh at 135*l*. He was sold some years afterwards to the late Hon. Lady Menzies, Rannoch Lodge. The late Duke of Hamilton bought a fine three-year-old dun heifer at 126*l*., which he afterwards sold to the late Duke of Athole, and which, as the property of the present Duke of Athole, stood the first prize cow at the Highland Society show at Inverness in 1865. A dun bull of the same line, belonging to the Duke of Athole, won the first prize as a three-year-old at the Centenary show in Edinburgh in 1884, and the first prize in the aged class in 1885 at the Highland Society's show at Aberdeen. The great feature in the Breadalbane herd was great weight combined with fineness of quality, and the representatives of that herd still show those characteristics in a marked degree. The present Marquis of Breadalbane has again established at Taymouth a very good fold, chiefly from the best Perthshire herds, and they always carry high honours at the national shows. The folds of Mr. John Stewart, Bochastle, Callander, and of Lord Kinnaird, at Rossie Priory, represent some of the best strains of Perthshire blood, and are much thought of.

It may here be mentioned that in 1882 Mr Robert Campbell, of Merchiston, Manitoba—a Glenlyon man, and for long a valued agent of the Hudson Bay Company—imported to Manitoba a number of Highland cattle selected from the Athole, Bochastle, and other good folds; and he reports that they have suited uncommonly well. They are entered now in the Highland Herd Book.

The management of Highland cattle varies considerably in different districts, and according to the size of the herd. In the larger and more reputed folds the cattle are at large summer and winter, the breeding cows only being placed in separate pens or sheds at calving time, which is usually from January to

March or April. In winter the stock generally get meadow hay or straw, and in many cases a few turnips in the open field, in addition to what rough grass they pick up in the woods and other sheltered places; and it is surprising how they maintain condition under such treatment during the most severe winters. For some time after calving and until the young grass comes on in May, the calves are kept separate from their dams, and let in to them to suckle three times a day; but when the cows are turned afield the calves are turned out along with them, and remain at foot until they are weaned, which is usually about the beginning of October. The experiment of allowing the cows to calve in the open field, and letting the calf follow the dam at will from birth, has been tried, but the result was that both cows and calves became very wild, and the cows were often dangerous to approach. In general, Highland cattle are gentle and good-tempered; but when left to roam at large in the woods or on the hills, where they seldom see the face of man, or at least of a stranger, they become shy, and, like all wild animals, guard their young with jealous care; and the means of offence and defence at the command of a Highland heifer are not to be lightly regarded by the most courageous. In some good folds, in Inverness-shire for example, the breeding cows are housed and milked like dairy cows, and the calves reared by hand; but this is done only on mixed farms, partly arable and partly pastoral. A Highland cow yields nothing like the quantity of milk that an Ayrshire does, but the quality is much richer. The age at which Highland cows calve is usually four years, because it is found that, unlike softer breeds, the heifers are not at maturity until they are three years old, and of course breeding at an earlier age stops their growth. The usual practice with Highland farmers is to draft off in October or November their old cows and surplus young stock, the latter generally at six quarters old. Prices of course vary with demand and quality but from 8*l.* to 12*l.* is the ordinary range of prices for the better sort of this class of young cattle.

The crossing of Highland heifers with Shorthorns is a subject which is often discussed, and generally viewed with great favour by good judges of both breeds of cattle, but the experiment does not seem to have been yet tried with such success as to have commanded much attention. There may be various reasons for this, but it occurs to us that a main cause is that the experiment has hitherto been chiefly, if not exclusively, tried by southern breeders crossing two-year-old heifers or aged cows with Shorthorn bulls, producing in either case a diminutive offspring. If three-year-old heifers were brought direct from the hills and crossed with a pure-bred Shorthorn, and afterwards maintained on their usual "sober" fare, there is every reason to expect that the result would be satisfactory; and no cross is so likely to be useful in upland districts as this, combining, as it should do, the "growthy" qualities of the Shorthorn with the hardiness of the Highlander.

This valuable breed of cattle does not receive the attention which it deserves, especially at the present time, as a means of utilising much waste pasturage and of improving the condition of the smaller class of tenants or crofters. It is undoubtedly the breed best adapted to the Highlands of Scotland; but no race of animals can withstand cold and hunger for many generations without marked deterioration in size and quality, for the young of all animals require both food and warmth for their due development, and there is no greater mistake than that of thinking, as many do, that exposure to cold and hunger tends to hardiness, which, however is truly the result of good keep and comfort. If the crofters were practically taught this lesson, say, by the providing for a township or commonalty of a well-bred bull to be kept by a responsible member of the township at the common cost; and if they were taught to provide a little natural hay or meadow grass and some shelter for their young cattle in winter, they would discover the advantage in the enhanced value of their stock, and the discovery would lead to a general adoption of the improved system.

CHAPTER XVI.

THE GLAMORGAN BREED OF CATTLE.

By MORGAN EVANS.

SOME of our old breeds of cattle are rapidly disappearing, and it is well to note their existence ere it is too late, and to record their merits and failings whilst any trace of them remains. The exertions of a few eminent breeders have raised some of our indigenous cattle into world-wide repute. Other once celebrated breeds have not been so fortunate, and have been all but crushed out in rivalry with contemporary animals that have been cultivated with more care and further developed towards perfection in shape, size, and quality. The tendency of modern agriculture is to obliterate local breeds of farm stock. Improved farmyards, improved systems of cropping and manuring land, gradually lead to the adoption throughout the kingdom of improved breeds of cattle to the exclusion of purely local strains. The Shorthorns, Herefords, and Devons are ever extending the boundary of their influence, and the counties from which they originated are not now, as at one time, the sole homes of these cattle. Where high farming is practised, one of our fashionable breeds is generally adopted. Great hardihood may be dispensed with when cattle are never exposed to cold and rain. Early maturity and rapidity in fattening when in the stalls are the qualities most sought for. Wherever a country is in a highly advanced state of cultivation, the hardy native oxen of the district become abandoned for breeds more suited to the commercial interests of the farmer in a time when quick returns on his capital are of vital importance. Rough waste pastures give way to broad acres of wheat and turnips; active grazing cattle are consequently replaced by large docile animals, which accumulate great weight of flesh when kindly treated in stalls or boxes. The plough is no longer drawn by oxen, but by stalwart horses, or by the more powerful aid of steam; and the modern farmer, instead of selling all his cattle from the summer grass, to be turned into profit by the grazier, manufactures meat himself by the aid of cake and corn, and sells the produce direct to the butcher.

Among the breeds doomed to extinction is the once so well known and highly prized Glamorgan breed. The Glamorgans are of ancient lineage, and their origin is hidden in the past. They belong to the class called middle-horns, and in character and antiquity of descent they rank with the Herefords, Devons, the Welsh black cattle, and other allied breeds. As far back as the twelfth century, it is said that a Norman knight, Robert Fitzhammond, who had seized a great portion of Glamorgan, introduced some Normandy cattle into the county, which are supposed to have been crossed with the native cattle. The swelling crest of the Glamorgan ox is by some traced to the influence of this admixture of foreign blood. Youatt says that the influence of Devon blood could not be mistaken at the end of the last century, and attributes it to the importation of Devons into the district by Sir Richard de Grenaville, one of the knights who at one time divided the lordship of Neath. It is certain, however, from all legends and historical accounts, that the Glamorgan cattle are a very old breed, and were the native cattle of the district from a very early date, and that their principal characteristics remain unchanged, Norman and other breeds notwithstanding. In more modern times the Glamorgan farmers were particularly careful of their breed, and we are told that in the last century they prided themselves greatly on the fact that they admitted no admixture of foreign blood into their cattle. The Glamorgan cattle soon became famous. Stock in England at that time were fed at grass. There was no stall feeding and no improved Shorthorn. There was land to plough, and active strong oxen did most of the work. The ideal of a good breed consisted in the females being hardy and profitable milkers, and the males active, docile, and strong workers in plough and cart, and beasts that, when their allotted period of farm labour was done, would, at six or seven years old, fatten into brave oxen on the broad English pastures, on their way up to London and other great centres of the beef-eating population. The Glamorgan breed was celebrated for these desired qualities, and about the commencement of the present century they were highly prized and much sought for by the great English graziers and feeders in the counties of Northampton, Warwick, Wilts, and Leicester. George III., who has been dignified by Youatt with the character of being a "good judge of cattle," was very partial to this breed. He stocked his farm at Windsor with them, and periodically recruited the herd with fresh blood from the Welsh country fairs.

Notwithstanding the high patronage of a king, and other circumstances which might be thought favourable to their development, they have gradually declined in character and in numbers, until at the present time there is no pure herd of these cattle to be found in the county where they were so long held supreme. The Glamorgans are almost extinct. A cow here and there of the old type might be found, but they have greatly degenerated in size and quality; and a pure-bred bull of the true sort it would be difficult, if not impossible, to find. The reason of the decline of these once famous cattle is popularly attributed to the high price of corn during the French Revolution and the succeeding wars of Napoleon, which were followed by the breaking up of the old fine pastures of Glamorgan for the purpose of growing greater breadths of grain crops. Why the Glamorgans should succumb under such influences more than many other well-known breeds I cannot say. It is, however, certain that from that time less care was bestowed on them, and they diminished in number. Consequently their fame became more circumscribed, and when the farmers of the county once more turned to breeding cattle, they took advantage of the fashionable improved breeds that had already gone so far in advance of their native stock.

The Shorthorns, Herefords, and Devons had stolen a march on the Glamorgans. The native breed still held its own for a long time in the dairy, being much superior to either of its rivals in milking qualities. A few energetic breeders now rose to do battle for the cattle of their forefathers, and, although very great progress was made, and a fair standard of perfection at certain points was attained, it became impossible to stem the tide against the invaders. Three or four local breeders were prominent to the last, but death and other changes caused the last strongholds to give way, and we might almost say the end has come.

The breeders of greatest note in late times were Mr. David, of Radyr, and Messrs. Edward and Christopher Bradley, of Treguff. Respecting what was commonly known as the "Treguff breed," Mr. Edward Bradley—who, in 1872, at the age of eighty-six enjoyed lively and pleasurable reminiscences of days when he championed his favourite breed—kindly recounted its history to me in a letter which lies before me. When a boy he had often heard conversations by practical men on the superiority of the Glamorgans, and a regret expressed that they had not been taken more care of for the purpose of the shambles, instead of being yoked to the plough in the farmer's team; and he found himself, some seventy years ago, in a position, with the aid of a younger brother, to enter into an undertaking towards the restoration of the breed of Glamorgan cattle. "The origin of the Treguff breed" (he says) "was purchased from a mountainous district in this county, and she possessed great valuable points to be admired as desirable towards establishing a herd likely to become valuable. This cow was selected as having a frame to be desired for the dairy as well as for breeding. She was comparatively a small animal, showing a capacious udder and a perfectly formed body, straight over the back and loins, deep chested, wide hips, short legs, and a particularly small bone. Her hips and shoulder joints were round—such as are seen generally in the Hereford, Shorthorn, &c.—showing the greatest aptitude to fatten quickly. Her colour was an admixture of brown and bay or red. This cow was in calf, and was the founder of the Treguff breed, which in a few years established its celebrity at the provincial shows. My brother, who knew well the characteristics of the different points requisite towards furthering the value of the breed, took advantage of every opportunity to purchase cows having qualifications likely to improve the stock. The Glamorgans for several years became great favourites. They were valuable in the dairy, and no meat in the butcher's stalls showed more quality, nor would any other breed vie with them in thick marbling throughout the carcase. Many of them were fattened at three years old for the butcher, and the Treguff cattle have taken upwards of sixty prizes at the Tredegar and other shows against Herefords, Durhams, Devons, &c. A prize given at one of the Tredegar shows for 'the best three-year-old beast of any breed' was awarded to my three-year-old, weighing nineteen score a quarter, and he was sold to a Newport butcher for 53*l*. After my brother's death, and at the expiration of the lease of Treguff farm in 1850, the stock were all disposed of by auction; the breeding portion became crossed with such description of animals as were at the time in the purchaser's possession."

The Glamorgan cattle produced a rare quality of meat, highly prized in the metropolitan and provincial markets. They were profitable to the butchers, being well lined inside with tallow, and their meat, from its first-rate quality, always commanded the highest price. The average weight of cows, says one who wrote in 1814, was at that time from eight to ten scores, and oxen from twelve to fourteen scores per quarter. Youatt thus describes the breed: "They were of a dark brown, with white bellies, and a streak of white along the back from the shoulder to the tail. They had clean heads, tapering from the neck and shoulders; long white horns, turning upwards; and a lively countenance. Their dewlaps were small, the hair short, and the coat silky. If there was any fault, it was that the rump, or setting on of the tail, was too high above the level of the back to accord with modern notions of true symmetry."

Martin says they were a superior breed, "generally of a red or brown-red colour, often with white faces, and otherwise varied with white. The head was small, the aspect lively, the neck inclined to be arched, the carcase round and well turned, the back rising to the root of the tail, which was peculiarly elevated." That the Glamorgans had white faces is incorrect. Youatt also errs in a similar way, for he illustrates his description of them with a drawing of a white-faced cow from the

Royal farm at Windsor. No pure-bred Glamorgan ever had a white face. The most fashionable colour for a Glamorgan cow was an admixture of a rich brown with red. The bulls were invariably black, with, of course, the usual white markings, and many of the cows were of the same colour.

The Glamorgan breed at one time extended through the counties of Monmouth and Gloucester. There can be little doubt that the old Gloucester cattle were descended from the Welsh stock, and they may therefore claim a notice here. They also, like their progenitors, are almost extinct. The only existing herd of pure Gloucesters is that kept by the Duke of Beaufort at Badminton, where they were first established nearly a century ago. Fifty years ago there were two other herds in Gloucestershire, the one at Leonard Stanley, and the other owned by the late Colonel Kingscote. The former was sold off in 1843, and the Kingscote stock in 1852. The Badminton herd now alone remains. Several cows and heifers were bought at the sales of the Stanley and Kingscoté herds to introduce fresh blood into the Badminton strain; since that time no change of blood from the old sort has taken place, but many years ago four heifers of the best Glamorgans that could be obtained were purchased, and bulls bred from them. The cross did not effect any change in their character or colour, but reduced the size materially. The true Gloucester cow shows a good deal of character, being a lengthy, good-looking animal, light fore, but deep hind quarters, with good milking points. The body is brown; head, nose, and legs black; well-shaped white horns with black tips; tail and top of rump (or ridge of tail) white; white udder with black teats; the upper side and end of tongue is also black.

The great peculiarity is the white mark, which extends from the loin along the ridge of the tail, and down between the hind legs to the fore part of the udder. Only a sufficient number of calves are reared at Badminton to keep up the herd, which numbers about fifty head; consequently the draft cows are sold to the butcher. I am indebted to Mr. John Thompson—who has so ably managed the Badminton estate since 1842—for much of the above information. In 1872 he wrote to me:

Although good, hardy animals, they fatten slowly, which I think is the chief cause of the breed not having extended; but when fat the meat is very good, having a larger portion of lean meat to fat than most animals, and are considered by the butchers "good cutters;" but I have no doubt a well-bred Shorthorn would make 3lb. of meat from the same food that a Gloucester would require to make 2lb.

To the foregoing information I am happy to add the following letter, which I have just received from Mr. John Thompson as the sheet is going to press:

I am happy to inform you that the Duke of Beaufort's herd of "Old Gloucester" cows continues to go on very well. About two years ago we were fortunate in getting a little fresh blood introduced, which I think will do great good, without altering the fixed type of the breed in the least. Having heard that Dr. Price, Llantrissant (generally known in Wales as "The Druid"), had preserved some of the old breed in its purity, we managed to get him to part with four of them (two bulls, one cow, and one heifer). The three-year-old bull and the cow are as true specimens of the breed as I have ever seen. If they had been bred at Badminton they could not have matched the herd better. . . . . We have now a young bull by his old one out of one of his grace's cows, also one by his grace's bull out of Dr. Price's cow, which will soon disseminate the fresh blood throughout the herd. I may mention that although we have gone on so long without any change of blood whatever, there was no falling off in the character or constitution of the animals, the only noticeable feature being a barrenness in the heifers; for some years past not more than a moiety of them have bred, a failing which I hope the fresh blood now introduced will remedy.

Several attempts have been made to improve the Glamorgans by crossing with Ayrshires, Herefords, and Devons, but without success. A breed so old, and of such a fixed type, was not likely to blend harmoniously with any other. Improvement could only come from within; all attemps to improve by crossing must and did end in failure.

CHAPTER XVII.

PEMBROKESHIRE OR CASTLEMARTIN CATTLE.

By MORGAN EVANS.

IF WE LOOK for the oldest breed of domesticated cattle in Great Britain, we must evidently turn to the western coast. With the descendants of the ancient British people will be found the descendants of the ancient British cattle. The successive invasions on the east and south coasts of our country of conquering Roman, Saxon, Dane, and Norman, drove the aboriginal inhabitants with their cattle into the west and far north of the island. There can be no doubt that the Welsh and the Scotch Highlands are the oldest of all our existing breeds of cattle; not even excepting the white cattle in Chillingham Park. The oldest colour I believe to be black, notwithstanding much popular tradition to the contrary. From the mention of white cattle in certain ancient records, it has been too readily assumed that the oldest native stock of the country was white. The laws of Howel Dda, or Howel the Good, said to be written in the tenth century, certainly speak of white cattle; and the Dimetian code says that "the privilege of the Lord of Dinevwr is to have for his *saraad* [fine or tribute] as many white cattle with red ears as shall extend in close succession from Argoel to the palace of Dinevwr, with a bull of the same colour along with each score of them." Youatt says that the "same records that describe the white cattle with red ears speak also of the dark or black-coloured breed." I do not know the passages in the records to which he alludes. Indeed, it appears that there was not in early times any uniformity in colour or characteristics of our herds and flocks, much more than now, for among the sayings attributed to Catwg Ddoeth, or Catwg the Wise, we find:

> The best of the ewes are the polled,
> The best of the cows the spotted [or grey, *brithion*]
> The best of the horses the gentlest,
> The best of the goats the white.

The distinct specification of colour in the Dimetian code supposes the presence of cattle other than white. It may even be assumed, as far as the above proves to the contrary, that white cattle were not the common stock of the country at the time, although they appear to be the sort most highly prized. The great favour in which they were held might even be because of their rarity; possibly also because of real or traditional value attached to them; or, as has been conjectured, they may have been larger than the blacks. The tribute, however, appeared to demand a special rather than a common kind of beast. There need have been no selection of a particular colour, unless cattle of another kind co-existed with them, and, indeed, prevailed to a greater extent. There is no doubt of the great antiquity of a race of wild white forest beasts. They are alluded to very distinctly, in 1598, by John Leslie, Bishop of Ross, and also by Hector Boece, in his "History and Chronicles of Scotland," quoted by Professor David Low in his work on the "Domestic Animals of the British Islands." And, according to Speed, King John received from Maud de Breos 400 cows and a bull, all white and with red ears, as a present to his Queen, in order to appease his Majesty, whom her husband had offended. This variety, hitherto considered the oldest, is of very early origin, yet still possibly an offshoot developed from the older blacks, and perpetuated in certain centres by natural selection. Their present descendants, the white cattle of the parks, frequently throw back calves, appearing to revert to the earlier type. Mr. Darwin evidently doubts the great antiquity generally attributed to them, for the following passage occurs in his work on "The Variation of Animals and Plants in Domestication," p. 85: "The cattle in all the parks are white; but, from the occasional appearance of dark-coloured calves, it is extremely doubtful whether the aboriginal *Bos primigenius* was white;" and he also adds that certain facts which he enumerates show "that there is a strong, though not invariable, tendency in wild or escaped cattle, under widely different conditions of life, to become white with coloured ears." There is a general belief in Wales, as in England, that the old breed of the country was white, and many specimens of this sort might have been seen very recently in the Principality. Indeed, Low, who published his superbly illustrated book in 1842, selects as a specimen of the wild or white forest breed a drawing of a "cow eight years old from Haverfordwest, in the county of Pembroke." Cattle of this kind are now very rare in Pembrokeshire. The breed of the country is black, and known sometimes as the Castlemartin, but now more generally as the Pembrokeshire, breed. There can be little question of the great antiquity of this breed. "The Pembroke race in England," says Mr. Darwin, "closely resemble in essential structure *B. primigenius*, and, no doubt, are its descendants."

Youatt says, "Great Britain does not afford a more useful animal than the Pembroke cow or ox." There is no breed which for general usefulness can successfully compete with the Pembrokeshire cattle in their native district, so perfectly adapted are they to the climatic and physical characteristics of the country, and to the system of farming generally practised there. The country is hilly, and some of the best pastures are greatly exposed to the storms of wind and rain so frequent in the autumn and winter months. Farmyards with sufficient accommodation for breeding and feeding, after the English fashion, are extremely rare. The landlords are somewhat to blame in this matter for not providing the farmer with better accommodation for man and beast. As a set-off to severe criticism of the landlords, it may be urged that the system of agriculture practised in South-western Wales is generally such as does not demand very extensive outbuildings. To the entreaties of a tenant who asked for cattle sheds, such as were built on other farms on the estate, one landlord is reported to have replied, "I'll build you a new barn and cart house, John; and when I see you grow turnips I will build you a cattle shed." This answer appeared very pertinent to the individual holder, but I think a very self-evident retort might also have been made; at least, no better inducement to growing green crops and stall-feeding cattle can be given than by building the appropriate number of sheds, which the farmer always readily turns to proper use. Improved farming can only follow, and never precede, improved farm buildings, in Wales, as everywhere else.

Many reasons might be adduced to show that the breeding and rearing of stock must long remain the principal feature in the agriculture of the district, and that under existing conditions the native cattle are those most profitable to the majority of the farmers. The counties of Pembroke, Cardigan, and Carmarthen are well adapted for breeding cattle, sheep, and horses. The humidity of the climate is favourable to the growth of grass, whilst the soil is firm and dry. Foot rot among sheep is almost unknown; the hoofs of the colts are well formed and hard, and very different from the spongy, flat-footed animals bred in the fens on heavy clay soils. The bed of outlying stock is firm, and never becomes sloppy in the wettest weather, the undulating country allowing any excess of rainfall to run off freely to the swollen rivulets. The grass is not rank and coarse, but short and sweet. It is not a corn-growing country. The spring season is late, and the autumn weather commences early. Crops that are grown profitably elsewhere are liable to great damage, in consequence of the lateness of the harvest and the wet weather which frequently sets in at that time. In consequence of the humidity of the climate, it is more suited to growing oats than wheat. Beans and peas are a most uncertain crop. Many attempts at growing large quantities of both have been made; the result is that beans have been entirely abandoned, whilst a few acres of peas may still occasionally be seen. But the uncertainty of harvesting this crop in proper condition will be understood when it is proverbially said in Pembrokeshire that no man grows peas more than five years in succession. High farming as in England can never become the rule here; whole parishes may be found in which there is no more than a single field that could possibly be tilled by steam. The late Mr. Mechi's advice for treating thin-skinned land—viz., to plough deeper—would result in imbedding tons of ploughshares fast in the eternal rocks, to be philosophised over by confused antiquarians in the next century, or the plough would turn up the "yellow rab" in such quantities as would ruin any farmer less opulent than a City alderman. In fact, all attempts at high farming—growing crops and harvesting them after the fashion in the midland counties, or in the Lothians—have been delusive. All Englishmen and Scotchmen who have migrated to this part of Wales, have either signally failed or succeeded only by adopting the peculiar farm practice of the country. The late Mr. Rees, bailiff to the late Lord Cawdor at Stackpole Court, having been asked at an agricultural dinner at Carmarthen to give his opinion on Scotch farming, said he knew little about it, as there was "no Scotch farmer's grave in Pembrokeshire—none of them ever remained long enough to be buried there." And I remember many years ago being asked whether I ever knew a Scotch farmer who was able to hold his own in Pembrokeshire for more than seven years. I was unable at the time to reply in the affirmative.

The above remarks, will, I trust, be found a digression more apparent than real. I wish to show that, notwithstanding the undoubted pre-eminence in many respects of some of our more widely-spread and improved breeds of cattle, there is still some ground for supposing that in this peculiar locality the ancestral blacks are cultivated for sound practical reasons, and that their improvement is an object worthy of attention. Taking into account the climate, soil, and average homestead accommodation in the country, the Pembrokeshire cattle can be bred and fed cheaper than Shorthorns or Herefords. Surely an ungenial climate must tend to increase the expense of keeping a beast Wintering cattle is dearer than letting them run the fields in summer. The more cultivated and delicate breeds are under the disadvantage in Pembrokeshire of having to be housed a fortnight or three weeks earlier than the blacks, and they must be kept in later for about the same period in the spring. This makes a material difference in the estimate of cost for the year, where there is a mixed system of dairying, breeding, and feeding carried on. There can be little doubt that, in the district under notice, a herd of black cows can be kept fifteen per cent. cheaper than an equal number of Shorthorns, and still yield as much butter or cheese—two articles that form an important item in the rent-producing power of the Welsh farmer. Capitalists holding sheltered and luxuriant pastures, having extensive farm buildings, and who aim at producing large, prime fat beasts, may there, as elsewhere, keep Shorthorns to greater advantage than any other breeds; but persons of limited means, living on poor land, and with small farmyards, cannot do better, I think, than retain and cultivate the indigenous breed of the country. I hold there is no middle course: either blacks or Shorthorns. The Shorthorns are undoubtedly the grandest cattle in existence for early maturity, size, aptitude to fatten, and do well with proper care in any part of Wales, as in almost every part of the globe. The arguments in favour of introducing any other improved breed into the district must be futile, when even the suitability of the Shorthorn, with all its signal merits, is a question open to dispute.

The character of the Pembrokeshire breed may be better shown by comparison or contrast than by repeating the hackneyed phrases usually adopted to describe the ideal features of all oxen—whether Shorthorns, Herefords, Devons, or Welsh. The main requisites in every animal producing flesh for consumption are depth and breadth. A mention of peculiar features in virtues and failings is preferable to vague generalisation. The colour of the Pembrokeshire breed is black. The horns are long and white tipped with black points, widespreading and curving upwards. The head is of medium length, longer than the West Highlands, and somewhat longer than the Devons, approaching the Herefords or the improved Sussex in form. The nose is small, and the neck fine, with little tendency to the "throatiness" observable in some breeds. The eyes are prominent, but without the untameable gleam of the West Highland or Chillingham cattle, domestication having removed any special traits of wildness and of ferocity. The coat is long, not straight like the Highland cattle, but wavy, or sometimes curly. The forehead is broad, and the tail of good length. These may be said to be some of the chief characteristics of the Pembrokeshire breed in contra-distinction to other well-known cattle, although it does not very correctly represent the type aimed at by the breeders generally. For instance, in

Wales no more than elsewhere is a white horn considered the best, but a yellow, mellow, and oily-looking horn, having the unction mark of a predisposition to fatten; a horn in which the black extends more than a few inches below the tip, or one that has a hard blue colour throughout, is to be condemned. Several writers have remarked on the colour of the skin as being of an orange yellow, and the coat on the barest parts of the body as being of a brownish hue. Some of the best breeders in Pembrokeshire are careful to maintain these characteristics in their herds. These, along with a yellow horn and a wavy coat, almost invariably indicate a beast that will feed well either at grass or in the stall. A short, crisp, coal-black coat is much inferior to one that is long and wavy. The outer covering of hair put on in the winter months should, with outlying cattle, at the end of spring and during the early summer months be of a russet brown. One frequently sees cattle of this breed whose coats are one mass of ringlets; but experience, I think, shows that they are not the most easily fattened, and I do not know to what source to attribute this peculiarity. The hair on the forehead of bulls is often very much curled, and it is rather to be admired than otherwise for the sake of its picturesqueness, as well as that it indicates hardihood and masculinity.

The meat produced by these cattle is excellent, and not to be surpassed in texture and quality. The milking properties of the cows are certainly equal, if not superior, to those of most modern improved breeds. I have the authority of eminent London dairymen for stating that Welsh black cows are on the average equal to any class of cows in milk-producing capabilities. The only objection to them at dairy farms around the metropolis is their colour. The admixture of black with red, and white, and roan in the herd is not thought fashionable, neither is it pleasant to the eye.

There is a tradition in Pembrokeshire that the Castlemartins were improved by the importation of Devon bulls, but I do not know on what basis such tradition rests. I do not place much reliance on it, although the red colour of the Devons easily merges into black when crossed with them. The late Mr. Thomas Lewis, of Norchard, Pembroke, many years ago, assured me that in all the best strains of Castlemartins there was a cross of the Herefords. A Mr. Hitchings, of Ermytage, he said, had migrated to the north of Pembrokeshire to the home farm of a Mr. Barham, of Trecwn, and brought back with him to Ermytage, in Castlemartin parish, on his father's death, a lot of half-bred Herefords, and black cows in calf to Mr. Barham's bull. The best breeders, according to Mr. Lewis, adopted the infusion of strange blood, and he says that the most celebrated stock in Castlemartin were soon after impregnated with it.

I have seen and closely watched several attempts to improve the breed by crossing. One effort, within my recollection, to introduce Hereford blood ended in utter failure. The white face was difficult to wipe out, and the progeny of the cross never appeared to work into the blacks. The North Devons amalgamated more easily, and the produce was more of the proper type, but did not on the whole improve its general character. The West Highland cross resulted in small, useful beasts, and, though they sold very readily at the fairs, were not as profitable as the native breed of the district, for there was a loss in size and in milking qualities, with no corresponding advantage gained. I speak of all these crosses from practical knowledge and home experiment.

The late Mr. Richard Harvey, of Haverfordwest, was the first person who made any serious attempt to improve the breed by introducing a cross with the North Wales breed of black cattle, and he produced some remarkably fine stock. The Earl of Cawdor, who has recently become an enthusiastic patron of the black cattle of Wales, has also committed himself to this experiment, and is one of the most successful exhibitors of the day. Notwithstanding all this, although the cross leads to an increase of size, it destroys the original characteristics of the Pembrokeshire cattle. On the whole, this cross was unadvisable in its conception, and unfortunate in its result. The two breeds should have been kept distinct.

The quickest and best way to improve the breed is by making a judicious selection of the old stock of the country. By breeding from the best and weeding the bad, considerable progress might be made in developing it to a perfection hitherto unattained. It is not impossible that a superior class of animal might be produced by judicious crossing; but the path is one beset with difficulties when we have to deal with a breed so distinct and of such a unique character as the Pembrokes. Holding this view, I, in 1867, proposed the formation of a herd book as the only safeguard and guide to improved breeding. The opportunity given was but lightly esteemed, for no more than three or four breeders put forward any claim to making an entry in the book. I also proposed the formation, on joint-stock and co-operative principles, of a herd of no less than twenty of the best cows that could be obtained (with, of course, the requisite number of bulls) to found a herd farm, where scientific principles in breeding were to be adopted, and where an annual sale of the surplus stock was to be made. The estimation in which joint-stock companies were held at that time was fatal to the project; but I still think it to be the most important and feasible ever made to develop the Pembrokeshire cattle. It must be understood that no single breeder was to be found who would attempt the tactics of a Bakewell or a Colling. I therefore proposed a combination of effort, which, had it met with proper support, would in one decade have raised the Pembrokeshire breed to a position they are not likely to reach for some time to come.

The entries I received for the "Black Cattle Herd Book" and subscriptions thereto, I placed in the hands of Mr. Harvey when I left the county in 1870. Soon after, he, in conjunction with Mr. J. B. Bowen, of Llwyngwair, revived the project, and in 1874 the first volume was published. The third and last issue is dated 1883, and it is proposed to publish a fourth this year (1887). Up to the present time there are 324 cows and heifers, and 127 bulls registered as pedigree animals in the "Welsh Black Cattle Herd Book," which bears the unfortunate sub-title of "Pedigrees of Animals of the Castlemartin, Dewsland, and Anglesea breeds."

A few errors current in popular descriptions of the Pembrokeshire cattle have to be corrected. For instance, Youatt says that a "few have white faces, or a little white about the tail or udders," and that the "Pembrokeshire cow is usually black, with occasionally a dark brown, or less frequently a white face, or a white line along the back." Mr. W. C. L. Martin, in his work on "The Ox," following Youatt, commits similar blunders. No white is admissible, except, perhaps, on the udder or scrotum; any other markings of white obviously denote strange blood. Even a white udder or scrotum is not to be admired, and is exceptional. A coat of a brownish colour is not uncommon; indeed, a brown tinge many breeders consider an indication of aptitude to fatten, and as denoting rapid growth in their young stock. I lean very decidedly to this opinion. It is necessary to state, however, that this brown is of a peculiar hue, and the slightest tendency to red must be emphatically condemned.

The old white cattle are nearly becoming extinct in Wales. It may not be uninteresting to record a late attempt to form a herd of these white cattle. I remember a neighbour of mine, an English gentleman, who took a farm in Pembrokeshire (Mr. Tebbitt, of Castlecenlas), collected a few white cows of the old breed common in the country at one time, and, after obtaining a white bull, continued for several years breeding cattle of this character. The experiment was not long persisted in, for he afterwards resorted to Shorthorn bulls, and crossed his stock. But some of the white cattle he produced, especially many of the cows, were remarkably fine. When he inherited a great fortune from his London ancestors, he wisely enough gave up farming in Wales. Of late years Lord Cawdor has made a similar experiment, and has resuscitated a small but unique herd representative of what is commonly supposed to be the aboriginal breed of Britain.

In one instance only have I witnessed the appearance of albinos in a herd of black Welsh cattle. They were a dun fawn, with light eyes of a pinkish hue; some of them were very short-sighted—almost blind. These animals cropped up in a herd of the purest blood in the country. They could all be traced to be descendants of a certain black cow, likewise of the purest strain. The albinos appeared in the herd with great frequency. They were principally females, but the albino cows gave birth to calves of the usual black colour. As their offspring was always black, it did not greatly interfere with the course of breeding pure stock on the farm. It was, however, very annoying to find the tribe that had the best blood in the herd throwing albinos in every direction. The herd was that of my father, the late Mr. John Evans, of Mabus, Pembrokeshire, to a portion of which I succeeded. I have no delicacy at this date in mentioning the fact, for he was the greatest prize-winner of his day, and his cow Ruth, that took eight prizes in the shows of Pembroke and Carmarthen, is remembered to this day. Some of her blood has impregnated, with improving and characteristic features, the best of the existing herds of Pembrokeshire black cattle. *Fiat justitia ruat cœlum.*

The Pembrokeshire black cattle are the principal stock of the counties of Pembroke, Cardigan, and Carmarthen. The county of Pembroke is divided into two sections, the English and the Welsh speaking population occupying separate districts. The hundred of Castlemartin, the south-western part of Pembrokeshire, is, along with the hundred of Roose, entirely English. The black cattle of Castlemartin were fifty years ago pre-eminent in the county. From their geographical position, and from the greater enterprise of Castlemartin farmers, the Castlemartin cattle had become famous before the other stock of the country. The farmers of North Pembrokeshire, living in the hundreds of Dewsland and Kemes, until recently always resorted to Castlemartin for their bulls, and advertisements of sales of black stock invariably held out as an inducement to purchasers that certain bulls and cows were of the pure Castlemartin breed. At the present time these cattle are characterised as of the "Pembrokeshire breed" in all such documents, and also in the catalogues of local shows. Careful breeding commenced with the farmers of Castlemartin, and the stock of the country has been much influenced by the use of Castlemartin bulls. The cattle in the hundred of Dewsland are, however, larger than those of Castlemartin, and at the present time are of equal if not of greater merit than the strain so long considered representative of the Pembrokeshire cattle, and which gave them the distinctive name of Castlemartins.

CHAPTER XVIII.

THE ANGLESEA CATTLE.

By MORGAN EVANS.

ISLANDS always had a peculiar fascination to aggressive people bent on conquest; islands also generally breed heroic men, and possess a distinct fauna. The Isle of Anglesea is famous as the scene of frequent invasions, and as the home of an ancient breed of cattle. Soon after the Roman general Suetonius Paulinus had supreme authority in Britain, A.D. 58, he pushed his forces on to subjugate this little island spot of but 173,000 acres, or about 270 square miles. His troops, it is said, swam across the straits to Mona, as the black cattle imported therefrom periodically swam hitherward during nearly eighteen centuries after. The island was again attacked in the same century under the direction of Julius Agricola, who was sent by the Emperor Vespasian to command the forces in Britain in 78. Besides the numerous affrays between Welsh princes, who appear to have had a faculty for fighting with their kith and kin, and amongst other incursions of its enemies, the Danes in 900, and again in 969, landed there and made great havoc; and in 913 and 966 the Irish, with their peculiar instinct of practising home rule by entering into quarrels with their neighbours, laid the place waste with great cruelty. In 1096 it fell a prey to English troops under the Earl of Chester and the Earl of Shrewsbury. Henry III. invaded Wales in 1245, and made a tool of his judiciary in Ireland to attack Anglesea, but, not coming quickly to the support of his minion, the Irish forces were assailed and driven back to their ships. Edward I., also, in 1277 reduced the island with a fleet from the Cinque Ports. After the inhabitants submitted to English laws and government, the natives became patriotic—perhaps to a fault; for in 1648 they had a council of war, and issued a general declaration in favour of King Charles. With the blind pluck inherent in islanders, they thought their little strike for the king would change the aspect of affairs and avert his impending certain fate. But the last humiliation of the inhabitants was at hand, when on the 2nd of October in the same year they capitulated to the Parliamentary forces under General Mytton, and, like peaceable, good-natured people in novels, they have lived happy ever after.

The island of Iona, one of the western islands of Scotland, was remarkable in ancient times for its want of cows; but the island of Mona, in Wales, has long prided itself on having a good stock of these animals. St. Columbus prohibited cows from grazing on the slopes of Iona, for he said, "Where there is a cow there will be a woman, and where there is a woman there will be mischief"; but saints and sinners agree that Anglesea was the special home of herds of prolific cows, whose progeny were transported in great numbers into the adjacent land to become food for the people of North Wales, and in later times to penetrate into the large pastures in English counties. The ancient British called their island Mon, a name the Romans Latinised into Mona. The Welsh characterised it as "Mon Mam Gymru," or Mon the mother of Wales—supposed by some to refer to its general productiveness, by others to its being possibly at one time the principal seat of learning and Druidical lore. I imagine the term "Mother of Wales" arose from its maternal capabilities in supplying food to the mainland by its corn and hordes of cattle. An old proverb says that "As Mona could supply corn for all the inhabitants of Wales, so could the Eryri mountains afford sufficient pasture for all its herds if gathered together." The late Rev. Robert Ellis, of Carnarvon, celebrated as a Welsh scholar and literary antiquarian, suggested, in answer to some queries put to him by me, a derivation for the ancient name of the isle as pertinent as it is original. The word "môn" signifies cow; thus Pontfôn becomes Cowbridge, "Henfona" a place to keep cows; the Isle of Môn is therefore the isle of cows. However correct this derivation may be, Anglesea has certainly ever been conspicuous as a cattle-breeding island.

The productiveness of Anglesea in cattle has always been great for such a limited area. Roberts's Map of Commerce, published in 1649, gave 3000 as the number of cattle annually exported and swum across the Straits of Menai. The losses by this mode of transit were few; cattle are good swimmers. This fashion of swimming beasts has been known elsewhere along our coasts in equally difficult places. The Rev. Walter Davies, in his "General View of the Agriculture and Economy of North Wales," says that Mr. Lewis Morris, in his "Book of Charts," in 1747, puts the number of animals exported in his time as 15,000, although this must be an evidently exaggerated account. Mr. Davis, in 1810, gives the average export as "not above 8000, from one to four years old." Youatt, writing after the erection of Menai Bridge, considers he does not exaggerate when he estimates the annual export at 10,000 head, of an aggregate value of 50,000*l.*

The black cattle of Anglesea are nearly allied in character and race to those of South Wales. Mr. Darwin definitely pronounces the Pembrokes as descendants of *Bos primigenius*, but thinks, with Professor Richard Owen, that the blacks of North Wales have their origin in *Bos longifrons*. I am not tempted to discuss here the osteological distinctions on which he bases his inferences. History and the geographical situation of black cattle on the western and northern coast of Britain seem to favour the idea of one common origin, and that they are the oldest breeds in the country. They are found at intervals in a line from St. David's Head in Pembrokeshire to as far north as Iceland.

The Anglesea, or North Wales black cattle as they are now generally known—for they are found in Carnarvon, Denbigh, Merioneth, and other counties of North Wales—are very like the Pembrokes. The coat, as with the Castlemartins, should be long and wavy. This generally denotes good quality, and a growing beast easily fattened. In colour they are generally darker than those of South Wales, being a pure black. A little more white is allowed than in the Pembrokes, the scrotum of the bulls and the udders of the cows being very frequently white—indeed some eminent authorities insist on this as being a "characteristic and desirable feature," denoting purity of breed. A white streak is sometimes found along the chine, but this feature cannot be commended. The horns, which may be broadly described as white with black tips, curving gracefully upwards in cows and oxen, are usually much darker-coloured than in the Pembrokes, and the white portion not so mellow and creamy in appearance. They are perhaps a little larger than the Castlemartins, and stand on short strong legs; but they are not so good in the head or shoulder as the Pembrokes. The head of the ox is very frequently heavy and bull-like. Davies, in his time, attributed the "bull-like features in the head and dewlap" of the Anglesea ox to the fact that the calves were not weaned in Anglesea until "double the time at which they are weaned in other counties," together with their not being "gelt until they be about a year old;" but this will hardly account for the persistency of this feature in stock not thus treated. The shoulder is often coarse, and they are inclined to be flat behind the bladebone. In short, comparing them once more with the Pembrokes, they are altogether coarser in the fore part, but have better hind-quarters and broader loins than their southern rivals.

As breeding and rearing cattle has from time immemorial been the pride of Anglesea, the development of good dairy qualities in the cows was for ages much neglected. In the beginning of the present century the island but barely produced enough butter and cheese for home consumption. It is not therefore to be expected that the Anglesea cattle, under such a course of treatment and selection, should have inherited great milking powers, and consequently they are in this respect surpassed by many other breeds.

Mr. T. Congreve, at the Christmas Meeting of the London Farmers' Club a few years ago, speaking as a grazier, deeply lamented that year by year it became more difficult to buy a lot of good beasts to graze. The demand for good grazing beasts is increasing. Our large dairies have relinquished breeding; our stall-feeders are many of them entirely dependent on the fair and market for their stock. The production of good grazing young beasts and those ready for the feeder's stalls will probably for some years prove to be a profitable pursuit—as times go—for farmers in our breeding districts, whatever the breed of cattle they raise, provided they be of the requisite quality. Few districts are so suitable as Anglesea for stock raising; few cattle so highly prized for the quality of their meat, or thrive so well on good English pastures as the black cattle of North Wales. They are hardy, and may be reared at little expense. For some time to come—at least as long as the present aspect of things remains—the farmers of North and South Wales who breed the native cattle will not have cause to regret any attention they may bestow on improving the native stock. And although their farmyards are generally ill-constructed, and deficient in requisite accommodation for feeding purposes, the farmers may become somewhat more reconciled to their position with the increased demand for their cattle consequent on the relinquishment of breeding elsewhere, and the scarcity of healthy animals throughout the country.

Black stock in Wales are always readily bought up by the drovers who frequent the fairs in the districts where black cattle are a speciality. They can be disposed of, half fat, at a pound or two a head more than coloured beasts of the same weight. Whilst this continues, it is some inducement to the local breeders to keep to the type and colour of their beasts—especially for those who have not the proper farm buildings for stall-feeding all their oxen and selling them only as prime fat.

The Anglesea cattle are now cultivated to equal perfection in Carnarvonshire and some parts of the adjoining counties, as in the "mother" isle; and diminutives of this breed are the principal stock of the mountainous districts of Carnarvon and Merioneth. They offer good sound material for development. Bakewell, it is said, thought highly of them in this respect. The Rev. Walter Davies naïvely remarks that this eminent breeder thought that "in some points they were nearer his idea of perfection in shape than any other he ever saw, *his own* improved breed excepted." But they cannot be improved by crossing with English breeds. They will not blend with foreign blood; the colour becomes destroyed and the type broken, and the produce cannot be reduced to an uniform standard. Endeavours in this direction have been fruitless. An improved type of Angleseas must be evolved from themselves, as must also be the case with the Pembrokeshire breed. But, as the only possible cross of the latter not evidently retrogressive is with the Angleseas, so the cattle of North Wales allow of no fresh blood except perhaps that of their kindred in South Wales; and a writer in Morton's "Cyclopædia of Agriculture" suggests this cross as the best, and one calculated to improve them in many important points.

Of late years great efforts have been made to improve the breed, and with considerable success. They have been honoured with a "North Wales Black Cattle Herd-book," the second volume of which was published last year (1886). The number of bulls registered is 138, and of cows 411. The most energetic and untiring of their advocates and the most eminent of their breeders is Colonel Henry Platt, of Gorddinog, Bangor. The gallant colonel in a letter which lies before me writes, "I find a well-reared Welsh beast will make as much or more weight than most of the English breeds. At Islington every year the Welsh runts are amongst the heaviest cattle exhibited. I have taken three first prizes and one second at the Minnesota State Fair in America this autumn [1886], and have succeeded in making the Americans who saw them most anxious to possess some bulls to cross with their grade cattle. Welsh is the coming breed; they have everything to commend them." He dolefully adds, "Is it not sad to think that when a demand arises from the other side we shall not be able to meet it? Our farmers, most sad to say, are selling their two-year-old heifers by hundreds to the butcher!"

Like the black cattle of South Wales those of North Wales out of Anglesea are good milkers. A small herd of dairy cows of the breed has been established by Captain Ross, of St. Alban's, Herts, with great success. One of this herd a few years ago took first prize at the dairy show of the British Dairy Farmers' Association held at Islington.

Cattle like the Angleseas and the Pembrokes should require no apology for their existence or cultivation into an improved type. The peculiar character in climate, soil, and geological contour of the districts in which they are bred, and the system of farming commonly pursued in these parts, are a sufficient excuse to warrant their patrons in adhering to them with tenacity.

CHAPTER XIX.

THE KERRY BREED OF CATTLE.

By the late R. O. PRINGLE,
AUTHOR OF "A REVIEW OF IRISH AGRICULTURE."

FROM the earliest period of which we have any record, the rearing of cattle has been a principal feature in the industrial pursuits of the people of Ireland. This is evident from the frequent mention made of cattle in the early annals of the country; and it is recorded that among the tributes paid to the King of Cashel in the fifth century were about nine thousand head of cattle. At a still earlier period the "inhabitants not only fed upon the flesh of oxen, but were clothed in their skins, formed weapons (pins and fasteners) out of their bones, used their sinews and intestines for strings, and employed different parts of these animals in ministering to clothing and decorative arts."

Such are the statements made by Sir William R. Wilde, in a most interesting paper "On the Ancient and Modern Races of Oxen in Ireland," which he read a few years ago at a meeting of the Royal Irish Academy. Sir William informs us that "cattle formed not only in early times the chief wealth and produce of the country, but were also employed as a means of barter. Thus we read of ransoms being paid with oxen, and as many as 140 milch cows being given for a manuscript." The names of many places in Ireland are derived from circumstances connected with the rearing of cattle, and most of these names have come down to the present day unaltered, or with a very slight variation from the original.

Sir William's paper was founded upon an examination of the

heads and other remains of ancient oxen found in bogs in Ireland, and these relics, he remarks, "are exceedingly curious in an historical point of view, as they afford undeniable evidence that, so far back as the eighth or tenth century at the latest, we had in Ireland a breed of cattle which, for beauty of head and shortness of horn, might vie with some of the best modern improved races so much admired by stockmasters, and which are now being re-introduced from England." The heads of ancient oxen show that four breeds or races of cattle existed in Ireland in early times. These, according to Sir William R. Wilde, were "the straight-horned, the curved or middle-horned, the short-horned, and the hornless, or Maol," or Moyle.

Of the latter, living representatives may still occasionally be met with in remote parts of the country; but the Kerry breed, which belongs to "the curved or middle-horned" race, must be considered the sole modern representative of the ancient breeds of Irish cattle.

The Kerry cow is a handsome animal, and small in point of size. The late Earl of Eglinton, when Lord Lieutenant of Ireland, described the Kerry breed as "the thoroughbreds of cattle." The points of the true Kerry are described as follows in the "Review of Irish Agriculture," published in the *Journal of the Royal Agricultural Society of England*. "The former," that is, the true Kerry, "is a light, neat, active animal, with fine and rather long limbs, narrow rump, fine small head, lively projecting eye, full of fire and animation, with a fine white cocked horn tipped with black, and in colour either black or red." The woodcut of the Kerry cow given in Youatt on Cattle, while correct in most points, gives the idea that the breed is of a roan colour, which is never the case.

We have used the term "true Kerry" on account of certain crosses or sub-varieties which exist of the breed. The first of these is the "Dexter," which, it is said, was introduced about seventy years ago by a Mr. Dexter, who was land agent on Lord Hawarden's estates. It is doubtful whether the "Dexter" variety is the result of a cross with another breed or of selection. If it arose from a cross, it is very difficult to say what breed was used, and we are rather inclined to believe that the "Dexter" variety originated in a selection of males and females having certain distinguishing characteristics, which were perpetuated in after generations by attention to the same principle in the selection of animals retained for breeding. Certainly the points of the "Dexter" variety are sufficiently marked to enable anyone to pick them out after a little experience. "The 'Dexter,'" as stated in the "Review" above quoted, "has a round, plump body, square behind; legs short and thick, with the hoofs inclined to turn in; the head is heavy, and wanting in that fineness and life which the head of the true Kerry possesses; and the horns of the 'Dexter' are inclined to be long and straight." It must not be supposed, however, that the "Dexter" is to be considered a spurious Kerry, although not an original Kerry, which is the meaning we wish to convey by the term "true Kerry." This fact is recognised at all the Irish royal shows, and, what is perhaps of more importance with reference to this point, at the shows of the County of Kerry Agricultural Society, the only difference being that there is no separate classification at the royal shows, whereas the prize sheet of the county society has separate sections for the original Kerry and the "Dexter." The latter is, indeed, a great favourite with many persons, and we have frequently observed that the majority of the prizes at shows where there was no separate classification were awarded to "Dexters."

A cross of the West Highland Kyloe blood has been introduced into the Kerry breed. This, however, has not been done recently, so far as we can learn, nor to any great extent. The traces of the cross are sufficiently evident to those who are familiar with the particular characteristics of the two breeds. The generality of people, however, either do not detect the slight traces of the Kyloe cross which exist in some Kerries, or do not attach any importance to them one way or other.

The greatest drawback the Kerry breed has had to encounter is neglect on the part of breeders, and the result is that it is often difficult to pick ten really nice animals for breeding out of a lot of fifty or sixty heifers as exposed for sale in any of the fairs in Kerry. The remainder would be good grazing beasts, but not such as a fancier would select as breeders. Still, very nice specimens are often picked up amongst such lots, and we have reason to believe that prizes have frequently been awarded to heifers picked up at fairs, and of the breeding of which nothing was known. In fact, entries in royal show catalogues frequently run as follows: "Kerry heifer, the property of Mr. A. B.; pedigree and breeder unknown." These animals had been bought by good judges simply because they had bred back to that character which the old breed possessed. This tendency to breed back to the old type is a feature in the Kerry which proves it to be an original or native breed of cattle, and should be taken advantage of in all attempts to improve the breed.

The Kerry cow is very docile, and is always a great pet when kept by the owners of suburban villas. She has also the property of being able to bear confinement; for instance, we have known a Kerry cow kept for five years in a stable in Dublin; she had only two calves during that time, and was scarcely ever dry, keeping up a full supply of milk for a large family. We have stated that the breed is small in size, heifers, and sometimes even cows, not exceeding 40in. in height at the shoulder. This was the height of Alderman Purdon's very neat first prize heifer, exhibited at the late show of the Royal Dublin Society. She had a calf just before the show, although only two years old, and her milk vessels were as large as those of many cows three times her weight. The dimensions of a fat Kerry cow, which was awarded a prize at a show of the Royal Dublin Society, are given in the article on Irish Agriculture already referred to—namely, "38in. in height at the shoulder, 70in. in girth, and 42in. in length from the top of the shoulder to the tail head," indicating "a weight of about thirty imperial stones." Of course, many Kerries run larger, or rather stand higher.

The Kerry cow is very easily kept, and this characteristic, combined with her milk-producing qualities, well entitles her to the appellation of "the poor man's cow," which has been bestowed upon her. The yield of milk and butter depends, of course, very much on the keep; but we happen to know that the average yield of milk produced by the Kerry cows belonging to a gentleman who has for many years paid great attention to the breed is about twelve quarts, or three gallons, daily, and the average yield of butter from 6lb. to 7lb. per week. Some of his cows have produced more, but the quantities stated have been about the average. These cows are no doubt well kept; but it is a large yield, considering the size of the animals and the comparatively small amount of food which they consume.

The Kerry breed is, however, not only suited for supplying the dairy, but cattle of this breed also fatten rapidly on even middling pasture, and their beef is exceedingly fine and well-flavoured. This is a feature of all the varieties of the Kerry, even of the most neglected specimens of the breed, and hence it is that cattle which would not be selected for breeding are quickly bought up by graziers when driven to other parts of the country.

Considering the merits of the Kerry breed as a really useful as well as a fancy breed, it is gratifying to know that breeders in Kerry have of late been paying more attention to it than they did. The County of Kerry Agricultural Society is rendering valuable assistance in this work of improvement, and we confess we look to that Society for doing more good in this way than either the Royal Agricultural Society of Ireland or the Royal Dublin Society. Of the gentlemen residing in Kerry who have evinced most interest in the improvement of their native breed of cattle we may name Richard Mahony, Esq., Dromore Castle, Kenmare; James Butler, Esq., Waterville, which is a post town; and the Knight of Kerry, Glanleam, Valencia. Several gentlemen in other parts of Ireland have also small herds of Kerry cattle, to the breeding of which they have given careful attention. Among these are Earl Fitzwilliam, Coollattin Park, co. Wicklow; C. Brinsley Marley, Esq., Belvedere House, Mullingar, Westmeath; Capt. Bayley, Friarstown, Tallaght, co. Dublin; and Mr. James Brady, Raheny, co. Dublin, who has for a long period been a careful breeder and successful exhibitor of Kerry cattle. The blood which Mr. Brady has chiefly bred from has been that of the herd belonging to the Knight of Kerry. It was amusing to see Mr. Marley's first prize three-year-old Kerry bull, Rory of the Hills, bred by Mr. Brady, turned into the ring at one of the spring shows of the Royal Dublin Society, to compete with Shorthorn, Hereford, and Devon bulls for the Chaloner Plate, valued at 150 guineas, as "the best bull of any breed over two years and under six years of age." For the first time the judges declined to award the cup to the Shorthorn bull, and gave it to the Devon; but when it was observed that the judges were going to set the Shorthorn aside, many thought, and very reasonably, that the Kerry bull would have carried off the greatest honour of the show.

Sir Robert Peel and the Rev. J. C. Macdona have of late years introduced the Kerry into England; the latter gentleman, after carrying off the prize at the Royal Agricultural Show of 1871, sold his herd to Mr. J. H. Murchison, who possessed the original of our engraving.

The prices of Kerry cattle vary considerably. Ordinary cattle, suitable chiefly for grazing, may be bought at moderate prices; but if a heifer is likely to prove a fancy animal, her value is increased a hundredfold. Prize cows and heifers frequently bring as much as 15 to 25 guineas, and we have known even larger sums refused for choice specimens. The following may be taken as ordinary rates: Common Kerry heifers, well selected, from 6*l.* to 7*l.* each; common cows, well selected, 10*l.* to 11*l.* 10*s.* Mr. James Bogue, Passage West, co. Cork, who has been long resident in Kerry, advertises that he buys Kerry cattle on commission; but, as we have already said, it is not easy to pick up superior animals, unless one travels a good deal through the county, and takes time to look about him. We may mention, however, that the principal fairs in Kerry where Kerry cattle are to be met with are those held at Killarney, Killorglin, Castlemaine, Cahirciveen, Sneem, and Dingle.

CHAPTER XX.

THE JERSEY BREED OF CATTLE.

By JOHN M. HALL.

JERSEY, the largest of the Channel Islands group, whose entire cultivated area, 20,000 acres, is not one-fourth of Rutlandshire, supported in 1885 12,217 head of cattle, of which 6843 were animals in milk. The annual export of cattle from the island amounts to about 2000 head. Of the origin of the breed little is known; but since 1763, when the first of a series of Acts prohibiting importation of cattle from France was passed, the island blood has been kept practically pure. No systematic effort had been made to raise the general standard of the breed, either as regards appearance or profit, prior to 1833-34, when the Royal Agricultural Society of Jersey was founded, and its first show held. From this date commences the visible upward movement of the breed amongst dairy tribes, a movement which in subsequent years was accelerated by the foundation of farmers' clubs in the various parishes (1852-60), by the formation of the Herd Book Society (1866), and recently by the establishment of the Jersey Dairy Farmers' Association, which held its first show in 1884.

In England the early history of the breed is obscured by the perverse misapplication of the term "Alderneys." The island of Alderney exports 100 head of cattle where Jersey exports 2000. Yet to this day Jerseys and Guernseys are still popularly included in the named "Alderneys." Channel Islands cattle were imported to this country in the first half of the eighteenth century. Fourteen animals delivered at Goodwood in 1747 cost the Duke of Richmond au average of 3*l*. 17*s*. 6*d*. per head. In 1811 Lord Braybrooke paid an average of 18*l*. 15*s*. 8*d*. for twenty-four Channel Islands cattle delivered at Audley End.

During the year 1811 (about forty years previous to the practical recognition of the breed in America) Channel Islands cattle began to be systematically distributed throughout this country, chiefly through the agency of Mr. Michael Fowler, of Little Bushey. Of private breeders who since that day have contributed to advance the breed it must suffice to name Mr. Philip Dauncey, of Horwood, whose herd, dispersed in 1869, has made "Dauncey" blood famous on both sides the Atlantic; Mr. Duncan, of Bradwell, the breeder of Marjoram—the head of the Stoke Pogis line of Jerseys; Mr. Walter Gilbey, the breeder of the bull Banboy; Lord Chesham; Mr. G. Simpson, of Reigate; and many others.

In 1871 the Royal Agricultural Society of England separated the classes of Jerseys and Guernseys at their Southampton Show; and eight years later the English Jersey Herd Book had been founded, and its first volume was published.

Jersey cattle having been bred for generations mainly with a view to develop dairy qualities—especially butter production—there is an antecedent probability of a divergence in type from the beef-making animal. This exists in fact. A good specimen of the Jersey cow viewed either from the front or side presents a wedge-shaped appearance: that is, she is wider at the hips than at the shoulders, and is lighter at the front than towards the rear parts. Reviewing her points in detail, we note that her head is long and clean cut; her eyes placid; her horns fine, waxy, and of rich colour; her neck is long, clean at the throat, and lightly set on flat, sloping shoulders; her withers are fine, and her chest broad and deep; her chine is thin, and her back even, uniform and strong over the loins. The spinal processes are somewhat open, giving the backbone a ridgy character. Her rump is long and level; her hips wide, and fine in bone. Her ribs are large and full; her body long and deep in girth, especially in front of the udder, but not too capacious. Her tail is fine from root to tip, hanging neatly over the rump-ends. Her hind-quarters are wide apart; her legs small below knee and hock, and her hind legs stand square, neither "cow-hocked" from a near, nor arched from a side view. Her hide is thin and mellow, moving freely to the touch over ribs and hip-bones. Her skin has a golden tinge inside the ears, at the arm-pits, and between thighs and udder. Her milk veins are prominent on the abdomen, and ramify over the udder. Her udder is capacious and shapely, but not fleshy; when distended with milk, it stretches in form like the arc of a circle, from a point high up between the thighs to, and well forward under, the belly. The teats are moderately large, of equal size, wide apart, and evenly placed. The skin of her udder is soft, rich coloured, and devoid of harsh wiry hair.

Finally, the cow has a general look of *healthy quality*, and is what has been described as "a loose, open-made animal, rather pointed, or sharp and well-defined; in fact, the contrary of what we should look for in a flesh or beef-producing animal."

In the selection of a bull it is of primary importance that his dam should be of a dairy type such as we have attempted to describe. She should be a rich and continuous milker, giving fifteen quarts per day in her flush, and difficult to dry. These points conceded, it is desirable that the bull should be a long, level, shapely, rich-coloured animal, with a general look of quality, and inclining to definition of form rather than to its concealment, amid fat and muscle. His horns and hoofs should be small, throat clean, head heifer-like, loins and hips wide, and flanks deep. If, further, his rudimentary teats be rightly placed (not on the scrotum), wide, well formed and level on both sides, he is the sort of bull to beget stock of high dairy merit.*

The Jersey is a continuous and—relatively to her size—deep milker, but it is her butter qualities that decide her value. The claim advanced on behalf of the breed by its advocates is that the Jersey cow, properly treated, will produce *a larger quantity of butter from a given amount of food than any other animal.*

* Indications as to milking capacity and duration of yield are supplied by the form and development of the escutcheon, and although certain growth of hair may not invariably predicate yield, yet we believe it is true that a bad milker never had good escutcheon marks. In America it is greatly believed in, and also in Jersey, though possibly less now than formerly. Those who would study the subject will find an interesting and able article in the R. A. S. E. Journal, Vol. XXI., 2nd Series, by Willis P. Hazard, Secretary of the Pennsylvania Guenon Commission.

This is a matter of importance to a nation which pays thirteen millions a year to the foreigner for butter and butter substitutes.

In order to ascertain the value of a herd, individually or collectively, *the churn is the simplest and best test*. The creamometer, which measures cream by volume, is inconclusive, owing to the varying densities of cream. Analysis is likewise inconclusive, because the butter value of milk is partly dependent upon physical causes, such as the size of the fat globules, which are not determined by chemical analysis.

In order to establish a practical standard by which to measure the efficiency of their herds, it is not necessary for breeders to have recourse to the island, or to America, for their data; nor to base their conclusions upon phenomenal results whencesoever obtained. Existing records supply abundant evidence as to what has been accomplished by English herds treated on simple practical lines.

The introductory articles, written by Mr. John Thornton, and printed in Vols. I. and II. of the English Jersey Herd Book, contain a mass of information on this point. We quote therefrom a few facts relating to the herd owned by Mr. P. Dauncey: "The herd at Horwood, as a rule, was kept up to fifty cows. . . . . . Careful measurement has often shown fourteen pounds weekly from one cow; indeed, in one instance sixteen pounds was obtained. The greatest yield was in June, 1867, when the entire herd of fifty cows made $10\frac{1}{2}$lb. each cow, and $9\frac{1}{4}$lb. over. The average produce the same year from the whole was 'within the slightest fraction of seven pounds per head per week, dry or milking.' Twenty-two quarts of milk was the highest record from any one cow in one day."—Vol. I., E. J. H. B., pp. 75, 76.

This is essentially a *grass* record, and the yield of a herd of dairy cows treated on Mr. Dauncey's system would necesssarily fluctuate, to some extent, according to the season. But in unfavourable seasons, it is possible for the butter dairyman to use artificial foods advantageously and economically, and in this manner to correct the inequalities of our variable climate.

Bearing in mind our conclusion upon this evidence, as well as upon other sources of information, we assert that it is not unreasonable to expect a herd of Jersey cows to show an individual annual record of 600 to 700 gallons of such milk as will produce one pound of butter for each two gallons of milk set for cream. In other words, in a standard herd each cow should produce her calf each year, and should yield, besides, 650 gallons of milk, churning 325lb. of butter.

Imported Jerseys require extra care until they have passed a winter in England. They should be kept in at night until the summer has fairly set in, and directly the autumn evenings begin to turn chilly. In cold, wet rainy weather, whether spring or autumnal, they are best under shelter.

Home-bred animals are, as a rule, hardier; but *even these* will not yield their best returns unless they receive a different treatment from that which is ordinarily given to a dairy cow. The province of the Jersey is to convert her food into butter. If she secreted fat, instead of giving it away in her milk, she would be better provided against emergencies of food and climate. But her habit of body being such as it is, two conditions are especially requisite to her welfare: warmth, and a sufficiency of regular diet. Both are fulfilled in a favourable season during eight to ten weeks of early summer, at which time an average herd of Jerseys, on grass alone, ought to give in the dairy a return of seven pounds of butter per head per week. But if this figure is to be maintained during periods of drought, or when pasturage is short, or when the autumn grass becomes thin and watery, the employment of some form of artificial food will be found necessary.

Calves.—Sucking calves may be removed from the dam on the third day, when they readily learn to drink from the pail. From three to four quarts, one-third of it being sweet skim milk, will be a sufficient daily allowance for the first four weeks; it may be subsequently increased to eight quarts a day, half new and half skim milk. Should diarrhœa appear, it must be checked at once. First reduce the milk, or altogether suspend it for twelve hours; and apply remedies. Mr. Simpson recommends the following treatment in cases of white scour (diarrhœa): "Immediately the symptoms occur, give one or two tablespoonsful of linseed oil, followed by a tablespoonful of bicarbonate of potash dissolved in water. Keep the calf warm and the shed clean."

The calf will begin to eat a little sweet hay at an early age; and at about eight weeks old will take daily small quantities of bean and pea-meal ground coarse, damped, and mixed with chaffed hay. At twelve weeks old the milk may be gradually diluted to clear water. The main point to be borne in mind in connection with weaned calves—say, four to six months old—is that their food must be easily digestible, and must be sufficient to maintain healthy growth. A mixture of linseed cake two parts, bean-meal, pea-meal, and lentils one part each, will be found to answer, regulating the quantity by the age and capacity of the calf. After nine months old the lentils, bean and pea-meal may be replaced by a mixture of equal parts decorticated cotton cake and maize-meal, being careful to gradually change the food. If the calf feeds badly, a little locust bean-meal sprinkled over the chaff and meal often stimulates the appetite. In fine weather calves may be let out to grass from about six months old, but their feed should be maintained for six months longer. At fourteen months (some breeders advise earlier) heifers can be taken to the bull, and during succeeding summer months, if grass be plentiful, will need no extra food. During autumn and winter they will be better for receiving two to four pounds daily of ground oats and bran, mixed with their chaff and roots.

Cows and heifers in milk.—Summer. If, in addition to grass, any artificial food is given, about four pounds daily will suffice, which may consist of equal parts of decorticated cotton cake and maize-meal. In autumn and winter the daily ration should be gradually increased to seven pounds—say, decorticated cotton cake, linseed cake, and maize-meal, two pounds each; ground oats or bran, one pound. Cotton cake must be of *good quality*, and should be ground to the size of peas to facilitate digestion. This artificial food may, of course, be varied in composition, regard being had to feeding value; its use should be altogether discontinued six weeks before calving, when the cow is dried off.

Silage.—If part of the grass crop is converted into silage, it will improve the colour of the butter in the autumn and winter months. Twenty-five pounds daily is an ample allowance for a cow.

Roots.—Parsnips and carrots are the best for butter production; mangolds, though increasing milk flow, are too cold and watery to be of much service. A few of them, however, kept till spring, may be used mixed with chaff. Cows fed with artificial food, and receiving a daily proportion of silage and parsnips or carrots, will not require much hay. About seven pounds cut into chaff, with four pounds oat straw, will be a fair daily allowance in winter.

In very cold weather the water given to the cows should be slightly warmed.

Before quitting the subject of feeding, it may be remarked that the Jersey, being a small cow, should be fed in proportion to her size. It is very common to overfeed them. If she receives more than is needed to develope her butter production to its full normal extent, the surplus will be to all intents wasted. [For information upon this important and interesting point, viz., the economic use of food, the reader may be referred to the Report on the Vernon Experiments, published by the British Dairy Farmers' Association, pp. 60-63, 1886.]

Milk fever is undoubtedly the scourge of the breed, and its victims include some of the most noted cows, only brief remarks tending to its *prevention* can be offered here. This disease does not commonly overtake heifers or old cows. The greatest danger appears to threaten animals with third and fourth calf, especially those in high condition, or constitutionally deep milkers. The months of July and August are marked as very fatal to cows of this stamp. In all such cases watchful care must be exercised. If the cow be dried off (some deep-milking Jerseys are never allowed to dry) her food must be restricted—artificial food entirely suspended, and green food administered sparingly. She must have fresh air and exercise; and for this purpose may be kept in a close yard, or a pasture eaten close by sheep. Nine days before due to calve a dose of linseed oil—say, a half-pint to a pint—should be given, and repeated every other day if *any tendency to costiveness appear*. If, in spite of these precautions, the symptoms before calving are unusual and such as to cause apprehension, send at once for the veterinary surgeon.

CHAPTER XXI.

THE GUERNSEY BREED OF CATTLE.

By A NATIVE.

FROM time immemorial the island of Guernsey has been famous for its breed of cattle, and a very just reputation it is, for there are very few localities in Europe, and certainly none in Her Majesty's dominions, where a more jealous care has been observed to prevent the mixture of foreign elements. Of course, the isolated position of the island has greatly aided the inhabitants in their endeavours; in fact, we doubt if any but a locality so situated could for so long a period have preserved a breed so intact. The cattle are larger and more valued than even those of Alderney, the name of which is so familiar throughout England. They are exquisitely delicate in form; colours varying from light red to fawn and dun, with a few black, each generally with white intermixed. The head is long and handsome, eye large and prominent, horns gracefully formed. For flesh-giving qualities they are profitable, and for dairy stock they are truly excellent, yielding on the average (if properly fed and cared for) 1lb. of the finest butter per day throughout the year. The size is a fair average, and doubtless the breed would be much larger were it not for the peculiar treatment they have ever been subject to. The two beautiful animals represented in our drawing were Cloth of Gold, No. 1, and Portia, No. 2, the property of the Rev. J. R. Watson, of La Favorita, Guernsey, who has taken great pains to improve the island breed during his residence in that locality. The bull and cow carried the first prize at the show of the Royal Agricultural Society of England at Bedford.

The farms of the island being limited in size, it is found necessary to tether the cattle, whereby they lose much of that exercise and freedom which would tend to larger growth. They are also by this means too frequently exposed to excessive heat or cold without the possibility of choosing the necessary shelter. Notwithstanding these drawbacks, it is really remarkable how well the animals have always thriven. So great is the demand for this breed that, on an average, 700 cows and heifers, with about a dozen bulls, are annually exported.

It is very essential that purchasers of Guernsey cattle should know the character and reputation of those through whom they purchase, inasmuch as the demand has at times induced fraudulent exporters to resort to practices by which other breeds have been only too successfully palmed off on the unwary. The pure Guernseys are chiefly exported through the Messrs. Fowler.

It is interesting to trace from the annals of the island how extremely jealous the native inhabitants have ever been of the reputation of their cattle, and how, when the importation of foreign cattle for breeding purposes has been suggested, the farmers have been alive to the mischief that would ensue, and have not rested until an Act of the Royal Court has been passed to prohibit the introduction of such element, and impose heavy penalties on those who may attempt it.

Early in the present century a feeling was prevalent among certain commercial inhabitants of the town of St. Peter Port (not natives of the island) that the importation of cattle from France and other neighbouring countries would tend to improve the local markets by reducing the price of butcher's meat; also that it would enable the islanders to increase their export trade, since the Guernsey breed was in such good repute, and the limited admixture of foreign breeds would but very slightly, if at all, deteriorate the purity, and, while Guernsey blood preponderated, would certainly not detract from the quality. These individuals, therefore, petitioned the Royal Court to repeal the stringent laws which had for many years been in force regarding the importation of foreign cattle, and allow, under certain conditions, the introduction of bulls, cows, and heifers from other countries. Immediately on this movement becoming known, a counter-agitation was set on foot, and we find the inhabitants of nine parishes combining to send up a petition, wherein they set forth that whatever might be the temporary advantage gained to commerce, it would be an ultimate loss to the agricultural interest; it would, moreover, be an unwarrantable innovation regarding one of the distinguishing characteristics of the island, and perpetrate a mischief never to be repaired. Regarding these petitions, the records show that the Royal Court, after having heard the various arguments *pro* and *con*, and the conclusions of the Crown officers, passed an Act confirming and strengthening the time-honoured custom, forbidding the importation of bulls, cows, and heifers from foreign ports, under a penalty of 2*l.* sterling for each animal, and its entire confiscation, the said fine to be applied one quarter to the Crown, one quarter to the poor, and the half to the informer. It was further ordered that the said Act should be published in the market, and affixed to the other market laws, that no person might pretend ignorance of the same.

A few years later it would appear that a suspicion or fear was aroused on account of cattle from France having been permitted to be imported for slaughtering purposes, the said cattle being kept alive too long, and allowed to associate with the Guernseys. The inhabitants, therefore, to prevent the possibility of the foreigners being used for breeding, again petitioned the Royal Court, praying that, unless stringent measures were adopted to prevent the said mixture of bulls, cows, and heifers of other countries with those of the island, Guernsey cattle would not long be superior to other breeds. In response to this petition we find the Court avowing its appreciation of the value set upon the purity of the Guernsey breed, and their concurrence in the opinion that it was mainly attributable to the unremitting attention and jealous care of the inhabitants that Guernsey cattle were so renowned, and therefore decreed:

1. That all individuals possessing cows, heifers, calves, or bulls imported from France, should, within eight days of the passing of said decree, make a declaration in writing to the high constables of their respective parishes, under a penalty of the confiscation of the said animals, also a fine, at the discretion of justice, not exceeding 20*l.* sterling; and that a copy of said declaration be deposited at the registrar's office.

2. That it be forbidden henceforth to import from France, without special permission of the Court, any bull or bulls, under a penalty of confiscation, and a fine of 20*l*. sterling for each animal imported or otherwise got into possession.

3. That from and after the passing of this decree all cows, heifers, and calves imported into the island direct or indirect shall be killed within four months after their arrival, and foreign calves born in the island killed within eight weeks after their birth, all under penalty of confiscation of the said animals, and a fine not exceeding 20*l*. sterling.

4. That in the exportation of cows, heifers, calves, and bulls, a declaration shall be made on oath by the proprietor or other person possessing the knowledge that the animals are not only of the island, but that they have not mixed with foreign cattle in any way.

By the published ordinances or Acts of the Royal Court of the island we learn that from time to time the aforesaid decrees have been repeated with stronger and yet stronger force, as, for instance, in 1823, lest by chance permission of the Court for the importation of foreign cattle might at any time have been unduly or too readily granted, it is ordained that henceforth the said permission be valid only when sanctioned by the president and at least seven of the magistrates; also that foreign cows be not kept alive for more than the allotted four months, even though they be in calf, under the penalty aforesaid, to be imposed on the person who imported them, or in his default the master of the ship that brought them, or in his default the person who had them in his possession.

Frequent attempts have been made at different periods to contravene these laws, but in every instance they have been detected and checked. To the credit of the native inhabitants, however, be it said that such attempts have never been made by them, even though at times some pecuniary advantages might have been gained thereby. The attempted contraventions have always been made by those from other parts who had made Guernsey for the time being their place of residence.

On one occasion it was very plausibly pleaded that certain heifers which had been, and were still being, imported into Guernsey from France were by no means fit to kill in the four months allotted by the law. Guernseymen, however, were not to be done in this way. They knew that such an excuse meant one of two things—either that the owners wished to keep them for cows, or had an idea of passing them off in England as Guernsey cattle, and they were too proud of their reputation to allow themselves to be thus robbed of it by interlopers. They, therefore, once again appealed to the Royal Court, not only to turn a deaf ear to such excuses, but to forbid for a time the importation altogether, lest a fatal blow should be struck at this their favourite branch of industry. And, in ready compliance, the Court decreed that it is now forbidden to import from France or elsewhere any heifer whatever under the penalty of confiscation and a fine not exceeding 10*l.* for each heifer; also that all masters of vessels carrying cattle are bound, within twenty-four hours of their arrival, to furnish a list to the constables of the parish in which they are landed, under a penalty not exceeding 5*l.* sterling.

We have said at the beginning of our article that Guernseymen will not tolerate admixture into their breed of cattle even from the neighbouring island. In this respect the Guernsey people are much more exclusive than the inhabitants of the larger island of Jersey; and it is this exclusiveness which is their boast and pride. It may be, and indeed is the case, that the breeds of the other islands derive advantage from their admixture with Guernseys—for instance, the old and well-known breed of Alderneys, which is now nearing extinction, have by this means become assimilated to the Guernseys. But, like the Arabs with their horses, Guernsey has ever kept, and boasts of her determination still to keep, her breed of cattle distinct and separate; and hence the law is made equally binding on the importation of cattle from the sister island as from foreign ports.

Acts of the Royal Court have, as we have said, been enforced from time to time, and in no case have the measures been repealed or made less forcible. On the contrary, the system at the present time adopted is even more decided than any we have quoted. For instance, it is now forbidden to import bulls under any pretence whatever; and, moreover, it has become an impossibility for spurious breeds to be exported from the island as Guernsey cattle, since every cow, heifer, and calf imported at Guernsey is immediately branded with the letter F, to signify foreign, and the importer has to deposit the sum of 1*l*., which is forfeited if the animal is not killed within the given time. If, therefore, a purchaser of Guernsey cattle ever supposes himself to have been duped, he has only to ascertain positively whether the animals have ever stood on Guernsey soil. If they have, and are branded with the letter F, they are originally from other places; but if they are free from brand, the purchaser may rely upon it they are the genuine Guernsey breed. We repeat, however, that it is absolutely necessary to ascertain for certain that they have stood on Guernsey soil; for tricks have been practised to avoid the branding, by shipping cattle from France to Guernsey, and thence to England, without landing them at Guernsey at all—or, at least, only landing to pass from one vessel to another.

The most satisfactory method for gentlemen desirous of being possessed of Guernsey cattle is to visit the island for themselves. A splendid choice would be afforded them at the time of the show, which is held annually on Whit-Tuesday, when from 170 to 200 head of pure Guernseys may usually be seen; and no one could witness the sight without being impressed with the beauty and quality of the breed.

There is another show held on the 20th of December in each year for fat cattle. A good idea of the class of animals to be seen at this exhibition may be gathered from the following list of weights of beef at the 1872 show, it being borne in mind that the quotations given are Guernsey weight, 102lb. of which are equal to 112lb. English. It is, moreover, worthy of note that from the age of two years till within six months of their death the oxen are doing the work of the horse at the cart and the plough.

By whom fed.	Age.	Weight.
Thomas Hocast	8¼ years	1199lb.
Samuel Best	4¼ years	966lb.
P. Hocast	7 years	1152lb.
J. Le Page	6¼ years	1257lb.
John Corbin	5 years	916lb.
J. Le Page	5¾ years	830lb.
Col. Feilden	6 years	908lb.

As our article may possibly lead to the Guernsey cattle being even more than ever in request, we will just observe that purchasers in England must not be too hasty in condemning us or the Guernseys if they do not find the cattle at once turning out all that we have described. In this matter, like all others, a fair trial is necessary. With cows and heifers especially, the change of climate and pasture may have a checking effect at first; but, as they become naturalised to their new home, they will soon prove themselves true to the character they have so long and so justly borne.

In the management of Guernsey cows the islanders make it a special study to supply that kind of food, and otherwise subject the animals to that method of treatment, which tends to promote good milking qualities. Flesh-making is quite a secondary consideration with them, so long as good quantity and quality of milk and butter are forthcoming.

They have their special treatment for special seasons. For instance, in spring and summer the cows are fed on clover, lucerne, and grass, care being taken that neither is of too mature a growth, and that only a given quantity of each is supplied. Clover and grass, when not too ripe, produce abundance of milk; lucerne also is beneficial, but not in too great quantity, otherwise it is apt to kill the delicate flavour of the butter. If these foods are supplied in too ripe condition they tend to make flesh instead of increasing the milk.

Those who take pains with their cattle are careful that the cows feed when the sun is shining, and place them in the shade as soon as a sufficient quantity of food has been taken, as allowing them to remain in the heat after feeding tends to reduce milk.

In winter the cows are fed on hay, straw, carrots, turnips, and mangold wurzel. The greatest quantity of milk is produced from food in the following ratio: first, carrots; second, common turnips; third, mangold wurzel. Other roots, such as parsnips and Swede turnips, are avoided where practicable, inasmuch as they produce flesh rather than milk. Care is also observed by those who take pride in their cattle to house them at night and in rough weather, allowing them open air and exercise only when the sky is clear.

Disease among cattle is a thing comparatively unknown in Guernsey. The majority of premature deaths occur from calving or milk fever, but these are very rare. Every precaution is taken to prevent diseased cattle being imported, and equal precaution is observed with regard to the native cattle coming in contact even with the healthy importations.

Guernsey heifers are frequently in calf at the age of two years, and, having calved, they yield milk to the age of fifteen, and often beyond that time, being dry at intervals to the extent of only from three to six weeks each time. It is sometimes urged that heifers are permitted to breed too young; but the experience of many, if not most, Guernsey breeders is that the cows are not a whit the worse for it.

In England there exists a strong prejudice against cow beef, but the method of feeding in Guernsey is such as to render the flesh of the cow as delicate as the heifer, that is, so far as the prime joints are concerned. Flesh-making is avoided until the animal has run dry, and its time has come for slaughter; then attention is turned to the clothing of bone and sinew with new flesh, which, when killed and brought to table, is found to be as tender and delicious as can be desired.

Since the former edition of this work the Guernsey cattle have advanced in favour more rapidly, perhaps, than any other breed. To start with their island home, larger prices have been realised than ever before. A cow has been sold to America for 400*l*., others have been shipped to the States at prices up to 150*l*.; 100*l*. for a cow is not at all an uncommon thing; but such prices as these have only been for the best, and the dairy history of whose ancestry was good, and whose pedigrees were traceable and reliable. In fact, the demand for trustworthy information as to pedigree had increased so much, that it led to the formation of a herd book by Mr. James James, of Les Vauxbelets, and other gentlemen, on the principle of selection. One volume was published, and shortly after this work was taken over by the Royal Guernsey Agricultural Society, who carried on the selection principle on the following lines: No animal was admitted that had not been adjudicated on by judges appointed by the society, and received at least a commendation. Such animals, even though pedigree was wanting, and all that had won prizes in previous years at island shows, were placed upon what was termed foundation stock. All animals in Mr. James's herd book were admissable in the Royal Society's first volume. Any animals in addition, whose sires and dams were on the foundation stock, were admitted as pedigree stock. On these principles this herd book is now carried on. Unfortunately, many breeders set their faces against the system of selection, and also thought there was no necessity for a herd book at all, as the island laws against the importation of cattle are so stringent that they preclude any idea of the possibility of a cross. However, seeing the feeling there was on the part of English and American buyers in favour of animals with a pedigree, they partly went with the stream, and started another herd book, and called it the General, for the simple registration of all stock, good, bad, and indifferent, with or without pedigree. Such a book as a guide, or recorder of pedigrees, is absolutely useless—in fact, is mischievous; as breeders and farmers who are careless or lazy can go to the General, and enter an animal without any sire or dam being named. An animal certainly can be admitted into the Royal Herd Book without a pedigree, but in this case it has to be judged first, and no animal is allowed to be brought forward till she has calved, so that the judges can form some idea of her dairy qualifications, and not pass her simply on good looks. (Such a happy-go-lucky way of business as the General Herd Book cannot last long, as both English and American buyers set great store by the Royal Guernsey Agricultural Society's certificates of pedigree, and will not take animals on the General unless they happen to know something about the family of the animal they are about to purchase.) The policy, if such it can be called, of two herd books on such a small island as Guernsey is suicidal and unpatriotic, and can only be sustained by those whose animals are not good enough to get on the "Royal" Herd Book by selection, or who do not take sufficient interest in the breeding of their cattle to take the little trouble necessary to get them entered on the "Royal."

Let us now turn to America. Here the breed makes rapid strides. A herd book has been started some years, and the rules as to pedigree are most stringent. The annexed form will show how particular they are, as one of these forms must be sent with every imported animal, if it is to be entered in the American Guernsey Herd Book. The form, as may be seen, is in blank, but must be shaded in, so as to show the distinctive markings of the animal to which it refers.

F, fawn; W, white; R, red; O, orange; Y F, yellow fawn; O F, orange fawn; L F, lemon fawn; L R, light red; B, brown; Br, brindle; or any other letters indicating colour. Sketch each side of the animal, the head, and the shape of the horns.

Sex_____   Name of Animal_____

Herd Register No._____

EXTRACTS FROM THE BYE-LAWS OF THE AMERICAN GUERNSEY CATTLE CLUB.

" 7. There must be certificates of sale and description of animals from the Island breeders, certified by the United States Consul, or by a Justice of the District. The original papers or certified copies of these, with the importer's indorsements, must be placed on file in the Secretary's office. After the 1st of July, 1882, these papers must be accompanied by a sketch of all white markings of each animal, made before shipment, on the outlines furnished by the Secretary of the American Guernsey Cattle Club, the same to be signed in duplicate by the breeder or seller, and by one witness who shall certify that he has compared the sketch with the animal known by him to be the one described, and that it is correct; one copy to be sent by mail, under the seal of the American Consul on the Island of Guernsey, to the Secretary of the American Club, to be filed and held by him for confirmation of the animals when imported.

8. No animal imported after the 1st of July, 1882, can be entered in the Herd Register of the American Guernsey Cattle Club, unless previously registered on the Island of Guernsey.

9. Breeders shall furnish a certificate of the service of dams, and the name and Herd Book number of bull, if served before shipping from the Island.

11. The names of the Island breeders of the animals shall be printed in the American Herd Register, and after the 1st of July, 1882, the names and Island Herd Book numbers of their sires and dams.

14. In the outline sketches on the back of each form the white markings of the animal must be drawn, and the coloured parts shaded or indicated by letters, and certified to by the applicant."

CERTIFICATE OF THE APPLICANT.

I hereby certify that this sketch represents the white and coloured markings of the within described animal.

Signature of Applicant.

CERTIFICATE OF WITNESS.

I have compared this sketch with the animal described, and find it to be correct.

Signature of Witness.

Date_____

Our American cousins find the Guernsey bulls most useful for crossing, as the "grades," if males, are steered, and grow to a large size; and, if heifers, are first class for dairy purposes.

The breed must be thought highly of, as large shipments from time to time are made from the island, and these in the face of high prices, as only the best are purchased, heavy expenses for transit, and a three months' quarantine on the other side.

One other fact about the island. Now that depression in agriculture has become so acute, people have given up looks and show points, and gone in for animals that are likely to pay their way. The Guernseymen, recognising this, have taken to making tests to show what their breed is capable of, and in this praiseworthy endeavour they are well assisted by the Royal Agricultural Society of Guernsey; and the *modus operandi* is as follows: Anyone wishing to have his cow tested, notifies the Royal Society of his wish; they appoint two persons to see the test is fairly carried out. These two attend at the farm, night and morning, for a week, see every milking, and see the milk carried away and locked up. The whole milk is churned, and at the end of the week the judges send in their report of the quantity of milk at each milking, and the amount of butter made therefrom. Such a test can be relied on as being absolutely exact. It is feasible in a small place like Guernsey, but it has not as yet been attempted in England; and in America, the country of magnificent distances, it would require some working.

These tests are only just commencing in the island, but it shows that the Guernseymen are becoming alive to the fact that now-a-days utility must be their watchword, and that they must prove what their animals are capable of. Now, with regard to England. Here the breed has made tremendous strides in public favour; and within the last few years, from being almost unknown, it has become universally known, and, what is more, thoroughly appreciated—not as a pet lawn cow, or to grace a park, but as a reliable rent-paying beast, either pure or for crossing purposes. The writer was visited by a farmer who makes butter in a large way, and had kept Jerseys for thirty years. He had seen and been struck by the appearance of the Guernseys at the Bath and West of England Show at Brighton, and determined to cross his Jerseys; and he came and bought three young Guernsey bulls for that purpose. The fact is getting slowly known amongst farmers that the Guernsey and Jersey are not at all the same animal, and that the cross with the *Guernsey* bull on almost any breed is bound, if a heifer, to be first class for dairy purposes, and if of the male sex, to make a fine large beast, easily fattened.

With regard to its increasing popularity, it is only necessary to note the fact that at all the leading shows the Guernsey now has classes to itself, where formerly it was put with the Jersey in classes for Channel Islands cattle, and also to glance at the entries; the Guernseys are only outnumbered by the Jerseys; and generally the Channel Islands cattle outnumber *all other breeds combined* by two to one.

As a proof that the Guernsey is a really good dairy cow—perhaps the best—I may point out that one of the largest dairy companies in London, the Express, has kept for the last five or six years a large herd of pure Guernseys, which they intend to increase and keep pure, so satisfied are they with it as a profitable dairy cow. Such a fact needs no comment.

Without being able to record such phenomenal milk and butter yields as occasionally startle the bucolic mind when they come from the other side of the Atlantic, still a few facts from carefully kept records may prove interesting as demonstrating the capabilities of the breed.

In the Royal Guernsey Society's Herd Book appears the following official test of a *heifer*, five months gone with second calf, therefore the great flush of milk was thoroughly past:—
" Vesta 6th, No. 625; born 20th November 1881. Official test dated 28th May 1885; gave 13lb. 15¼oz. of butter in one week."

There are numbers of cows who make from 14lb. to 20lb. of butter weekly, and this without forcing.

Sir J. F. Lennard, of Wickham Court, Beckenham, Kent, who has kept a pure herd for twenty years, occasionally importing a bull from the island, has built up a herd of great excellence, any of which appear, when *cows*, to make from 16lb. to 22lb. of butter a week. His cows are tested two following days every month, and the average struck from that. Some of his cows by this test have made over 3lb. of butter from one day's milking. As regards quantity of milk also they, the Guernseys, deservedly rank very high. I have a herd of from twenty-five to thirty animals; the milk of each one is measured night and morning, so that I have facts to go upon. Last year one cow in twelve months, between second and third calving, gave 825 gallons. A heifer, first prize winner at the Royal Show at Norwich this year, calved when two years two months and three weeks old, and gave fifteen quarts a day. I have had others giving with their first calves thirteen and fourteen quarts a day, and as much as 700 gallons in ten months, and the second calf *within* twelve months. I have several which have given from 500 to over 600 gallons between first and second calvings.

Within the last three years the English Guernsey Cattle Society has been started, which has given an impetus to the breed. The society numbers over fifty members, and has issued two volumes of a herd book, which is intended to be an annual affair. Its rules for registration of stock are stringent, and rightly so; and it appears impossible for any but a perfectly pure animal to be admitted. Appended is the scale of points drawn up and approved by the society, from which it will be seen what great importance is attached to the dairy properties and indications.

No.		COUNTS.
1.	Head fine and long, muzzle expanded, eyes large, quiet and gentle expression	4
2.	Horns yellow at base, curved, not coarse	3
3.	Nose free from black markings	1
4.	Throat clean, neck thin and rather long, not heavy at shoulders	7
5.	Back level to setting on of tail, broad and level across loins and hips. Rump long	10
6.	Withers thin, thighs long and thin	4
7.	Barrel long, well hooped, and deep at flank	10
8.	Tail fine, reaching hocks, good switch	1
9.	Legs short, arms full, fine below knee and hocks	2
10.	Hide mellow and flexible to the touch, well and closely covered with fine hair. Yellow inside the ears, at the end of tail, and on skin generally	12
11.	Fore udder large, and extending forward, and not fleshy; udder full in form, and well up behind, with flat sole. Teats rather large, wide apart, and squarely placed	25
12.	Milk veins prominent, long, and tortuous	6
13.	Escutcheon wide on thighs, high and broad, with thigh ovals	5
14.	Size, general appearance, and apparent constitution	10
		100

THE SHEEP OF GREAT BRITAIN.

CHAPTER I.

INTRODUCTORY.

SHEEP occupy a prominent place in the history of British Agriculture. To the products of these valuable animals, and especially to their wool, were our progenitors indebted for much of their national prosperity. Youatt, in the admirable preface to his work on Sheep, makes allusion to our woollen products, which became eminent soon after the subjugation of this island by the Romans. Dionysius Alexandrinus tells us that "The wool of Great Britain is often spun so fine that it is in a manner comparable to the spider's web." In later times the wool trade was by far the most important industry of our country; and wool frequently took the place of money. The Lord Chancellor of England sits on the woolsack; and maidens of all degrees were taught to spin—hence the name spinster, now so hateful. As population increased, the food supply became a question of importance, and eventually, as the carcase became more valuable than the fleece, the latter was to a certain extent sacrificed. It has been reserved to our own generation to bring both the yield of wool and the weight of flesh to perfection in the same animal; but this is undoubtedly done at some sacrifice of quality. We do not grow such fine wool as formerly, and for our own better class of goods have recourse to the German and Australian merinos. An extraordinary impetus was given to sheep culture by the introduction of root crops, which, with artificial grasses, secure winter food. The revolution effected by Bakewell and others would have been impossible but for the development of arable farming, which allowed of regular and progressive feeding from birth to death; and hence we find that a given result is produced in half the time formerly occupied.

Unfortunately for the historian, it is only of late years that the importance of statistical information has been acknowledged. Youatt, in 1837, estimated the sheep stock of the kingdom at 32,000,000 head; their wool at over seven millions of pounds sterling, employing nearly 350,000 individuals, and ultimately yielding at least 21 millions of pounds annually. Lavergne, who wrote in 1855, estimated the sheep stock of Great Britain at 35,000,000, of which nearly 30,000,000 are found in England alone, being proportionately three times more than in France. Our statistical information dates from 1868, and is admittedly defective, owing to the hostility evinced by many who ought to have known better, but, such as it is, reads thus:—

Years.	England.	Wales.	Scotland.	Ireland.	Isle of Man.	Channel Islands.	Total.
1868	20,930,779	2,668,505	7,112,112	4,822,444	72,116	1856	35,607,812
1869	19,821,863	2,720,941	6,995,337	4,648,158	62,198	1775	34,250,272
1870	18,940,256	2,706,479	6,750,854	4,333,984	53,565	1545	32,786,783
1871	17,530,407	2,706,415	6,882,747	4,228,721	53,565	1545	31,403,500
1872	17,912,904	2,867,144	7,141,459	4,262,117	61,683	1335	32,246,642
1873	19,169,851	2,966,862	7,290,922	4,486,453	66,728	1588	33,982,404
1874	19,859,778	3,064,696	7,389,487	4,437,613	84,424	1699	34,837,597
1875	19,114,634	2,951,810	7,100,994	4,248,158	74,239	2113	33,491,948
1876	18,320,091	2,863,141	6,989,719	4,007,518	70,566	1544	32,257,579
1877	18,330,377	2,862,013	6,968,774	3,989,178	Inc.Tot.	In.Tot.	32,220,067
1878	18,444,004	2,925,806	7,036,396	4,094,230	,,	,,	32,571,018
1879	18,445,522	2,873,460	6,838,098	4,017,889	,,	,,	32,237,958
1880	16,828,646	2,718,316	7,072,088	3,561,361	,,	,,	30,239,620
1881	15,382,856	2,466,945	6,731,252	3,258,583	,,	,,	27,896,273
1882	14,947,994	2,517,914	6,853,860	3,071,493	,,	,,	27,448,220
1883	15,594,660	2,581,250	6,892,361	3,219,098	,,	,,	28,347,560
1884	16,428,064	2,656,997	6,983,293	3,243,572	,,	,,	29,376,787
1885	16,809,778	2,767,659	6,957,198	3,477,840	,,	,,	30,086,200

Defective as the above undoubtedly is, it is pretty clear that Lavergne, and probably Youatt, estimated in excess of the actual number; because from 1855 to 1868 the general progress was very considerable, and it is tolerably certain that the sheep stock also developed. Between 1868 and 1872 there was a rapid and severe diminution. This was mainly attributable to the dry summers of 1868 and 1870, which caused great losses, lambs being destroyed wholesale in those districts which suffered most. It is also probable that the sheep stock of 1868 was increased on account of the losses of cattle by disease, which led to a larger number of ewes being kept; but it will be evident that in 1874 we had nearly reached the same numbers. In 1876 and to 1879 there was very little variation. Then came the wet years, with liver disease and foot-and-mouth complaint, and we saw an alarming decrease, which culminated in 1882, when in round numbers our sheep stock was lessened by 8,000,000 from the figures of 1868, or about $22\frac{1}{2}$ per cent. The last three years have shown a steady increase, mainly attributable to the comparative freedom from contagious disease which we have enjoyed. The fact that a large area of land has been laid to grass, whilst it may increase the resources for cattle feeding, would not help sheep stock, as it is well known that more stock can be kept on tillage land than grass; but the leaving down clover and grass seeds for two, three, and four years leys should afford more food for sheep, and this may partly account for the facts.

Without going into the vexed question of the origin of races, we find in the natural conditions a sufficient cause for the differences observable in our domestic animals. Let us compare, for example, the short and long woolled varieties, which appear to have existed in this country from the earliest times. In the former we have small bodies, short wool, great activity, and hardiness—just the features that would result from having to travel far, and feed close upon the scanty though sweet pasturage of the downs; on the other hand, the larger framed Longwools, though equally or probably of even slower growth, finding greater abundance of food on the rich pastures of the Midland districts, or in the uplands of the Cotswold ranges, had not to travel so wide, and consequently became more developed both as to wool and frame. We acquire this knowledge by analogy—viz., by seeing the effect of a change of circumstances. The small frames and close wool of the Sussex Down alters materially when placed upon rich lowlands, and so great is the change that it is often (as we believe wrongly) attributed to a cross of blood.

We would refer the reader to the different types of sheep

which were shown by the late Mr. Ellman, of Glynde, and Mr. Jonas Webb, or Lord Walsingham, for example, and which led the nephew of the former to protest against the decision of his colleagues at one of the meetings of the Royal Agricultural Society, on the ground that the sheep were not characteristic. But we can see, in the conditions to which each was subjected, sufficient cause for modification without the introduction of foreign blood. The Sussex Down of old time, and to a certain extent at the present day, is the manure carrier from the open downs to the arable land; consequently, after running out all day, the sheep were driven into a fold on the arable land, and there left for ten or twelve hours without food. Small frames and great hardiness were necessary for such a life. Contrast this with the high culture as carried out by the foremost lowland feeder—lambs allowed every advantage from birth, accustomed to pick at the best of artificial food, receiving frequent changes of diet, and having the supply at hand and not to be sought for. Regular feeding insures increased size.

Little, however, is known either of the type or date of the British sheep. Writers, who have assumed that because the oldest records contain no mention of them that therefore they did not exist, appear to forget that the same reasoning might apply to the country itself; for, during the barbaric ages, and before the Roman times, we search in vain for an authentic history of Britain. It is quite true that the Phœnicians traded with the Britons for lead and tin only, and no mention is made of sheep; but it does not therefore follow that they did not exist. The Phœnicians, it is believed, confined their dealings to Cornwall, and sought such articles as they were deficient in; whereas wool was grown by neighbouring countries, and undoubtedly used for clothing. It is, however, known that the skins of animals were exchanged, and there is no reason why these should not have included the sheep. If sheep were not in existence in these islands, it is a little remarkable that, within a short period of their subjugation by the Romans, a woollen manufactory should be established at Winchester, and goods of a very high class be furnished. Indeed, the finest and most superior robes used only on state occasions were supplied from Britain. It is not important to carry the speculation further,

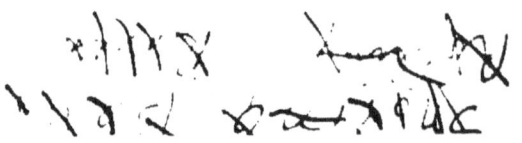

although it might be urged that the well-marked differences in our local breeds could only have been acquired by the influence of climate and soil through long periods of time. We believe that the sheep is a native animal, and that the wide distinction between the early Cotswold, Southdown, and mountain sheep is owing to the influences by which each was surrounded. These peculiarities rendered them specially suitable for their life, because they had arisen from external influences. And even at the present time, speaking generally, it will often be found more profitable to improve than supersede. Hence we find in all the great sheep-breeding districts a tolerably distinct type; and these, again, are only blended together on what may be described as intermediate land. It is probable that the pure breeds may be imperfect in form, slow to feed, and apparently very defective when compared with more improved sorts; but they possess qualifications which render them invaluable for the peculiar country they inhabit; and it is only when the conditions to which they are accustomed, and which have stamped their very peculiarities, are modified, that we can succeed either in improvement or replacement. As examples we may mention the Welsh mountain sheep, the black-faced sheep of Scotland, the hardy Herdwicks of Cumberland and Westmoreland, the Cheviots, and the denizens of the Romney Marshes in Kent. Improvements in farming, tending as they do to diminish the evil influences of climate, enable us to modify and improve the native breeds, and in some instances to supersede them; hence there have grown up sorts that result from the blending together of totally distinct types. These—and we particularly allude to the Shropshire, the Oxfordshire, and the Border Leicester—have, over large districts, taken the place of the original sheep, to the manifest advantage of the farmer. Having great facilities for improved management, and especially being able to provide, by a judicious succession of crops, for steady progressive development, our object is to combine plenty of wool, weight of carcase, aptitude to feed, and quality of flesh —points that are not united in any of the old distinct breeds. Thus the Longwools yield large fleeces and plenty of mutton, and have great aptitude to feed, but the quality of the flesh is unsuitable for any but the coarser consumers—a fatal drawback. The Southdown gives us good flesh, but there is too little of it, and the fleece is light. It is evident that something between these extremes is very desirable when we have favourable conditions, and it is a fact that the Shropshire sheep more especially, and the Oxfordshires to some extent, have pushed their way far from home, and have, in several instances, superseded the pure breeds. This substitution is possible just in proportion and to the degree that our treatment differs from the natural conditions under which the sheep originally existed. In order to do justice to an improved sort, great care must be exercised as to the selection and succession of food. Constant progress is absolutely necessary, and it will not do to fill one day and starve the next. Climate really decides the question as to the kind of sheep. On the Cotswold hills, which range from 400 to 600 feet above the level of the sea, where the soil is a thin limestone with stones, and where clover and sainfoin grow well, and where the root crops can be eaten on the land, the heavy-woolled and somewhat strong-boned Cotswolds are unrivalled. With such jackets they do not feel the wind, and, so long as the lair is dry, flourish excellently. The Lincoln is capable of holding its own on the strong grass of the fen districts; and the cross of Lincoln and Leicester is most approved of on the uplands. Even the Romney Marsh sheep, retained pure on the marsh district, have been considerably crossed on the upper lands, with manifest improvement. Nothing can live on the high lands like the Welsh or Scotch mountain sheep; but wherever the conditions are ameliorated we find a cross or another sort substituted with advantage. Modern farming has a tendency to mitigate extremes, and the farmer who is supplied with all the accessories of improved agriculture has means at his disposal which enable him to provide for his stock in a way conducive to their development, and which would have been quite impossible fifty years ago. Hence we are not surprised at the very rapid spread of the Shropshire sheep, which appear likely to monopolise the great west-central district of England.

CHAPTER II.

THE MANAGEMENT OF EWES UP TO LAMBING.

HOWEVER great are our present facilities for rearing stock, natural conditions, especially as regards soil, will define breeding from non-breeding land. We may lay down as a general rule (from which exceptions will occur) that light soils, and especially where these are derived from limestone strata, as the chalk, the oolite, &c., are the most suitable for breeding. Sandy or gravelly land is also healthy, but the natural food, probably from the insufficiency of lime, is less suitable for young stock, and care must be exercised to keep up the fertility by a sufficient addition of such elements as are deficient. Strong land, though growing under good management the best of food, is not, even when thoroughly drained, suitable for a breeding farm. There is always a risk of loss from disease, especially from attacks of the fluke. We do not mean to say that breeding sheep cannot be kept on strong land—for, as we hope to show, very profitable results may be obtained—but such land is most suitable for a fugitive flock. The ewes must be bought in each fall, and sold out with their produce during the ensuing summer. It is to the dry healthy limestone soils—which under the head of wolds and downs are largely distributed throughout the country—that we must look as the nurseries for sheep stock; and it is a question whether, considering the active demand that exists at the present day for sheep for wintering, the practice of the Hampshire and Wiltshire farmers as to selling their lambs might not with advantage be more generally followed, as it would allow of a heavier stock of ewes being kept, and thus our sheep might be increased.

It is the nature of the soil as to texture, and the presence or absence of lime and phosphoric acid, rather than climate, that determines suitability or otherwise for breeding. Nature provides against the latter by the character of the fleece; but if the food is deficient in phosphates, and if the ground is constantly damp and cold, it stands to reason that the young animals must suffer. There was a time, years ago, when the demand for feeding sheep ruled low, and when it was often more profitable to buy than to breed. Such times may recur again, for it is not easy to adjust demand and supply. At present breeding under favourable circumstances is a profitable business, and those who sell out lambs in the fall at from 30s. to 45s. probably realise more into pocket, regarding the largely increased numbers that can be bred, than those who make out their stock fat, even if they succeed in making of wool and carcass from 70s. to 80s. Sheep stock are highly profitable in themselves, and indispensable for the proper cultivation of the land. Much of the surface that was once in rolling downs is at considerable elevation, and far removed from the homestead. A good crop of turnips starts the rotation, and sheep with their golden hoofs do the rest. To them must be attributed the bulky crops of waving corn, which may well astonish the stranger, seeing that they are grown on soil often only a few inches thick. Nay, so rich does land become under sheep farming that it is often necessary, in order to obtain quality of corn, to remove the excess of richness by a crop, like oats, that loves fatness, after which only can prime malting barley be grown.

We will suppose the soil suitable for breeding to be principally arable, for it is an undoubted fact that light limestone soils grow more sheep food under the plough than if in grass, besides giving us very excellent crops of corn. The choice of breed will be decided by local circumstances, and of course the quantity of head that can be kept in a healthy state will depend upon the nature of the land and the capacity of the animal.

There is no doubt that in animals of equal cultivation the consumption of food bears pretty close relation to live weight and increase; so that in reality it does not much matter whether we keep three Southdowns or two Cotswolds, so long as each is properly placed. The consumption of food will be in proportion to the increase of weight; but, of course, for this comparison to be correct, it is most important to have the right sheep in the right place. Where sheep are the ruling live stock—as they should be on sheep land—the proportion will range from one ewe to every two acres, and two ewes for three acres, according to the sort, and according to whether the offspring are sold as hoggets, or finished off late in the spring. We have a very decided opinion that a considerably increased stock might be profitably maintained if a more economical system of feeding were introduced.

The first point we notice is, that, in order to be vigorous, the ewe must be young; and, except in the case of exceptionally valuable mothers, it is a sound practice to cull the ewes after they have lambed three or four times at most. They will then be either four or five years old. The draft ewes are thus good in the mouth, and capable of being rapidly fed, or of great service to the lowlander for another crop of lambs, of which more anon.

In a flock of two hundred ewes, from fifty to sixty shearlings should be introduced in each year, in order to replace the regular culls, and such as have proved themselves defective mothers. With the best management and good fortune a certain per-centage of the ewes are barren; these, without extra food, manage to get very fat, and should be sold, as it is not worth while running the risk of a second unprofitable season. Again, it is but too often the case that the mammary glands are affected by cold or fever, and though the ewe recovers under careful treatment, her milking properties are lost or injured. Such animals, it is needless to say, must be withdrawn. Hence, to make up for all sources of deficiency (amongst which we must not forget the ordinary mortality, which ranges from 5 to 10 per cent.) sixty shearling ewes should be reserved to replenish the flock; and it is almost needless to observe that they should be the best we grow. Only the culled shearling ewes must be sold, and these find a ready market, either fat or for breeding. The lowlander, again, comes to the fore, as he is precluded from breeding his own; he is only too happy to meet with healthy stock, and so it happens that our discarded ewes frequently realise from 50s. to 70s. a head.

The best time for lambing will depend upon the climate and our resources in the way of early spring food. If we are favoured as to situation, and can insure a good succession of suitable food, it is a great advantage to have our lambs early, especially if we breed rams, or sell our lambs. But mistakes are made by trying for early lambs where the farm is exposed and the climate backward. It is easier, safer, and cheaper to keep the lamb inside rather than out of the ewe. The duration of pregnancy is about five months, or 152 days, with less variation than is observed in many other animals. Mr. Tessier, in his report to the Academy of Sciences in Paris, gives the result of his observations on 912 ewes. The shortest period was 146 days, and the longest 161 days—a difference of fifteen days; but more than three-fourths yeaned between the 150th and 154th day after impregnation, bringing the average as nearly as possible to 152 days, or twenty-one weeks and five days.

By cultivating the habit of early maturity, it has been found practicable under certain conditions to breed from ewe lambs; and, though experience is as yet too recent to justify a very decided opinion as to whether such a system can ever be largely introduced, or be carried on profitably, we are quite satisfied that if due care is exercised as to feeding, there need be no check to growth—nay, more, we believe that the aftergrowth will be better. Mr. Alfred de Mornay, of Col d'Arbres, Wallingford, who was one of the first to illustrate the precocity of Hampshire wether lambs—and whose extra stock lambs at Smithfield led to the giving of prizes for lambs—is experimenting in this direction with every prospect of success. His idea is to develope early breeding properties by cultivation; he commences by selecting the forwardest ewe lambs, and mating them with a ram lamb equally precocious; from the offspring he uses only the forwardest specimens, and so very soon he expects to educate lambs to a habit of early breeding. No doubt, where the growth of the frame and the maintenance of the fœtus are going on simultaneously, the food must be plentiful and nutritious, but as we gain a year in time, we can afford better food. Mr. De Mornay always drafts when the ewes are turned four years old. According to the old system they will have produced three crops of lambs; by this plan they will go away at the same age, but will have yielded a fourth crop. Of course, climate and soil should be favourable for early and rapid development. That this is rather specially the case at Col d'Arbres may be gathered from the fact that the wether lambs are sold for the butcher when from seven to nine months old, averaging fully 24lb. a quarter.

The ewes should be in fair condition and mending whilst with the ram. If they are low in condition the result will be a large proportion of singles; whereas, if fresh, doubles are more numerous. The nature of the soil as affecting the quality of the food influences the result. The Shropshires are, probably, the most prolific breed we have; yet we have known instances where they lost their great fecundity on a change of situation, and therefore we are inclined to believe that soil has a good deal to do with the produce. Great mistakes are sometimes made by forcing on poor ewes just at the time they are placed with the ram; in such cases blood is made so fast that the ewes are almost certain to turn. An unlimited supply of rape is said to have this effect; although an occasional fill of this valuable crop answers well if the ewes have been properly prepared.

The ordinary practice of allowing the ram to run with the ewes may be altered in the case of very valuable or heavy sires; for such the ewes should be stocked, the ram being kept up in a shed. In this way old sheep are found useful, and more certain in their work. During the day the ewes are allowed to range over the seeds which are to be ploughed, and on the grass land; and at night they are folded on rape, or mustard, or early turnips. In some cases the stubbles, where grass occurs, afford a healthy change. After about eight weeks the rams are removed, and the ewes are placed behind the fatting or stock sheep, as the case may be, in order to clear up the food that might otherwise be wasted; very few turnips suffice, and much good sweet straw will be eaten with advantage.

The folly of the practice, formerly so common, of stuffing

breeding ewes with turnips cannot be too strongly impressed upon the reader. We have no hesitation in condemning this as most extravagant and unwholesome. Indeed, a good turnip year in Norfolk was, in old times, invariably followed by a bad lambing season. When the serious cost at which the root crop is produced is taken into consideration, the folly of those who persist in consuming it, not only unprofitably, but prejudicially to the stock, becomes strikingly apparent; and we are persuaded that the true secret of successful management is to use the minimum quantity of roots with the maximum of straw, chaff, &c., adding, if necessary, a small quantity of artificial food. A ewe supplied with turnips *ad libitum* will eat a quantity equal to one-third or one-fourth of her live weight (*i.e.*, from 20lb. to 30lb.) daily. Of this nine-tenths is water, the temperature of which water in the winter is seldom many degrees above the freezing point. How much of the food of the animal must be burned away, so to speak, in order to raise this mass to the temperature of the body! Such unnatural feeding may be comparatively harmless in fine dry weather; but if the carcase is chilled by lying on damp ground, as must often be the case, then the combined effect of water outside and inside tells fatally upon the fœtus, and so we have dead, potbellied, and weakly lambs. Nor is this to be wondered at, considering. In 1863 the writer had the honour to deliver a short lecture at Hanover-square on this subject, and the late Lord Berners, who was in the chair, confirmed the statements made in the lecture by his own experience with cattle. "I can," he said, "state that I have for many years carried out what has been recommended with respect to dry food both for cattle and sheep, and I have found that when I reduced the quantity of turnips given to bullocks and sheep, and supplied them with a certain proportion of cut straw, they have done a great deal better than they did before. An ignorant common labourer will often give his bullocks as much turnips as they will eat, whatever be their condition at the time. One day I found in a yard twenty or thirty bullocks tied up and shivering dreadfully. I asked the man in attendance what was the cause of this, and he replied, 'O yes, they always be so after eating so many turnips.' I at once ordered the quantity of turnips to be reduced, and gave the animals dry food, and there was no more shivering." Since 1863 we have had several winters in which the roots were so deficient that flocks had to be kept without such food.

The winter of 1864, when the roots were generally destroyed by the severity of the weather, afforded valuable experience as to the possibility of keeping ewes alive and flourishing upon a minimum quantity of roots. We inspected a flock of Hampshire Down ewes that did not get a root before lambing. They ran at large on grass land during the day, and were folded at nights, so as to improve the pasture. Morning and night they got trough food, consisting of straw and hay chaff—two-thirds of the former to one-third of the latter—with a mixture of bruised oats and palm-nut meal; the outlay for artificial food reaching $2\frac{3}{4}d.$ a head weekly. We inspected the flock previous to lambing, and we never saw animals in a more promising condition: not one died during the winter. Again, in 1868-9, we had no turnips, and good managers learnt to keep their flocks in healthy condition. Last year (1886) in the southern counties the turnip crop was a failure, yet sheep were kept in a thriving state by straw and hay, with a moderate allowance of artificial food.

Mr. J. C. Morton, in an interesting paper published in the 27th volume of the "Journal of the Royal Agricultural Society," entitled "Some of the Agricultural Lessons of 1868," details the experience of many leading breeders as to the possibility of wintering ewes without root crops; for in that disastrous season there were none to use. And it is hoped that the lessons then learnt have not been forgotten.

One cannot, however, travel through the country without noticing the want of systematic management which too frequently prevails, and therefore, at the risk of being styled monotonous, we reproduce the arguments and evidence then introduced into the lecture which has been alluded to, in reference to the value and economy of using dry food, and especially straw, for breeding ewes. The two points we propose to illustrate are the effect on the manure and on the animal. First, as to the land. An acre of turnips will, on an average, weigh about 15 tons, yielding in the manure made therefrom 74lb. of nitrogen and 300lb. of mineral matter. The latter is composed as follows:

Per Acre. lb.		Per Cent.
110·94	Potash	36·98
20·28	Soda	6·76
1·77	Chloride of potassium	·59
23·55	Common Salt	7·85
10·83	Magnesia	3·61
33·42	Lime	11·14
29·22	Phosphoric Acid	9·74
37·29	Sulphuric Acid	12·43
10·29	Silica	3·43
3·27	Iron	1·09
19·14	Carbonic Acid	6·38
300·00		100·00

Supposing that the ewes consume at the rate of 20lb. a day when fed on turnips alone, and that we reduce the turnips by one-half and substitute 1½lb. of straw, it follows that with each acre of turnips 2 tons 4 cwt. of straw will be eaten. Straw, on an average, yields ½ per cent. of nitrogen; therefore we have 24·6lb. of nitrogen, of which probably 21·6lb. is left in the land as manure, the remaining one-eighth being exhaled from the animal's skin, &c. The same quantity of straw will yield 221lb. of mineral matter, of which about 48 per cent. is silica; the remainder contains minerals which are valuable as plant food. The addition to the soil of this large amount of silica—viz., about 107lb.—in a condition available for plant food would act beneficially on the barley crop, which requires for grain and straw more silica than any of the other cereals. The following table shows the composition of oat straw, and the quantity returned per acre if used as above described:

	An Analysis of the Ash of Oat Straw. Per Cwt.	Quantity returned per acre as Manure. lb.
Potash	19·14	42·29
Soda	9·69	21·41
Magnesia	3·78	8·35
Lime	8·07	17·83
Phosphoric Acid	2·56	5·68
Sulphuric Acid	3·26	7·20
Silica	48·42	107·00
Peroxide of Iron	1·83	4·04
Common Salt	3·25	7·18
	100·00	220·95

Assuming that the silica is of the value we have supposed, it is fair to calculate that the consumption of this quantity of straw would increase the value of the manure from one-third to one-half, which, especially on land remote from the buildings, is a matter of considerable importance. So much for the manure question.

Let us see, in the second place, what the feeding value of the straw is, and how far it compensates the sheep for the loss of half the turnips. Fifteen tons of turnips will yield nearly 3024lb. of dry matter; 2 tons 4 cwt. of straw will yield 4233lb. The proportion of this latter that would be digested is a point on which, unfortunately, we have not very precise data. The value of straw depends upon soil and climate, and especially upon the period of ripeness at which it is harvested. This is evident from the following analyses of straw grown on the Royal Agricultural College Farm at Cirencester, where neither land nor climate was particularly favourable:

ANALYSIS OF OAT STRAW, CALCULATED DRY.

	Fairly ripe.	Over ripe.
Oil	1·25	1·49
Soluble Protein compounds	3·13	1·54
Insoluble Protein compounds	1·74	2·79
Sugar, gum, mucilage, &c.	12·59	3·78
Digestible fibre	35·92	33·04
Indigestible woody fibre	37·84	49·80
Soluble mineral matter	4·31	2·70
Insoluble mineral matter	3·22	4·86

The reader will carefully notice the great difference in the proportions of sugar, and digestible and indigestible fibre, in the two analyses. The totals of these three items are practically identical; but how much more valuable is the fairly ripe straw —little, if at all, inferior to much of the over-ripe hay which is so frequently made. We are understating the actual facts when we say that 1 ton of No. 1 would go as far as $1\frac{1}{2}$ tons of No. 2. What a lesson have we here as to the time for cutting our crops! For, be it understood, the grain was not injured by too early cutting; on the contrary, the sample was fully developed, reached a great weight, and, being thin in the skin, was more valuable than if it had been left to a later period. It is necessary to explain that what the chemist distinguishes as digestible woody fibre is that portion which is soluble in dilute acids and alkalies (similar reagents to the gastric juice and biliary secretions), and which he therefore concludes would be converted by the digestive processes into food; and all that resists such tests will pass through the system, being insoluble. Sir John Lawes has shown that the power to thrive upon straw depends upon the proportion between the intestines and stomachs; that cattle possess greater power than sheep, and sheep than pigs. This affects the proportion in which such food can be used.

We have reason to believe that the animal laboratory is even more perfect than the chemist's, and that we may safely accept the figures under this head. Supposing, however, that only one-half the dry matter of straw is available, whilst the whole of the dry matter in the turnips can be made use of (which is never absolutely the fact, since in full-grown turnips there is always about 3 per cent. of woody fibre, and much more when the plant is over ripe, especially towards spring, when the flower stem is shooting), we have $2116\frac{1}{2}$lb. of available food against 3024lb. in the turnips; and it therefore follows that sheep eating turnips and straw, and thereby economising the consumption of the root crop by one-half, will consume in a given time 5140lb. of dry matter against 6048lb. when feeding solely on turnips. It is only fair to conclude that this difference will be partly compensated for by the healthier condition of the sheep, and by the lesser amount of fuel required to maintain the temperature of the body.

Ewes fed with an unlimited quantity of turnips are apt to scour, particularly in wet weather. We believe that it pays exceedingly well, especially towards the period of lambing, to give in addition to the straw a small quantity of artificial food, selecting such materials as are cheap and nourishing. It is of such vital importance that the ewe be strong, vigorous, and in condition to produce a full flow of milk, that such an outlay will be abundantly repaid afterwards. Straw may be used either given long in cages or as chaff. On sheep farms, and where the ewes are on arable land, we prefer the former plan, for several reasons—principally because the ewe has a power of selection, and can take only the more nourishing portions. Chaff is liable to blow away and be injured by moisture.

Where the turnip fields are far from home (as is so often the case on sheep farms), and where, from the nature of the soil, the surface treads in wet weather, it is an excellent plan to stack so much corn, either in the most sheltered corner of the field or in one adjoining, as will suffice for the requirements of the ewes whilst eating the turnips—not only for fodder, but also to litter a fold in which they may lie at night. We thus secure healthy lairs, and give the ewes a dry warm bed instead of their having to remain often in water; and at the same time we also make a lot of good manure, which is very handy for application. If forethought is exercised, we can so stack the straw as to form one side at least of our fold (that which is most exposed), and the shepherd has only to cut down as he wants it. We have seen on the Cotswold hills this plan carried out most successfully; and even when the fodder was only barley straw (the least nutritious of straws), the ewes will assemble at the corner of the fold nearest the desired haven long before the time they are to be admitted, and require no driving, but hasten along, and are soon busily engaged at the racks, where they quickly fill themselves with dry if not very nutritious food; and later on it was a goodly sight to see their woolly forms covering the ground, in evident comfort and content. In the mornings, and before they are let out, they have a fresh supply of straw.

Order must be observed in the consumption of straw. Barley and bean straw may be used in the earlier stages; then oats, which are more nourishing; and, lastly, well harvested pea straw, if that crop is grown. The latter, which is often little inferior to fine hay, is reserved for the time when the ewes are lambing, or shortly before. We think it is an excellent practice to carry this straw as it is thrashed, and build the stack rather long and narrow in the centre of the space intended for the lambing pen; it makes a shelter and helps to divide the pen. The ewes can to a certain extent help themselves, and the shepherd supplies the racks with great facility. When sainfoin hay is grown, and can be spared, a portion is most valuable for lambing ewes. Our own practice, when managing the Royal Agricultural College Farm at Cirencester, was to supply hay once a day, and pea straw once; in addition to which a moderate allowance of roots was given, and water *ad libitum*. The ewes were very healthy, and had abundance of milk. We cannot too strongly urge the great importance of a regular and abundant supply of water. We are aware that this was not recognised in former days, and there are a few flockmasters who adhere to old-fashioned ways; but science and common sense indicate that access to water is most desirable. Then, if required, it will be taken; and especially is it absolutely necessary during the lambing period and in summer time.

With proper care and attention on the part of the shepherd ewes can be wintered very reasonably on regular breeding farms. One of the most practical and successful Cotswold hill farmers we know managed as follows: Three hundred ewe tegs took the top of the turnips, gnawing them, and having hay. Six hundred ewes followed, eating the remains of the turnips and straw. These lay back at nights in a fold-yard kept well littered by the refuse straw removed from the racks. It is surprising what a quantity of straw may be thus consumed, how small a weight of turnips suffices, and what a good manure heap results, which, with proper management, is in excellent condition for the young seeds the next year, if required.

Where ewes are wintered on sound grass land (a practice which prevails on the land less adapted for sheep), we prefer to give pulped roots with chop, composed of hay and straw, over which a little meal is dusted. Here, again, it is surprising how well repaid we are for care and attention. Nothing can do better than ewes thus treated in the autumn; and from then to Christmas, if the weather is open, the grass, if plentiful, with a few turnip tops or an occasional load of turnips, suffices to keep the ewes thriving; then a supply of pulp, &c., once a day, makes up what they cannot find for themselves; and later on, as they become heavy with lamb, the pulp may be given twice a day.

CHAPTER III.

PREPARATIONS FOR AND ATTENTION DURING LAMBING.

THE lambing time is a most important period; and in proportion to the care and attention preparatory to and during its progress, so will be the result. We are not believers in good or bad luck. Weather and peculiar circumstances may tend to a greater mortality at one season than another, and such matters are to some extent beyond our control; but, making due allowance for such disturbing elements, we repeat that it is a question of intelligent, thoughtful management, as opposed to careless, ignorant, and thoughtless treatment. Take the one question of food and lodgings. During November, 1875, we had an unusual and almost unprecedented fall of rain, and the sheeps' backs were seldom dry for long together. Now contrast the condition of two flocks—the one hurdled on turnips, with the land trodden into a thick puddle of mud, and the sheep allowed to gnaw what they could, getting no dry meat, and lying always on wet ground; the other, also in the turnip field, but having only a regular allowance of turnips, and taken each night into a sheltered and well littered fold, where they are supplied with plenty of straw. Is it a cause of surprise if, in the one case, there are many dead and rotten lambs, while those that are born alive are weakly, pot-bellied animals; and, in the other, we have healthy yellow lambs that go on and thrive well? Nevertheless, we commonly hear the difference of results in the two flocks ascribed to good or bad luck.

Animals, in order to make a profitable return, must be treated considerately. The master and man must both love the animals, otherwise things will not go well. We have indicated in general terms how feeding should go on up to the lambing period; but we cannot farm by rule, and it may be that the experienced eye of the shepherd sees something wrong. The skin ceases to look healthy, the eye is dull, and then it is that the judicious stock-owner will try a change. If we hope or expect for a successful lambing there are certain things that must be attended to. First and foremost the ewes must be strong and healthy, which can only be when their food has been sufficient to grow the foetus without drawing upon the mother. We must endeavour to steer a mean between having them fat—in which case there is a risk of fever—and poor, when they cannot nurse well.

The value of and necessity for a ewe pen depends upon the climate and period of the year at which the ewes lamb, and the hardy or delicate character of the sheep themselves. When the situation is exposed (as must be the case on the chalk and limestone hills) we are greatly in favour of making ample provision for comfort, remembering that warmth and shelter are most important aids to success. In selecting the site for a ewe pen the following points should be borne in mind: To obtain all the natural shelter, especially from the more boisterous winds; to have the pen near the food; to select dry ground. The lee side of a good thick plantation is a capital situation; but if we have not any natural shelter, we must make it artificially: this can be done by stacking a wall of straw, and weighting the top so as to resist wind; then under this shelter we make the covered pens, in which the ewes and produce are placed for three or four days. We prefer to have these pens partly open and partly covered. They are constructed with hurdles, one on the top being covered with straw; thus we get warm shelter, where the dam and produce can be fed according to their wants. The open portion of the ewe pen should be divided into two parts, the division being made by a stack of pea straw, as has been mentioned before. The walls, except on the more exposed side, may be made by fixing double hurdles about a foot apart, and filling the interval with straw closely trodden in; this breaks the force of the wind, and renders the pen very snug. It is a good plan to house the ewes at night for a few days before the process commences, and when the time arrives all the forwardest ewes should be put together on one side adjoining the shelter pens; and for these, and the ewes with lambs, food nearest the pen should be preserved, so that they can be brought home without much trouble. By housing the ewes thus early they become accustomed to the movements of the shepherd amongst them, who can thus move about at night with his lantern without disturbing the ewes. It is a pretty sight to see the sheep all lying down in their comfortable quarters, so tame and gentle that the shepherd's movements are scarcely noted; and it is pleasant to see the care with which a good man goes about his work—not hurrying nature, but ready to assist when help is required.

As soon as the lamb is born, the ewe and her offspring are placed in one of the pens, and supplied with suitable food; a little gruel should always be at hand, together with a few simple medicines, of which laudanum is one of the most important. We believe that the medicine chests supplied by Messrs. Day, Son, and Hewitt, are very valuable. There was formerly a difference of opinion as to the advantages or otherwise of giving the ewes water; our own experience is decidedly in favour of the practice; possibly the objection may have arisen from its abuse. When animals are feverish, and have not been allowed to drink, they take a large quantity at once, which often causes scour; but if water is always to be had, they drink a little at a time with decided benefit. Just at first, after lambing, a little sweet hay, a handful of bruised oats, and a small quantity of sliced turnips may be given, care being taken not to give anything in such quantities as to induce fever.

Whilst the shelter of the ewe pen is most valuable, we must be careful not to render either the mother or offspring delicate by keeping them too closely housed; therefore, after the third day, if all goes well and the weather is favourable, they may run out in the day time, and after two or three weeks may be removed altogether, and left to shift for themselves with the aid of shelter hurdles, *i.e.*, thatched hurdles set up crosswise, so that whatever the direction of the wind the lambs can shelter. These are very necessary and valuable defences, and well repay some care in the thatching. It is a good plan to have a few made beforehand, so as to be ready for use when wanted.

We do not here enter into details as to the shepherd's duties, we would refer our readers to the excellent directions given by Youatt; but this much may be said that false presentations are rare except the ewes have been frightened. Hence the importance of gentle treatment; rough handling or hunting by a dog may suffice to cause the evil. Nature should be assisted, not anticipated; usually a little help at the proper time is all that is necessary. The following remarks by Mr. Cleeve, in his prize essay, published in the first volume of the "Journal of the Royal Agricultural Society," are so sensible that we reproduce them:

"The shepherd must receive it as a general maxim, to be most attentively observed, that *Nature is the best midwife*. He must not be led by the appearances of uneasiness and pain to interfere prematurely; he must watch the ewe closely, and so long as she rises at his approach he may be assured that whatever uneasiness she may exhibit all is well. Much uneasiness is generally apparent; she will repeatedly lie down and rise again with seeming distress. If this occurs when driving her to the fold he must be very cautious and gentle in urging her. These symptoms ought to be continued for two or three hours, or even more, before he feels imperatively called on to interfere, except the lamb is in such a position as to warrant fears of losing it. In cold weather particularly the labour is likely to be protracted. Should the ewe appear exhausted, and gradually sinking under her labour, it will be right to give her some oatmeal gruel, with a little linseed, in the proportion of a spoonful of the latter to two of the former. (A note appended, by Mr. J. W. Childers, recommends oatmeal gruel, with treacle and one gill of ale; and after a difficult time of lambing, when inflammation is to be apprehended, rye meal gruel, with a good proportion of treacle, without the ale.) When the ewe feels that she is unable of herself to expel the lamb she will quietly submit to the shepherd's assistance. In giving her this assistance, his first duty is to ascertain whether the *presentation* is natural. The natural presentation is with the muzzle foremost, and a foot on each side of it. Should all be right in this respect, he must proceed to disengage the lamb, first *very gently drawing down the legs, and with all possible tenderness smoothing and facilitating the passing of the head with his fingers rather than forcibly extracting it*—the particular attention of the shepherd being given to these points. This may be effected by passing the fingers up the rectum until he feels the back of the lamb's head, and then urging it forwards at the same time that he gently pulls the legs. Sometimes the head is sufficiently advanced, but the legs are too backward. In this case the head must be gently pushed back, and the hand being well oiled must be introduced into the vagina and applied to the legs so as to place them in their natural position, equal to the head. Should the forefeet, on the other hand, protrude, they must in like manner be returned, and the same assistance given to advance the head. If the hind quarters present themselves first, the hand must be applied to get hold of both the hind legs together, and draw them gently but firmly: the lamb may often be easily removed in this position. It is no uncommon occurrence to find the head of the lamb protruding, and much swollen; but still, by patience and gentle manipulation, it may often be gradually brought forward; or even nature, not unduly interfered with, will complete her work if the pelvis is not very much deformed. Should, however, the strength of the mother be rapidly wasting, the head may be taken away, and then the operator, pushing back the lamb, may introduce his hand, and, laying hold of the fore legs, effect the delivery. It also happens that the legs are thrust out to the shoulder, and from the throes of the animal it is not possible to replace them so as to get up the head of the lamb; by partially skinning the legs you may disunite them from the shoulder-joint, there will then be room for the introduction of the hand, and by laying hold of the head you can deliver the ewe. A single season of practice will do more than volumes of writing to prepare the farmer for the preceding and some other cases of difficult labour. But let him bear in mind that, as a general rule, the fœtus should, if possible, be placed in its natural position previously to any attempt to extricate it by force. When *force* must be used, it should be as gently as is consistent with the object of delivery."

We have ventured upon this lengthened quotation because the writer puts his experience and advice in remarkably clear language. We entirely endorse his remarks, and again draw attention to the importance of quiet treatment. The shepherd must know his sheep, and be known by them. Kind, gentle ways will enable him to exercise his power and skill to the greatest advantage. Unfortunately we cannot always meet with servants who really take a keen interest in their work; hence it is well that the master should keep a watchful eye over affairs, not omitting occasionally to take a turn at night work. We have known sad mistakes made from overtasking the shepherd, who cannot properly do his work when overpowered by fatigue; for a month or six weeks he must be up several times during the night, and occasionally, in a busy time, which usually happens when the weather is wet and stormy, he gets no rest at all. At such times assistance should be provided, so that the man can rest during the day. The hut should be warm, and have a stove and bed, with appliances for heating water, gruel, &c. Beer is undesirable, as tending to induce sleep, but good coffee and nourishing food should be provided abundantly.

CHAPTER IV.

MANAGEMENT FROM BIRTH TO WEANING.

IN our remarks we suppose a sheep farm, *i.e.*, a light limestone soil, principally arable, if with water meadow so much the better, as the latter adds much to the food resources, and renders the farmer more independent of seasons. When dry healthy pastures exist, the ewes and lambs may be removed thence when the lambs are from ten to fourteen days old, and have a portion of roots drawn there; but, as a rule, stocking grass at this season is very injurious to its summer prospects, and it is better to keep the sheep on the turnips—a certain area having been planted for this purpose. A hardy winter variety, such as the Green Round turnip, or White Swede, is sown late, so as to stand the winter and make good seed stalks in the spring. These young shoots are readily eaten by the lambs, which run forward through creeps or lamb hurdles, of which the best we have seen is made by Messrs. Carson and Toone, of Warminster, Wilts.

The lamb commences to eat when about a month old. Now is the time to begin artificial feeding, if our object be to force forward for early sale. This will, of course, depend upon circumstances; but on the exposed downs, where the majority of breeding sheep are kept, it does seem a good plan not to winter the wethers, as their room may be so usefully occupied by ewes. The climate is often rather exposed for young sheep, and we can frequently get more money, in proportion, in the autumn than in the spring, even if we have no such object. The outlay in artificial food is repaid by superior growth, and in the relief to the ewe—another advantageous point, as she sooner comes into condition for renewed breeding. Therefore, provided we have a fair chance of maintaining the growth of the animal, we advocate preparing suitable food for the lambs, and placing the same in small sheltered troughs in advance of the ewes. If asked what kinds of food we recommend, we say something that combines flesh-formers with fat-producers. The following mixture will be found advantageous: Dust linseed and decorticated cotton seed cake, bean meal, and palm-nut meal; these may be used in equal proportions, and given with fine hay chaff; until weaning two ounces per head daily will be ample; afterwards, and until the lamb goes to turnips, a quarter of a pound, which may be increased to half a pound during the winter.

Probably the management of the Hampshire and Wiltshire down lambs offers the very best example of judicious feeding that can be found. Here the primary object is to get the lamb to market in the autumn, and no expense is spared to provide frequent changes of food. As soon as the lambs can eat they are allowed corn and cake in troughs in front. After the turnips are eaten, the flock is placed in water meadows by day, and on late swedes, if any remain, at night; then on rye and winter oats, Italian rye-grass, &c. If there are no water meadows, a portion of the clover layer forms excellent food; this should always be folded off, the lambs having the front pen. At this time they will eat a considerable quantity of food. There are two plans, either to keep the ewes in close quarters, having a lamb pen ahead, and shifting often twice a day, or else to let them lie back on the ground they have already cleared. The first plan is, on the whole, preferable, and may be safely carried out when the sheep have a night change. If, however, we are necessitated to keep them entirely on the seeds (which, of course, is very undesirable), more room is necessary; but even then we must not keep them on the ground long, otherwise the lambs eat the young shoots, and scour. If we can go rapidly over the surface—that is, eat the crop down close—and then pass on, it grows again evenly, and in due time we secure another crop either for the ewes or for hay. It is at this time, when the sheep are on clover, that sliced or pulped mangolds prove of great value. They contain at this season much sugar, and correct the laxative tendency of the clover. Mr. Rawlence's practice, as described by Mr. Jenkins in his report of the Bulbridge Farm, may be cited as an instance of successful treatment. As a breeder of rams, a forcing system was adopted:

"After lambing, the ewes get mangolds with hay chaff for about ten days in the lambing pens; and, in addition to this food, the ewes with tup lambs or with couples get either one pint of oats or one pound of cake; but, unless roots are scarce, the remaining ewes are denied artificial food. At the expiration of ten days or a fortnight the ewes and lambs go on turnips, and remain there until March 20th. About this date the ewes and lambs go into the water meadows by day, and are folded at night on swedes, for the first fortnight or so, and afterwards on Italian rye-grass, or occasionally on rye and winter oats which have been sown where rye-grass has failed. This treatment is continued until the middle of May, when the lambs are weaned."

This practice is excellent; but we hear no mention of vetches, which appear to us most valuable food for sheep on arable land. By sowing successive crops in the autumn and early spring we can secure continuous food from May to August; and, according to our experience, lambs thrive very well indeed on such food when taught to eat it. We have frequently weaned Cotswold lambs on vetches, removing the ewes to a short pasture, or, as is the practice in Hampshire, leaving them to follow, only doubling the line of hurdles that separates. It is astonishing how soon the maternal instinct disappears, especially if the food is properly stinted. There is little difficulty about the milk; an abundant supply of rock salt should be provided, and the ewes supplied with dry food. When the vetches are young they are best eaten on the ground: in this case it will help matters if tracks are cut, as less food is spoilt. Very soon the crop grows to the stage when it should be cut and placed in racks. The lambs may be educated to eat thus artificially by suspending portions of the crop to the hurdles. The tender leaves are readily eaten, and soon nearly the whole stem.

The period for weaning varies according to the climate,

description of sheep, &c. The most cultivated breeds remain, as a rule, the shortest time, whereas with mountain sheep it is often a perfectly natural process. The sooner the lambs are separated, provided the system is capable of digesting vegetable food, the more rapid the progress of the lamb. Hence we find the process of weaning ranges from May to July.

CHAPTER V.

FROM WEANING TO MARKET.

OUR remarks apply to the wether sheep destined for the butcher. These have either been emasculated according to the old-fashioned process of drawing when a few days old—a practice which is a serious trial to the lamb, and, in the event of a sudden change of weather, sometimes followed by casualties: or are left to be castrated towards the beginning of June—a plan which, where skilfully performed, we think decidedly preferable. There are various processes. We prefer searing and the use of blue ointment and lard. In the sheep districts of the West of England there are properly qualified experts, who travel at this season from farm to farm and are wonderfully skilful; so that under ordinary circumstances, there is very little risk. One advantage of this matured plan is that the purses are much larger and fuller, so that the fat sheep, when turned, show to greater advantage. After searing it is necessary to watch the sheep, and keep them moving to prevent the settling of blood and stiffness. As a rule they never suffer, and go on without a check.

Great attention is necessary at this stage to keep the lambs in a healthy progressive condition. Food suitable, varied, and supplied with discretion, are the requisites. One shepherd will starve in the midst of abundance, while another will make the best of difficulties, and keep the sheep thriving during a drought. Suitable food and frequent change are the great requisites. Vetches often form the main food, but this is not well too long; and if the night fold is thus provided, we must, if possible, give a change to sainfoin or clover during the day.

Since the first edition appeared, lamb classes have been introduced at Smithfield, in consequence of remarkable exhibits of Hampshire lambs as extra stock. So far the greatest excellence and the earliest maturity has been about equally shared, by the Hampshire and the cross-bred lambs—of which the Hampshire and Cotswold first cross have proved most successful. Mr. W. Parsons of West Stratton, near Micheldever, in an interesting paper read at the London Farmers Club, Nov. 1884, gives statistics as to the comparative weight and numbers of lambs and shearlings, shown at Islington—he divides the nine years from 1875 to 1883, inclusive, into three periods. In the first, 1875 to 1877, there were 76 pens of lambs which, at 9 months 1 week, averaged 4cwt. 1qr. 16lb., whilst 187 pens of wether sheep, 21 months old, averaged 6cwt. 3qr. 2lb. For 1878 to 1880, entries of lambs had increased to 90; age same; average weight 4cwt. 1qr. 17lb; whilst 187 pens of wethers, give an average of 6cwt. 2qr. 27lb. In the third and last cycle, 1881 to 1883, the pens of lambs had increased to 162, yielding 4cwt. 2qr. 12lb. average weight; whilst 204 pens of wethers averaged 6cwt. 3qr. 5lb., and thus, as Mr. Parsons points out, whilst the wethers showed no material advance, the lambs under a year old had gained 24lb. live weight, and the number of entries had more than doubled. Mr. Parsons states that the most striking instance of success in early feeding in his experience relates to 200 lambs belonging to Mr. G. Judd, of Barton Stacy, Micheldever. These lambs were shown at Winchester Fair on Oct. 23, 1882, in two lots, and besides taking 70*l.* prizes, were sold within a fraction of 84*s.* a head. These lambs were born in January or early in February, and were weaned as soon as the water meadows were fed off—about May 13. They were fed on tares, sainfoin, and oilcake till July, when rape took the place of tares, and the amount of cake was gradually increased, until the daily allowance reached about $1\frac{1}{2}$lb. oilcake with $\frac{1}{3}$lb. of split beans, and, during the last six weeks immediately preceding the show, turnips were substituted for rape. The average weight of the 200 lambs was supposed to be fully 12st. or 96lb. Mr. Judd states that his system of feeding young lambs is not quite in accordance with the modern idea of pushing them on as fast as possible from birth.

He has found it better to keep them in healthy stock order—giving plenty of green food, and a small quantity only of oilcake, or cake and split beans mixed, until the end of June or beginning of July, by which time they will have developed some bone and size. Very forcing feeding in an early season is apt to make the lambs go wrong later on.

Sheep frequently suffer in hot dry weather from want of water. An unreasonable prejudice formerly prevailed, and possibly still exists in the minds of some shepherds, against the use of water, under the belief that it induces scouring—an erroneous opinion, arising from its improper use. If, as is too often the case, the lambs are kept on dry food (*i.e.*, food that is getting old) without free access to water, and then are allowed to drink without stint, scouring is inevitable, and sometimes is attended with fatal results; but if the animal is always provided with water, and can drink when it likes, the imbibition will be with discretion and eminently advantageous. It is only man who is excessive in his drinking. We therefore advocate the providing water in iron troughs during summer. When the sheep are on the turnips it is no longer necessary.

In August the careful manager will have some early rape ready; and he must be careful not to put the lambs on during wet weather, or when the dew is heavy, or there will assuredly be mischief. It should be remembered that rape is very succulent and blood-making food, consequently there is danger of scouring and considerable loss from improper use; indeed, this applies to all changes: they should be as gradual as possible. Now it is a good plan to mix rape and vetches as an introduction, and in the same way to grow a small area of rape and turnips, and thus bring the lambs gradually to their main food. But the rawness of unripe food may be modified by the use of artificial food. We believe that decorticated cotton-seed cake, when good, is very valuable, and better suited to counteract the effect of green food than linseed cake, which is itself slightly purgative. Malt dust is also valuable when it can be easily procured. A quarter of a pound of each would be quite ample at this stage. Serious losses frequently occur when tegs are first placed upon swedes, entirely arising from the want of due preparation. Sudden changes of food are always to be avoided. The animals eat too much, blood is made fast, and a black scour sets in. We have known great mortality in this way, which we are convinced might have been avoided had the food been introduced more gradually. Supposing that the food is being cut (and we can hardly suppose that sheep for the butcher will be allowed to waste their forces in gnawing), nothing is easier than to mix a few loads of swedes with the last acre or so of turnips, even if it is not practicable to grow a few together. Such a precaution will be most advantageous, preventing that check which may be anticipated when food is changed. It is very possibly owing to such check that the mischief occurs; for when the food is taken to, the animals eat more than is good. Of course to some extent we can regulate the supply, but we cannot muzzle the greedy animals. The use of sweet hay, and especially good sainfoin, has a tendency to promote health; but few can spare such material, which is wanted for the ewes after lambing. We believe that hay supplied in covered racks helps the sheep at this critical time, but it is not an economical way of using such food. The chaff-cutter allows of a due admixture of straw; and if finely cut and sifted, a good deal of dry food may be coaxed down, especially if the artificial is mixed with it.

The progress of the sheep depends not only upon the quality of the food but upon the nature of the land and the character of the season. A dry lair is very important, and in wet weather, if a damp lair is added to a moist jacket, much of the food is required to maintain the temperature of the body, and progress is out of the question. We are satisfied that in such weather on all but the driest land sheep would do far better housed. And it is a question worth deciding by experiment whether house feeding might not be more frequently adopted with advantage than is the case. It is neither necessary nor desirable to have the building entirely closed: one side might be open so as to insure plenty of air.

Great care must be taken of the feet, otherwise the walls of the hoof grow long and cause serious lameness. All this is matter of detail. In a wet winter sheep on good-bodied land suffer terribly from the state of the ground, and it is only as the days lengthen and the surface dries up that decided progress is visible and the fattening goes on well. Of course in many cases the sheepfold is a necessity, and we must manage as well as we can. This is especially the case on light-land farms, or where the buildings are far off. Occasionally, when the weather is unusually wet, it may answer to litter the fold with straw, which helps to keep the poor animals out of the dirt; but in these days the article is too scarce for such use. Sometimes things come to such a pass that the removal of the sheep to a piece of dry grass or seeds that are to be broken up is absolutely necessary. Here there should be a provision of roots, so that no serious loss of condition occurs; but, as a rule, sheep always suffer, and therefore it is best to keep them where they are if it can be schemed. An increase of dry and nourishing food helps to counteract the evil, and it is really astonishing how, when well tended, they manage to hold on, in spite of counteracting influences. We have found a liberal supply of rock salt, which should be placed in covered boxes about a foot from the ground, tends to promote healthy digestion.

Towards spring Long-woolled sheep are often seriously inconvenienced by the clots or balls of hard mud, which hang to the wool, and not only make progression a labour, but chafe and irritate the skin. These should be removed, and some help should be supplied for this extra work.

Now it is that the sheep consume the maximum quantity of food, and require attention early and late. When we find them lying down, often stretched out at full length, and only up when feeding or at the change of food, we may rest assured they are thriving; but this is generally pretty evident from the appearance of the animal, and especially the condition of the wool; indeed, the latter is immediately affected by adverse circumstances, and regular feeding is just as necessary for the development of wool as of carcass.

In the above remarks we have taken it for granted that the best way of preparing the root crop was by the ordinary Gardner's turnip cutter, which reduces the root into strips three or four inches long by three-quarters in cube. We are not prepared, however, to maintain that this is so. Our own experience with ewes wintered on grass, and the experience of a friend who has given some attention to the subject, leads us to an opposite conclusion; and though, at present, there are difficulties in the way, we anticipate that some day the pulping machine will be quite as valuable for sheep fatting as it is a necessity in the treatment of cattle. There is no question as to the saving in roots or the satisfactory progress of the sheep, provided the materials are duly proportioned. The digestive system of the sheep renders a larger quantity of roots necessary. The difficulty is as to the additional labour. Pulping machines take a much heavier draught, and the material is produced more slowly. On a large scale we are of opinion that a portable horse-gear with pulper on wheels might be introduced with advantage, as well as a chaff-house. The subject is at least worth investigation and experiment.

The growth of cabbage as sheep food has not been alluded to, yet it forms upon some soils a most important item in our resources. Limestone soils, especially such as contain a fair proportion of clay, are eminently suitable. And we allude to the practice of such men as the late Mr. Rigden and Mr. Robert Russell as examples of what may be done with the cabbage crop. By the former great dependence was placed upon the cabbage for summer food. For this purpose a seed bed is prepared in the autumn, a sheltered spot being selected. The plants are set out at the end of February or early in March, and furnish food during July and August. Later plantings provide winter food. At the date of our visit the Jersey hundred-head and the drumheads were looking most promising. Mr. Russell's practice, as carried out in Kent, was the subject of an interesting paper read before the London Farmers' Club in the spring of 1876. The chief point was his advocacy of the thousand-headed kale, which appears most valuable. The subject was treated of under each month's work, starting with August; and we learn that twenty acres of this seed is sown to be fed off at the end of April or first week in May. In October two acres of drumhead cabbage seed is drilled for spring transplanting. "Towards the end of April," to quote Mr. Russell's own words, " we drill in the celebrated 'thousand-headed kale seed,' using four to five pounds an acre. This is the least known, and most desirable of any green crop I have ever seen. It is a plant that produces more feed per acre than any other, does not disagree with any stock, and does not impoverish the land; moreover, with us, it has never caused sheep or lambs to blow or scour. Eighteen perches of this plant per day, with a little oat straw, has kept 270 sheep for three months without the loss of one." Summer cabbage comes in for use in June. The trefoil and trifolium stubbles are ploughed up close behind the sheep, and drilled in or planted with sprouting broccoli or kale. We are unfortunately in the dark as to the area of this farm. Mr. Russell stated that at one period in the autumn they had 264 acres of cabbage, kale, and broccoli; 30 acres of swedes, and 30 acres of mangolds. We have alluded to Mr. Russell's paper in order to direct attention to the cabbage and kale, which evidently are particularly suitable to his requirements.

CHAPTER VI.

ON WOOL.

SHEEP are distinguished as Long or Short Wools, according to the character of the fleece. These, again, are frequently crossed, giving rise to varieties, such as the Oxfordshire, Shropshire, and Down Leicesters, in which the staple is not either short or long, open or close. The growth of wool is an important feature in sheep management. Careful and regular feeding is requisite in order to insure uniform quality and maximum weight. Cleanliness of skin and freedom from parasites are also important. When sheep are continually rubbing themselves we may be certain they are not healthy; and though free from the flockmaster's plague, "scab," they are probably alive with ticks. Under such circumstances wool cannot be grown in perfection; indeed, we may safely state that the subject would pay for more attention than it usually receives. Lord Cathcart contributed a remarkable essay to the "Royal Agricultural Society's Journal" (Vol. ii., Second Series, page 309), entitled, "Wool in relation to Science with Practice," which will repay careful perusal, as it abounds with suggestions and statistics. He therein points out the importance of the trade, and states that the general consumption of wool in England is said to be $4\frac{1}{2}$lb. per head of the population, as against about 3lb. per head in Germany. England and France stand at the head of the wool manufacture; in the years 1870 and 1871 England manufacturing 330,000,000lb. and France 300,000,000lb. of wool of all descriptions; and he gives the following tables of imports and exports:

	Imports. lb.	Exports. lb.
England	238,820,852	94,911,916
France	167,422,200	25,711,412
Belgium	147,092,128	66,543,920
Germany	90,000,000	25,000,000
Austria	21,680,900	16,392,700
Netherlands	16,991,972	13,906,260
Russia	2,648,700	28,558,577
North America	62,202,714	12,067,689

The above table indicates the importance of the wool manufacture in this country, and suggests the advantage of more attention in the growth and preparation of wool than usually prevails. It must, however, be borne in mind that climate and soil have much to do with results, and that no amount of care can compensate for naturally unfavourable conditions. Mr. William Brown appends to his little book on "British Sheep Farming" a map exhibiting the general distribution of prevailing breeds, which is particularly interesting as illustrating the influence of soil and climate. Experience has abundantly proved these facts. The peculiar lustre of the Lincolns, due to the habit of growth of the wool, is lost or greatly lessened when the sheep are cultivated out of their district or on less congenial soil. The Leicesters on particular soils also exhibit the same character. Certain tracts of land are noted for the superior quality of the wool. Thus wool grown in the Ripon and Thirsk district, in Yorkshire, always makes 1s. to 2s. a stone more than that from other parts. Whether this is due to the soil entirely, or to the combined influence of soil and climate, is a point worthy of investigation. Lord Cathcart introduces a note from a practical farmer of the locality, who refers to the dry healthy nature of the soil, a fine loam resting on porous substrata of red sandstone and freestone. This insures dry healthy lair, whilst the fine quality of the land produces nourishing crops. Although limestone soils are for many reasons peculiarly suited for rearing sheep, there is a tendency to harshness of wool which renders them less valuable for wool growing than good clays or gravels. Again, the quality of the water in which the sheep are washed has also an influence. The softer the water the better. Hard water—*i.e.*, such as contains much lime and magnesia in solution—is undesirable. Climate is responsible for determining the broad distinctions of wool. Thus the Leicesters, when cultivated north of the Tweed, lose both in quantity and quality. "North of Fife and south of the English Channel," Lord Cathcart, quoting the late Mr. Torr, remarks that "the quality of the wool falls off, and then becomes hair or moss. The valuable fine lustre is pretty nearly confined to a few degrees of latitude; so that, space being limited, there is little or no danger of wool ever glutting the market. Fine lustre wool will ever bear a great value." Possibly present experience might induce a somewhat modified view. At the time Lord Cathcart wrote lustre wools were in demand for ladies' dresses; fashions change, and shiny goods are not now in such demand.

Whilst, therefore, the improvement artificially is limited according to soil and climate, much may be done to promote healthy sustained growth; quality and quantity are not synonymous terms as regards wool. It is generally considered that our Long-wool has deteriorated with improved practice. Abundance of food, whilst it adds weight, tends to coarseness. The great Bakewell, who knew this fact, and looked upon form and aptitude to feed as of most importance, neglected the wool question altogether. The objects of the grower and stapler are not necessarily identical. The former looks to profit, and cares for bulk more than quality. Continual progress should be his watchword. Woe betide the fleece if, from adverse circumstances, the animal experiences a serious check. In that case the bulk of the nourishment is absorbed by the body. The yolk (that invaluable nutriment which appears to feed the wool) is deficient, and there, sure enough, will be found by and by a tender point in the lock, which breaks easily and renders the wool much less valuable.

One word here as to the important question of washing. The object of the farmer is to cleanse his wool without affecting its weight more than he can help. Yolk, which is a compound of fatty acids with potash, is soluble in water, and a great portion is so removed; but if the fleece is left on the sheep for a sufficient time fresh yolk forms, and to some extent the wool regains weight. The wool handles soft and tough—conditions which are esteemed by the buyer—and weight is regained. Therefore, as soon as the animals are washed they should be placed in a dry grass field, and well fed. According to the nature of the weather and the thriving state of the sheep, from two to three weeks suffices to reinstate the yolk. All know the difference between the dry harsh coat, when shorn immediately after washing, and the soft greasy condition that indicates a thriving state and a good weighing. Now, as the yolk removed comprises 20 per cent. of the whole, and yields 40 per cent. of mineral matter, containing from 59 to 84 per cent. of potash, it becomes an important consideration whether it cannot be recovered. To do this a different process of sheep washing is necessary; and the point to determine is whether the value of the recovered potash would pay for the increased labour. A running stream, especially if tolerably soft, affords an admirable medium for sheep washing. If the waste water could be utilised for irrigation, much advantage would arise; but it is seldom that the natural conditions allow of this economy.

As regards the price of the wool, it is most important that the washing should be well done. The amount of friction required depends on the soil. Where sheep are dirty, no mechanical contrivance equals the wringing and rubbing of the hand; but there is no occasion for the operator to stand up to his middle in water, thereby contracting rheumatic affection, despite oceans of liquor. Where there is running water a bricked sheep wash should be constructed, in which one or more operators can work. This should be so contrived that the water stands about four feet deep. The sheep are thrown in on one side, and seized by the washer, who stands in a water-tight box placed in one corner of the bath. This box is built of brick and cement, and should be at least six or nine inches above the water line. The sheep is allowed to swim about for a few seconds so as to get thoroughly soaked, then seized under the chin by the left hand, whilst the operator works his right up and down the back and sides; then, with a twist, the animal is placed on its back, the head just kept above water, and the belly, &c., brought under the action of the hand; it is here and on the legs that external dirt will be found. To do this properly with heavy Longwools requires time and strength. Each sheep will take from three to five minutes; but it is well spent, as the result is thoroughly clean wool. At intervals, depending upon the condition of the animals, the bath must be emptied, and fresh water admitted. It is at this point that the water should be utilised if practicable for irrigation. The sheep being constantly in hand, there is no fear of exhaustion; when finished, it should have one good plunge, and be allowed to swim out at the lower side of the bath, which should be shelved so as to assist egress.

The comfort to the animals from a clean skin is evidenced by their basking at full length in the sun, where they will remain for hours; and this brings us to another important consideration, viz., the value of sheep dipping, both as affecting health, removing vermin, and favouring wool growth. In order to understand this question it will be necessary, very briefly, to consider the principal external parasites to which the sheep is liable. Our remarks must necessarily be limited; those who wish to pursue the subject will find an exhaustive essay, by the late principal of the Royal Veterinary College, Mr. James Simonds, in the first volume of the "Royal Agricultural Society's Journal," new series, 1865.

The principal enemies of the sheep are the so-called tick (*Melophagus ovinus*); the louse of the sheep (*Trichodiatis ovis*); the origin of the hateful disease known as "scab," which is caused by a species of Acarus; and last, but not least, the maggot of the flesh fly, which frequently causes intense irritation in hot weather.

The tick is found, more or less, in all sheep, and if unchecked is sure to increase, though procreation is not so rapid as in many other insects. It appears first in the pupa state, in which condition it is expelled from the parent, and adheres to the skin of the sheep. How long the pupa is being formed has not been ascertained, but Professor Simonds believes that the after stages occupy about fourteen days; warmth and a certain amount of moisture are necessary, both being supplied by the skin. During winter the skin is tolerably free, but as warm weather approaches the parasites abound, and must be destroyed, otherwise they cause such irritation that the animal's health suffers.

The louse of the sheep is not seen where dipping is practised, and principally occurs in mountain sheep which have been neglected. Professor Simonds speaks of them as common in many parts of the western counties of England, where they are known as red lice; sulphuretted oil being used for their destruction. We may safely say that when proper care is exercised such plagues will not exist. The scab was formerly much more common than at present. When sheep were exposed in fairs, and travelled long distances, and when open grazing was more common than at present, this was a frightful scourge, being extremely contagious; now it is only when sheep are bought that there is dread of scab, and a proper dipping generally insures our safety. In former days it was very common to dress the sheep each autumn with mercurial ointment, with a view to ward off scab and parasites. Now the bath has superseded this plan, and we have a number of different materials that are available.

The arsenical compounds were for a long time greatly in request; these generally consisted of a preparation of arsenic and soft soap. They proved effective against parasites, but it is doubtful how far the heath of the animals suffered from the absorption of the mineral, and it is quite certain that very serious losses have occurred from the drippings of the liquid on to grass which was afterwards eaten. For these reasons there is a natural objection to arsenical dips. Tobacco, with various other vegetable products, mostly of a poisonous nature, is very largely used abroad, sometimes in conjunction with sulphur; there is not the same risk as with arsenic. Of late years the carbolic acid compounds, obtained from the distillation of tar, have taken a very important place in the sheep dipping category, and for many reasons we believe that they possess superior advantages. Not only are they thoroughly effectual in destroying parasites, but they cleanse the skin in a remarkable manner, possess extraordinary antiseptic properties, and for a considerable time render the sheep fly proof—the flesh fly abhorring the tarry smell which pervades the sheep for some considerable time after the operation. Mr. McDougall is well known as a successful manufacturer of carbolic dips; there was at one time a firm at Norwich who patented a glycerine dip, in which carbolic acid was associated with glycerine. With regard to the periods most suitable for dipping, the first should be after the ewes are shorn, and before the lambs are weaned; the second during the height of summer, when flies are troublesome; and the third in the autumn. The effect upon the health of the sheep is very apparent; the pores are opened, circulation is more healthy, and progress more rapid. The cost seldom exceeds 1½d per head. It is hardly necessary to add that where sheep are bought, either for breeding or feeding, the first step is to dip them before placing with the home-breds; this precaution should never be neglected. One outbreak of scab will cause more loss than the cost of dipping during a lifetime. And the worst is that when once a flock is affected there is no telling when the farm will be clear—it may hang about for years.

The process of dipping is too well known to need a detailed description. The carbolic mixtures, when their strength is regulated, are not unhealthy—the action is decidedly tonic, so that any absorption that may occur through the skin will do good rather than harm. The dipping tub with its drainer and rubbing board are useful apparatus on a sheep farm. Cages have been designed in which the sheep can be lowered into the liquid, but we prefer the offices of two men, one on each side of the tub; the one holds the fore legs and takes care that the ears are not submerged, the other the hind legs; and after keeping about half a minute in the bath the animal is rolled up and down on a grated surface, to allow of the liquid returning to the tub. The dipping is best done in a yard or bare ground where there is no possibility of the sheep finding any food.

CHAPTER VII.

LEICESTER SHEEP.

THE Native Breeds of Sheep may be classified under three groups, according to the length and character of their wool. Thus we have under one head *Long Wools*, of which we may name the Leicester, Lincoln, and Cotswold as types; the *Intermediate*, as seen in the Cheviot, Dorset, and Oxfordshire sheep; and *Short Wools*, embracing the Downs and Welsh. It is difficult to say whether long or short woolled sheep predominate at the present day; a few years since we should have had no difficulty in awarding the first place numerically to the long wools, but of late years the Shropshire sheep have made great strides in public favour. Looking, however, at the large number of long wools bred in the Border counties, and their general appearance along the north-eastern coast, as well as in the south-western counties, we think they would still poll well.

In our attempt to give a general account of the leading breeds, we commence with the Leicester, not because they are a more ancient race than either Lincoln or Cotswold, but because in the hands of Bakewell they were modelled into a type of animal that eventually impressed its qualities more or less upon every other variety of long-woolled sheep. Bakewell was so far fortunate in occupying the arena at a time when circumstances extraneous to himself enabled him to carry out his improvements. The successful cultivation of the turnip as a field crop provided a more regular supply of food. It was possible, by good management, to keep stock thriving from birth to death; hence the value of an animal that could make flesh rapidly. We have no record of the *modus operandi*. Bakewell was singularly and suspiciously reticent as to his operations; hence we have rumours of crossing in various directions, and there are not wanting those who would claim the honour of supplying the foundation stock for his wonderful success. We believe that he adhered entirely to the Leicester, and possibly did not at first go far from home, or even beyond his own flock, but trusted principally to selection. Observation would convince him that like would produce like; he went for quality rather than size, and, finding great jackets associated with strong bone and coarse offal, he was more careful as to frame, quality, and aptitude to feed than as to wool; hence one result of his operations was some diminution of fleece. Depending so much upon his own stock, and desiring to see what the males grew into before using them, the plan of letting, instead of, as was then the practice, selling the rams, would, of course, be very advantageous; hence Bakewell commenced a practice which has been pretty generally followed by his successors. At first the plan was not popular, and it is recorded that the first ram let in 1760, to a Mr. Wilmore, of Illston-on-the-Hill, only made 17s. 6d., the same price being taken for two others. His ideas were ridiculed more especially by his neighbours. He was not a man to be easily daunted, but held on his way, making sometimes a guinea, and occasionally double that sum, for the use of a sheep. It was not until 1780, or twenty years from his start, that he got a paying price, viz., 10 guineas, for the use of a ram. By this time his fame was extending, and prices rose rapidly, until in 1784 and 1785 his best animals were let for 100 guineas. Having gained this high position, Bakewell thought to protect himself by establishing a ram club, known as the Dishley Society. We take the following interesting account from an article by the late lamented H. H. Dixon in the fourth volume (N.S.) of the "Royal Agricultural Society's Journal": "Mr. John Breedon, of Rotherby, was the last survivor of the Bakewell Ram Club, whose rules bear date January 5, 1790, and pledged the twelve members (who paid 10 guineas each) to 'keep the transactions' secret upon their honour. Mr. Paget was the president of the club, which held its earlier meetings at the Bull's Head and the Anchor, at Loughborough, alternately, and fined each member a guinea for non-attendance. There were twelve members, and the rules were made and kept with Draconic strictness. No member might sell ewes and lambs to breed from, unless he sold his whole flock or dealt with members only. Only forty ewes could be taken in to keep, and those must be the property of one person; not more than two dozen rams could be shown to any person or company at one time; and even members could only show their rams to each other between the 1st and 8th of June, when the general show commenced. On July 8 they were bound to rigidly seal their pens for the space of two months. Certain flockmasters were not to pay less than 100 guineas in their first contract, and after that 30 guineas for wether-getters. Not more than thirty rams might be let by one member in one year; and it was further enacted that there were to be no dealings with flockmasters who showed rams in the market, and that the much-dreaded members of the Lincolnshire Society should not have a ram unless four joined and paid 200 guineas for him."

However absurd such rules may appear now, it is probable that they were of use in maintaining the purity of the breed. The Lincoln breeders, who were great rivals, had a somewhat similar code. The Dishley flock were distributed—after being in the hands of Mr. Smith of that place and Mr. Honeybourne —amongst Messrs. Stubbins, of Stone Barford; Paget, of Eleman; and Philip Skipworth the elder. The latter's purchase of ewes laid the foundation of the Aylesby flock, which, up to the death of Mr. Wm. Torr, ranked as one of the most valuable in the country; from Mr. Stubbins's purchase descended the once celebrated Holmepierrepoint sheep. In 1798, Mr. Nathaniel Stubbins realised 2176*l.* 18*s.* for the letting of thirty-one rams, and prices continued to rise until in 1805 the letting price was nominally nearly 100 guineas; the flock was in 1814 divided between his nephews, Joseph and Robert Burgess. Mr. Sanday, sen., succeeded the former at Holmepierrepoint in 1834. He was very successful in the showyard, having between 1847 (when he first won at Northampton with ewes) and 1863 (when he showed at Worcester for the last time) won fifteen firsts, seventeen seconds, and ten third prizes for rams; eleven first and ten seconds for shearling ewes, besides numerous commendations.

In thus briefly noticing the earlier history of the Leicesters, we must not omit the name of Sir Tatton Sykes, who, though he never showed, got together a most valuable flock, going in for quality. What his blood could do in the showyard has been exemplified by the late Mr. Borton, of Malton, and the late Lord Berners. He commenced operations by the purchase of ten ewes from Holmepierrepoint. He liked small, thrifty sheep, believing them best adapted for his wold country. The Sledmere sheep were not allowed any artificial food, consequently he did not make great prices, and probably his blood has been more valued since his death. At the present day the Leicester sheep are not in such repute as a pure breed, and even at Holmepierrepoint the Shropshires have succeeded them. Mr. Sanday, jun., sold off several years since, believing that the dark faces were more paying sheep.

It is to the influence of the Leicester in the improvement of other breeds that Bakewell's work has been most felt. Let us glance for a moment at their introduction into Scotland, which commenced at an early period. Several pure breeds were established about the Border, and these supplied rams for crossing. Cheviots and black faces have been in many cases improved into highly valuable sheep. Witness the Kelso sales of Border Leicester rams, and their influence in producing the "Barmshires," a cross from Cheviot, which are peculiar to the border counties of Roxburghshire, Berwickshire, and Northumberland. The cast ewes are sold in the autumn to dealers for Yorkshire, where recently they have made from 50s. to 80s. a head. They vary according as the influence of Leicester or Cheviot predominate; are hardy, with size, and aptitude to feed. They are frequently tupped once and sold off fat with their produce during the next summer. We have used upon them a Shropshire ram with success, always getting a heavy fall of lambs, from 60 to 80 per cent. of doubles. The Kelso public sales of rams are now quite a feature in sheep history; about 2300 rams are sold annually in the four sale rings, purchasers coming from far and near. Lord Polworth stands at the head of the list, his sheep always making a high average, come when they may. The order of sale is regulated by lot, and it is a great matter to be in the morning.

Yorkshire is divided between Leicesters pure and a cross with the Lincoln, which gives size and wool. On the moors a very useful class of sheep, known as Mug Leicesters, hardy, long-legged, and well adapted to run with the black-faced ewes on the moors, are kept; the produce, known as Masham lambs, are much liked on poor land in Yorkshire and the midland counties, and they exist upon bare pasture, and can be made fat the next summer after yielding a fair fleece of 5lb. to 7lb.

It would be difficult to say upon what breed the influence of Leicester quality has not been felt. How much of their present popularity do the Lincolns owe to it? It is commonly believed that the Shropshires acquired much of their aptitude to feed from a Leicester cross introduced by Mr. Meire; and possibly even the Southdowns are not clear of a strain. With both them and the Hampshires, the cross produces very muttony sheep, though it does not answer to prolong it. In the south-western counties, more especially Devonshire and Cornwall, Leicesters have long been the dominant sheep. We have several breeders of note, and a large infusion of the blood introduced into the original breeds. Go where it might, the result was always the same—a great increase of fattening properties, and early development. If we look at the Leicesters in their own country, we shall find them of moderate size, fine in bone, cutting a fair but not extraordinary fleece, averaging 7lb. to 8lb. round.

With regard to the principal points or characteristics, the form of a parallelogram on four legs is often adduced as the best definition of a correct outline in the sheep. This, however, does not apply to the Leicester, and would be scouted by breeders who desire the ovate form, and point to the partridge as the perfection of shape. The fore-quarter of the Leicester is remarkably well developed; the shoulders are wide and sloping, consequently there is no rigidness along the back; the bosom is deep and wide, and the fore flank very full. The animal stands close to the ground; the neck is short, so that the head is raised but little above the line of the back, and it is fairly muscular. In Youatt's language, "The neck full and broad at the base where it proceeds from the chest, but gradually tapering towards and being particularly fine at the junction of the head and neck; the neck seeming to project straight from the chest, so that there is with the slightest possible deviation one continued horizontal line from the rump to poll." The ribs are well sprung, and the carcass very true; the hips well covered, but not wide, and tapering to the rump, which is small; the back is covered with fat. The handling of the late Lord Berners' wethers at the Smithfield shows was quite remarkable, proving the extraordinary capacity for the accumulation of fat. In such over-fed specimens, however, the waste must be frightful, as, with the exception of a few joints, the carcass presents the appearance of a mass of luscious fat. With such a capacity of external and rapid development there is little inside fat; hence Leicesters are not favourites with the butcher; and, though looking much more wealthy, they do not scale heavier when arrived at a certain age (viz., as two-shears) than the Southdown. Their great point is early development and accumulation of weight on a given amount of food. The head is well set on the forehead flat and generally bare, or covered with short hair. Formerly a great point was made of bare heads, but now we believe breeders prefer to have close short wool, which protects from the fly; and this is certainly desirable, or otherwise the ewes must be capped during summer. The eye is full and prominent, indicating docility of disposition, and the head tolerably long and fine; the ears are thin and rather long. The muscular development is moderate; this is attributable to rapidity of growth. The legs of mutton are not large, and there is a deficiency of lean meat. The skin is thin and very supple, and the wool is fine and fairly long.

The Leicesters are not a prolific breed. In early days too many lambs were regarded as a great evil; if the breeder left off with an equal number with the ewes he was well content. In these days more fruitful sorts are desirable. In the grazing districts of Leicestershire it is the practice to run a few sheep in the feeding pastures; here they rapidly get fat, and appear well suited for the country. But, as we remarked before, it is for their value as a cross, and on account of the extraordinary influence they have exercised on most of our leading breeds, that they merit the first place in our notices of sheep stock.

CHAPTER VIII.

BORDER LEICESTERS.

By JOHN USHER, Stodrig, Kelso.

IN tracing the origin of the breed of sheep now commonly called Border Leicesters, it seems almost a work of supererogation to prove that they are descended from a flock known as the Bakewell or Dishley breed, and the more directly their lineage can be traced to that flock, and their exemption from the introduction of any other strain proved, the more they are generally allowed to be distinguished by symmetry of frame and purity of blood. The breed owed its existence to Mr. Robert Bakewell, of Dishley, in Leicestershire; by a course of systematic experiments, commenced about the year 1755, in crossing the old Leicesters—said to have been "large coarse animals, with an abundance of fleece, and a fair disposition to fatten"—with other long-woolled breeds, probably possessing smaller frames and more symmetrical proportions, he in the course of years worked them into a new breed. As the breeds he used, and the proportions in which he used them, are conjectural (his system having been carried on with much mystery), it seems vain to attempt to enumerate them. Bakewell must have had a good knowledge of animal physiology, and as his aim appears to have been, not so much to produce sheep of large size as of fine frame, and great aptitude to fatten, it is probable that he connected together animals of the purest blood, nearly allied to one another, thus producing sires which, in their turn, exerted a preponderating influence on their progeny. That he ultimately succeeded in establishing a distinct breed—their distinguishing feature being a capability of producing, compared with other breeds, the greatest quantity of fat with the smallest consumption of food in the shortest time—is an acknowledged fact, while as wool producers they yield to no breed of long-woolled sheep, Lincolns excepted. They gradually found their way into other localities; the first draft of them into the Border counties being introduced by the Messrs. Cully, who migrated thither from the county of Durham in 1767. The immediate followers of the Messrs. Cully were Messrs. Thompson, Chillingham; Jobson, Chillingham Newtown; Robertson, of Ladykirk; Smith, Learmouth; Compton, New Learmouth; Smith, Norham; Riddell, Timpendean, &c.

Whether some of the early breeders of Leicesters in the Border counties, in imitation of Bakewell's system, tried still further to improve them by crossing in with the Cheviot, a breed possessing fine style and quality; whether the change in their general appearance is due to selecting animals of the pure breed, high on the leg, with white faces and clean bone; and whether the soil and climate have had their influence—are questions that can never be satisfactorily answered. Certain it is that the distinguishing features of the Yorkshire and Border Leicesters, though sprung from the same source, have diverged considerably, the former now showing a blueness in their faces and a tuftiness in their legs, while the latter are white and clean in both, and, more, what are generally called *upstanding* sheep. As the Bakewell breed in early times are described as having white faces and legs, we leave readers to draw their own inference. Our hypothesis, that the Cheviot may have been used by the early breeders, is suggested by our having seen, within these few years, a lot of tups bought as pure Leicesters, which we happened to know were only the third cross from a very fine specimen of the Cheviot tup. The said sheep showed a style and conformation rarely equalled, and were particularly good in their necks and heads.

Our opinion, however, is, that the flocks tracing the closest lineal descent from the Dishley, untainted by any other strain of blood, selected and crossed with taste and judgment, tended with care, and "all appliances and means to boot," are still the best in the Border district. When so bred, they possess the following conformation: The head of fair size, with profile slightly aquiline, tapering to the muzzle, but with strength of jaw and wide nostril; the eyes full and bright, showing both docility and courage; the ears of fair size, and well set; the neck thick at the base, with good neck vein, and tapering gracefully to where it joins the head, which should stand well up; the chest broad, deep, and well forward, descending from the neck in a perpendicular line; the shoulders broad and open, but showing no coarse points: from where the neck and shoulders join, to the rump should describe a straight line, the latter being fully developed; in both arms and thighs the flesh well let down to the knees and hocks; the ribs well sprung from the backbone in a fine circular arch, and more distinguished by width than depth, showing a tendency to carry the mutton high, and with belly straight, significant of small offal; the legs straight, with a fair amount of bone, clean and fine, free from any tuftiness of wool, and of a uniform whiteness with the face and ears. They ought to be well clad all over, the belly not excepted, with wool of a medium texture, with an open *pirl*, as it is called, towards the end. In handling, the bones should be all covered; and particularly along the back and quarters (which should be lengthy) there should be a uniform covering of flesh, not pulpy, but firm and muscular. The wool, especially on the ribs, should fill the hand well. When the above conformation is attained, the animal generally moves with a graceful and elastic step, which, in the Leicester sheep, as well as in the human species, constitutes "the poetry of motion," and without which animals, even of high class in any breed, cannot now attain the chief honours in the showyard or the auction ring.

The foregoing may not suit the taste of all Leicester breeders. There has been a tendency in later times to attempt to improve the breed by crossing with sheep of looser frame, and wool of an opener and stronger staple. Such attempts have generally ended in failure, the strain of blood producing tender heads, weak necks and loins, and lack of constitution, and taking many years of careful and judicious management to eradicate. Our opinion is, that, in all such attempts, the coarseness, if any, should be on the dam's side, and that the sire should invariably be of symmetrical form and pure blood: nay, more, we think that where an apparent increase in the weight of fleece and frame has been attained it frequently proves fallacious when brought to the test of the scales, the extra open fleece weighing lighter than that of a medium texture, and the larger and looser frame, when stripped of the offal, being less heavy than the more compact, on the same principle as the bone of the thoroughbred horse exceeds in specific gravity the porous bone of the Clydesdale.

There is nothing in the general feeding and management of the Border Leicesters differing materially from that of other breeds. They require good land and good shelter, and, having these, will live and thrive on a small quantity of food. Having a strong tendency to fatten, they arrive at early maturity, and are capable of producing a greater quantity of wool and mutton in a given time than almost any other breed. Their mutton, however, does not stand high in mercantile value, being coarse in the grain and tallowy in the fat. Time was when it found a ready market among the pitmen in collieries. A story is still extant of one of them, when purchasing a portion of a cast ewe with several inches of fat on the rib, on being asked if it was not too fat for him, exclaimed, "Fat! I care na if it war as fat as atween Newcastle and the Scottish Border." Time has brought its changes to the pitmen, as to other members of society; higher wages and more leisure enable him to participate in the growing luxuries of the age, and while on gala days he relishes his leg of Cheviot or Southdown, his old titbit is only fit for melting into tallow.

The worth of the Leicester sheep does not, however, depend on its value as mutton. In all well-bred flocks the great bulk of the lambs on the male side are kept for tups, and in like manner the best on the female side for breeding purposes. Thus only a limited portion of each, the cast ewes, and tups of a certain age, find their way into the butcher market. Their intrinsic value consists in their crossing profitably with the Cheviot, Black-faced, Southdown, &c. The latter are not cultivated extensively in Scotland or the Border Counties, being generally considered too tender for the climate. The cross with the black-faced makes fine sheep at two years old, yielding mutton of fine flavour. That with the Cheviot also comes to fair maturity at the same age, getting to great weight with mutton of good quality. This cross also forms the foundation for another by breeding from half-bred ewes with the Leicester tup, and producing what are called three-parts bred sheep. For this purpose all the best of the half-bred ewe lambs are kept, and command a higher price than any other. On most lands of fair average quality, where a portion of turnips can be grown, half-bred ewes are kept. Their produce being a cross nearer the Leicester, their development is rapid; they are generally forced forward for the butcher market at one year old, or little over, and, in fact, form the great bulk of the mutton that now feeds our teeming population. Early maturity and quick returns are the order of the day; epicures in the middle and upper classes are fain to gratify their dainty appetites with mutton of two and three years old; while Southdown, Cheviot, and Black-faced wedders of four and five years, with the beautiful West Highland kyloe of similar age, are rarely found, unless in noblemen's and gentlemen's parks, where they are kept, regardless of profit, to tickle the palates of the aristocracy.

Flocks of pure-bred Leicesters are now not confined to the Border counties, but have found their way, wherever soil and climate suit their profitable cultivation, throughout Scotland, even to the "far north;" and auction marts for the sale of tups exist in many localities. In Caithness Sir George Dunbar has, by dint of high farming and selecting sires with great care and regardless of expense from the crack lots of the Border, raised a flock of rare excellence, and his annual sale of tups has reached a very high average. Edinburgh, for numbers, now treads closely on the heels of Kelso; but for sheep of first-class quality Kelso still bears the palm. There each September brings together upwards of 2000, and merchants from all parts of the United Kingdom. The position of the lots in four auction rings is arranged by ballot, and four auctioneers simultaneously sell single sheep at the rate of one in the minute for more than seven hours. The highest rate is generally attained by Mr. Penny, who sells about seventy in the hour, including stoppages and concise preliminary remarks, and finishes in his strong vernacular with a voice as clear as a bell. Lord Polwarth's (Mertoun), Miss Stark's (Mollondean), Rev. — Bosanquet's (Rock), and other crack lots always hold a *levée*, and thin other rings during their sale. The bidding seldom flags, there being customers for all sorts—tup breeders taking the choicest specimens—breeders of half and three-parts bred stock choosing sheep of large frame and open wool; of the black-faced cross, those with closer skins; fat lamb breeders, sheep of good quality, though lacking wool below; while some, contented with any quadruped if only cheap (alas for them!), fight it out with the butcher and local dealer. It is interesting to the close observer to note the change in the various lots from year to year, some from a bad cross, or untoward local circumstances, losing caste, while others come to the fore, showing the great difficulty of keeping the character of a flock at a uniform standard.

Since writing the above paragraph several years ago, a considerable diminution has taken place in the number of Border Leicester rams exposed for sale in Scotland and the Border counties. Their place is taken by half-bred rams—the first cross between Border Leicester and Cheviot. Thus breeders of half-bred sheep—which are considered the most paying in intermediate soils and climates—put half-bred tups to half-bred ewes, and keep up their flock less expensively than by buying yearly half-bred lambs of the first cross. A half-bred stock so managed does not lay on flesh so rapidly, but there is not a doubt of their attaining bigger bone, larger frames, and more robust constitution. We have a parallel to this in the system of breeding carriage horses in Yorkshire, by putting half-bred sires to half-bred mares, and thus producing horses with larger and stouter frames than can be attained by a cross with a thoroughbred.

Of all the Border flocks, there is none that has maintained such a uniform character as Lord Polwarth's, which deserves more than a mere passing notice. In 1872 his lot of tups showed to great advantage, the highest priced sheep reaching 170*l*., and the average about 37*l*., which had only once been exceeded in their history. The flock was formed about the beginning of the present century, being selected from the most direct followers of Bakewell. Our first recollections date back as far as 1835, when Tom Small, of immortal memory, was the presiding genius in their management, and no lover was ever more jealous of the honour of his mistress than Tom of his pet flock. We happened to know him well, and how he spurned the idea of using any strain of blood not strictly Bakewell, and well he could trace them till we got lost in a maze of g-g-g-g-g-g grand sires and dams. When Tom felt the infirmities of age creeping on, he was deeply solicitous as to how the flock was to be maintained in its purity, and, ere he " gave commandment concerning his bones," suggested his successor. His choice fell on Andrew Paterson, who had previously been instrumental in bringing a neighbouring flock into a state of great perfection, and, from personal knowledge, Tom knew that " he had the root of the matter in him." Andrew has amply justified his confidence, the success of the flock having been, in the sixteen years he has had them in charge, not merely uniform, but progressive.

Much conjecture exists as to how the perfection of the flock is kept up, and as no one ever hears of Lord Polwarth giving a long price for a tup, it is generally surmised that there must be a good deal of in-and-in, or what is called in Scotland " sib," breeding. We had lately an opportunity of seeing the ewes and gimmers. Their beautiful blood-like heads, deep chests, straight backs and bellies, uniform coating of wool, and family likeness, was a treat to look at. We fancied that we got some slight insight into the system of breeding, although Andrew, like Bakewell, is somewhat mysterious. Let it, however, be understood that our views are theoretic. From the circumstance of the ewes not being drafted at four or five years old, like the majority of flocks, but kept occasionally, if good breeders, till they enter their teens, it is evident that an opportunity is afforded, with Andrew's profound knowledge of pedigree (of which he is a walking dictionary), to preserve several distinct strains of blood, crossing them from time to time. A good strain is never lost sight of ; if rare, it is cherished as a miser would his gold, and animals of rare excellence are never parted with, without leaving their representatives. The procreative powers of nature are never taxed beyond certain limits, and not an ounce of muscular or physical energy is wasted. They are said to be "sib" bred; be it so. The student of animal physiology knows well it is the way to gain symmetry of form; and, so long as they keep up fair size and robust constitution along with it, we hold it to be the grand secret of their excellence—the accumulation of one blood, and that blood the purest, enabling them to make their mark wherever they are used, which is as palpable to the eye of a judge as the cross with a sheep of a totally different breed.

Since the above remarks were published, the Mertoun rams have had a hard struggle to maintain their supremacy at the Kelso annual ram sales, being closely pressed by the lots of Messrs. Clark, Old Hamstocks; Mr. Thompson, Baillieknow; Mr. Jack, Crichton Mains; Mr. Sampson, Courthill; Messrs. Bain, Legars, and others of scarcely less note, all of which, on analysis, are found to contain a large preponderance of Mertoun blood. In 1885 Messrs Clark's lot attained the highest average, while in 1886 the Mertoun lot once more gained its old position, but only by a mere fraction. Still, an infusion of Mertoun blood is almost universally resorted to, as an antidote to anything spurious being introduced into a flock, and is, in fact, in Border Leicesters what Bates' blood is in Shorthorns.

The custom of over-feeding tups so prevalent in the Border counties, although somewhat mitigated, is still the rule, not the exception. It is a great waste in many ways. Sheep so fed cannot be so active under any circumstances, and when taken, as they generally are, to a poorer soil and less genial climate, with the extra feeding entirely suspended, often succumb altogether under such barbarous treatment; at all events, their vital powers are weakened, and, instead of lasting two or three years, they get worn out in one or two. If ewes require to be in an improving condition during conception, why should the sire be in a declining one? and is he likely in such circumstances to impart a healthy constitution to his progeny? We hope to see the general adoption of a more natural and healthy system.

CHAPTER IX.

COTSWOLD SHEEP.

THESE sheep are natives of the Cotswold or Cotteswolde Hills, which run through the eastern side of Gloucestershire, in a direction from south-west to north-east. The name is derived from the practice in early times of protecting the sheep during winter in cotes or low sheds, which, according to Camden, were long ranges of buildings, frequently three or four storeys high, with low ceilings, and with an inclination at one end of each floor reaching to the next, by which the sheep were enabled to ascend to the topmost one. The antiquity of the Cotswold is established beyond contradiction. There is no record of sheep having existed in this country prior to the Roman era; indeed, the stupidity of the sheep would have insured their destruction by wild beasts. It was only when the Romans introduced something like a system of tillage that the conditions were suitable, and accordingly we soon hear of them, the best proof being the fact that both at Cirencester and Winchester cloth was manufactured, and the trade soon assumed considerable importance; indeed, Gloucester was an important settlement whilst London was only a burgh. Tacitus, writing on Cirencester, states that " great attention was paid to the good condition and cleanliness of the roads, which were repaired by the proprietors of the adjacent houses. Carriages with heavy burdens were prohibited, and nothing was to be thrown out before the shops, except a fuller hung his cloth out to dry." The late Mr. J. M. Reade, of Elkstone, to whose investigations we are mainly indebted for our knowledge of the antiquity and early history of these sheep, thinks that the use of the word " Cote," applied only to the

Cotswolds, indicates their eastern origin. The word is used in several places in the Old Testament: 1 Samuel, xxiv. 3, "David comes to the sheep cotes by the way, where was a cave;" 2 Samuel, vii. 8, "I took them from the sheep cote, from following the sheep." King Hezekiah had "stalls for all manner of beasts, and cotes for sheep." According to Goding, cloth-making was carried on at Gloucester during the Saxon heptarchy, for he relates that, when royalty visited Gloucester, the cottagers presented the king and the nobles with clothing of their own manufacture. Atkyns, in his history of Gloucestershire, states that Cirencester had two markets in the week—one for corn, &c., on Monday, which has been continued to this time, and one on Friday, chiefly for wool, for which commodity it is the greatest market in England. Both wool and sheep were exported, and this to such an extent that in 1425 an Act of Henry VI. was passed to prevent this; and it was therein provided that no sheep shall be exported without the king's licence. Later on, in 1468, King Edward IV. presented John, King of Aragon, with twenty Cotswold ewes and four rams; and a few years earlier the King of Portugal applied to Henry VI. for permission to export sixty sacks of Cotteswold wool, in order that he might manufacture certain cloths of gold at Florence for his private use.

There is, we think, little doubt that the original Cotswold sheep were, if not the earliest, at any rate one of the earliest, breeds of sheep in this country, and that they attained a position unrivalled for the production of wool. In the reign of Elizabeth they are described as long woolled and strong boned; and the poet Drayton speaks of their wool-bearing qualities when he says:

> T'whom Sarum's plain gives place, though famous for his flocks,
> Yet hardly does she tythe our Cotteswold's wealthy locks.

As we have said, these sheep in early days were valuable principally for their wool. They were large-framed, coarse, slow-feeding sheep; very hardy, and accustomed to travel in search of the short sweet herbage which invariably prevails on limestone hills. At first the wool was used for the manufacture of cloth, which in earlier days was very coarse. Fulling mills were established at Cirencester and in the neighbourhood.

In time finer sorts were necessary, possibly because, owing to improved management, long wool became longer and stronger; then, as arable cultivation improved, and the down land was broken up, the sheep would be kept in closer compass, and by degrees their outline improved and feeding properties increased. We think it probable that a Leicester cross has been introduced, and to this may be attributed, to some extent, the great aptitude for feeding which characterises the breed. As a pure breed Leicesters could not stand the severity of the winter; but at the time of Bakewell's marvellous success they were introduced upon the Cotswold hills, and though they speedily disappeared as a pure breed, they may have done good by increasing the tendency to feed, without materially altering the type of the sheep. Be this as it may—and it is a point that does not admit of proof—we find the Cotswold sheep of the present day remarkable for symmetry, early maturity, and weight. There can be no doubt that the establishment of the Royal and local agricultural societies did much for this breed, formerly so little known. It has never been the fate of the Cotswolds to be supported by great patrons. The farmers, however, on the hills were a wealthy and highly intelligent class, and do not lack enterprise. For many years, principally as a result of showing, the demand had greatly increased, and the best breeders made high averages. Indeed, the leading breeders experience such a lively demand for their sheep that showing is no longer necessary; and as it is a costly business, and results in more or less injury to the stock that are fed up, they have done wisely to withdraw from the arena, although the public lose the opportunity of seeing some of the finest specimens. We can well remember in years gone by, when Mr. W. Lane, of Broadfield, near Northleach, travelled up in the van with his sheep, and generally managed to win with the ewes. Messrs. Garnes were also great in the showyard. The late J. King Tombs, who bred largely from the Lanes and Garnes, took many prizes up to the time of his death. Recently the executors of Mr. Thomas Gillett, of Oxfordshire, have shown big sheep, at one time many prizes went into Norfolk, where the Cotswolds have proved very successful in the hands of Mr. Brown, of Marham, and formerly of Mr. H. Aylmer. Messrs. Gillett and Mr. Swanwick of the

R.A. College Farm are, at the present time, the most prominent prize taking exhibitors.

The thin soil and wide range of the Cotswold hills are particularly adapted for healthy sheep, and greater size is there attained than elsewhere; hence the lowlanders come to the hills to renew their blood. The sheep are used not only in a pure state, but as a cross on certain breeds. In this way they have developed in parts of Southern Wales, Hereford, Monmouth, &c. Some creditable specimens of Welsh-grown Cotswolds were shown at the Royal Meeting at Cardiff, especially the lengthy good ewes of Mr. Thomas Thomas, of St. Hilary, Cowbridge, and those of Mr. C. Spencer, of Gileston, which took second prize. Formerly the practice of having ram sales at the farms was general, and, as these were arranged in succession, and each breeder found it incumbent to put in an appearance at his neighbour's gathering, much time was wasted; moreover, the eating and drinking were often abused. Now a much sounder practice prevails. A great portion of the rams are brought to the fortnightly markets at Cirencester, and there disposed of by auction. The first sale takes place in the beginning of August, three auctioneers being often at work at the same time. A few of the principal breeders still hold home sales; Mr. W. Garne, of Aldworth, and Mr. William Lane, of Broadfield, may be named as instances. The practice of letting, so common amongst the Leicesters, is only followed in the case of particular sheep, and to a very limited extent. They are commonly sold out and out, and are at the purchaser's risk from the fall of the hammer; but it is generally understood that if a sheep proves useless another is lent.

We now attempt a description of these sheep, with the assistance of an illustration by Mr. Harrison Weir. They present a complete contrast to the Leicesters, and, if they are indebted to them for early maturity and tendency to feed, they do not take after them in external form. The Cotswolds are the largest breed of domesticated sheep in the world, and, standing rather high upon the legs and having very grand heads, they have a truly imposing appearance. The features are either white, grey, or mottled. The former predominates, but a little colour in no way detracts from appearance or indicates impurity. Some years ago, a flock of grey-faced sheep were bred near Bibury, by a Mr. Smith; a sale of rams took place annually, and every animal was more or less coloured. For purposes of crossing, with the Hampshire ewes, for example, grey-faced rams were esteemed, the produce coming darker in the face in consequence. Such sheep would be serviceable years since in forming the Oxfordshire breed. The head is rather large, wide across the forehead, the eyes full and prominent. It is considered a point of importance that the head should be well woolled, particularly the forehead and cranium, and that long locks hang down over the face; if the eyes and upper part of the nose are covered, so much the better. Of course we are now describing a shearling ram; but in all, whether male or female, bare heads are an abomination. The effect of the love-lock is striking, adding immensely to the style. Occasionally the profile is slightly Roman; but this is rather objectionable, as it gives a common character. The neck should be long and moderately thick, especially at the base, and where it joins the head. The setting-on of the latter is easy; and it is a great point when the head is carried high, as this adds grandeur to the general appearance. A ram should so carry his head as to be able to look over a hurdle. The carcass is long, level along the back, and the ribs well sprung; the under lines are not so true, and the flank is often weak. Indeed, the great defect in the *contour* of the Cotswold is the lightness under, and the short space between, hips and flanks. No animal fills the eye, however good upwards, except he represents the L's—long, low, and lusty. The Cotswolds are too often long, high, and lusty, and this height gives them a weak appearance. The thighs are moderately full, the legs of mutton being more developed than in the Leicester, and, though there is much external fat along the back, which gives a soft springy touch, lean meat is also abundant.

The wool should be long, open, and curly; the staple is coarser, and the weight of fleece is usually rather less than Leicester, and considerably under the Lincoln; neither is the quality equal to either. We have heard of instances of hoggets exceeding 14lb.; but a good average for flocks comprising half ewes is three fleeces to the tod of 28lb. Something, of course, depends upon management.

When size is a consideration the ewes are run thin on the land, one to two and a half or three acres; we seldom find more than a ewe to two acres, the produce being fed out; this gives about a sheep to the acre—not heavy stocking, but the size and weight of the sheep must be considered. Where it is not important to have individuals so large, a system of close hurdling is pursued, which, when properly managed, has many advantages. It consists in accustoming the sheep to graze on a limited area, and to have frequent change. For instance, instead of turning the ewes and lambs into a field of young seeds, when they can just pick out the more dainty parts and leave the grasses untouched, two folds are made, the sheep are turned into the first, which communicates with the second by a lamb creep. In the front pen are placed the troughs for holding artificial food for the lambs. According to the quantity of food the surface is cleared; sometimes a few days only suffice to clear a large field. More stock can be kept on a given quantity of food than when they range at large, and it is easier to provide a regular supply. In the young stage constant change is desirable, and the best flockmasters generally contrive two shifts a day. Sainfoin is the sheet anchor of the sheep farmer; the hay is of the greatest use during the lambing season, and the second growth affords a capital pasture for weak or sickly lambs. Indeed, a good sainfoin field may be regarded as the farmer's hospital. We hold that it is good practice to give the lambs a little extra food as soon as they can be induced to eat it; there are fewer losses, the constitution is strengthened to resist unfavourable conditions, and the sheep are sooner ready for the butcher. Commencing with a small allowance of dust cake, oats, beans, meal, &c., and continuing the same in increasing ratio up to the time when they go to the butcher (which should be when about a year old), we have frequently sold hoggets weighing 24lb. to 25lb. a quarter under a year old. The feeding qualities of the Cotswolds were subjected to careful experiments by Mr. Lawes, who found that in comparison with Downs, they consumed the least food to produce a given amount of increase, and made the greatest progress in a given time. The fat is principally external, and the flesh is coarse and open.

The Cotswold Hills, though formerly rough pasture, are now

chiefly under the plough; hence the sheep live on the arable land during the winter, and, owing to the large proportion of clay in the soil, become much draggled in wet weather. We have seen hogs with great balls of dry earth hanging from their breeches; and so hard do these accumulations become under the influence of March winds, that they have to be removed, otherwise, from constantly knocking against the poor animal's legs, injury would result.

Formerly, and too frequently even now, the ewes are allowed an almost unlimited supply of turnips, provided the crop is abundant. This is most mischievous treatment, and results in heavy losses. A few turnips, just the pickings up behind the fat sheep, and plenty of dry meat, will be found the best winter treatment until near lambing.

Should the weather be wet or the soil stiff, we prefer housing them at night in a sheltered straw yard, well supplied with litter. Not only have they the comfort and warmth, but we get in time a quantity of valuable manure. Owing to the severity of the climate, great protection is afforded during lambing time. A fold-yard with shelter-sheds is provided, which is made in the field where the roots are growing at the time. The forward ewes are brought up at nights a few days before lambing commences. The ewe and produce are removed to a small pen under cover, and carefully treated for three or four days; and, according to weather, the lambs are sheltered at night for a considerable time. Afterwards shelter hurdles are placed about the field. The lambs are delicate and require care, but soon grow out of harm's way. When a month old they are capable of eating, and should be supplied with ground oats, dust oil cake, and bran during the first six months; and, especially when first weaned, the lambs require great care. When once introduced to turnips, they go on at a great rate, and under liberal management are fit for market at from eleven to twelve months old, when they weigh from 22lb. to 25lb. a quarter. The draft ewes can be fed to immense weights; we have known instances of their reaching over 70lb. a quarter, dead weight.

CHAPTER X.

LONG-WOOLLED LINCOLN SHEEP.

THE improved Lincoln sheep, taking alone into consideration the weight of the fleece, is unrivalled as a wool producer; and for wool and mutton combined it has, in the writer's opinion, no equal as a rent-paying animal. It must be understood that he writes from a Lincoln breeder's point of view, but his aim will be to deal strictly with facts, and thus leave the reader to form his own conclusions.

Upwards of a century ago Lincolnshire possessed an established breed of sheep, with long bony legs and large carcasses, and, according to Ellis, they carried more wool than any sheep whatsoever. Milburn speaks of them as having large and coarse carcasses, the length from the head to the tail being in some cases four feet seven inches. The ribs were flattish, and not covered very thickly with flesh; the belly deep, and the shoulders so forward as almost to hide the breast; the neck thick and large, with a deep and flabby dewlap hanging from it; the skin thick and the flesh often grained; the hind quarters full and fat, the tendency being to lay on fat at the rump; and the legs fleshy and deep. The whole animal, continues Milburn, appears to be somewhat unshapely, taking the standard of a connoisseur as a criterion, but the valuable wool which covers it hides all imperfections.

There can be no question that the original Lincolnshire sheep were ungainly animals and gross feeders, possessing but little aptitude to fatten. When it was seen, however, what a vast improvement had been made in the Leicester sheep by Bakewell, the intelligent Lincolnshire breeders determined to follow in his footsteps; and, by a wise and judicious admixture of Leicester blood, they created an entirely new type of sheep, which retained the pre-eminent wool-bearing qualities of the old breed, and showed a marked improvement in form and in their aptitude to accumulate flesh. Foremost among these early improvers were the Kirkhams, Chaplins, Caswells, Duddings, and Clarkes, whose flocks continue to hold the highest rank up to the present day. The uneducated agriculturists of the past generation, however, for many years refused to take advantage of the skill displayed by the pioneers in this great work of improvement; but finally they began to perceive the vast superiority of the new type, which soon became the one established breed of the county.

Of course the great change which is now observable in the character of the sheep was not effected until after many disappointments, and until years of skill and patience had been devoted to the work. Fresh names were continually being added to the list of breeders—names which have since become famous — the Marshalls, Greethams, Davys, Wrights, Cartwrights, Howards, Pears, Paddisons, Battarsbys, Kemps, Havercrofts, Vesseys, Robinsons, Mayfields, Smith of Cropton Butler, and Mackinder; and the result of the competition has been the production of a sheep unequalled for wool and mutton combined.

It is somewhat singular that the Royal Agricultural Society of England should have refused a separate class to the improved Lincolns until the Battersea show in 1862. The slight was resented by many of the most noted breeders; hence the stewards reported that "the class was weak in numbers and in stamp, Mr. Marshall's first-prize ram being perhaps an exception." At Worcester, in the following year, Lincolns had again to compete as "Lincolns and other Long-wools," and this continued to be the case until 1870. At the Plymouth show, in 1865, the judges complained that the Lincolns "never come out in such numbers or form as to show their real character, the Lincoln flockmasters apparently thinking that the price obtained for them is a sufficient test of their excellence." In the interval between 1862 and 1870 the majority of the prizes in the Long-wool class were carried off by Lincoln sheep, the most successful exhibitors being Messrs. Marshall (Branston), Dudding (Panton), Wright (Nocton), and Cartwright (Dunstan). At Oxford and Wolverhampton Lincolns again stood on their own merits as a separate class. There were thirty-two entries at Oxford, the animals being generally so good that the judges commended the whole class.

It was not until about the year 1850 that the value of the improved Lincolns began to be acknowledged outside the county. The grand appearance of the sheep exhibited at the Royal and other shows impressed visitors from the colonies, who saw the benefit likely to be derived from a cross between them and their own short-woolled breeds. This was more particularly the case about 1864 and 1865, when wool became so valuable in the market. Enormous prices were made at the annual ram sales and lettings, one celebrated sheep belonging to Mr. Kirkham, of Biscathorpe, being let, in five successive years, at an aggregate of nearly 600 guineas. Naturally, however, the flocks of Mr. Marshall, the Messrs. Dudding, and other noted prize-takers, were best known to foreign customers, and the late Mr. T. B. Marshall opened a connection with Australia, New Zealand, the Cape of Good Hope, South America, and, indeed, all parts of the world, which was continued after his death by his brother, Mr. W. F. Marshall, the present owner of the Branston flock. The improved Lincolns are now spread over the whole of Lincolnshire and Rutland, and parts of Cambs, Notts, Yorkshire, and Norfolk, and in certain districts of Scotland and Ireland. Indeed, wherever Long-wools are grown, crosses are usually obtained from Lincolnshire.

As breeders the improved Lincolns are wonderfully prolific, and they arrive at maturity very early. The mode of management is exceedingly simple. The sheep, as a rule, being hardy and healthy, the lambing season commences the last week in February, and closes about the end of March. About one-third of the ewes produce pairs, triplets are frequent, and fours not uncommon. A week after birth it is the common practice to turn the lambs with their dams on the young seeds, there being very little grass on the heath. There they remain until they are weaned (about the second or third week in July), when they are placed on clover eddishes, cake and corn being allowed.

A grander show of sheep than those annually penned at Lincoln April fair it is impossible to conceive. The number of hogs ranges from 40,000 to 50,000, and when wool was high the best pens made from 85s. to 100s. each. These prices were obtained for hogs only fourteen months old. As these sheep are bred for both wool and mutton, it is necessary that they should not only have long and lustrous fleeces, but that they possess those qualities which indicate an abundance of meat—good necks, broad and even backs, with a firm touch—that they should be well sprung in the shoulders, and have good legs.

With regard to the fleeces, mention may be made of one flock of improved Lincolns, which may be taken as a fair sample of the best flocks in the county. Mr. Marshall, of Branston, has clipped no less than 26½lb. of wool from a shearling 14 months old, and his hogs have todded 160 twos and 40 threes. But even this average was exceeded by Mr. J. J. Clarke, of Welton-le-Wold, whose sheep have, it is said, the heaviest fleeces in the county.

Perhaps the fairest test of the estimation in which the sheep are held is the average obtained by the leading breeders at their annual sales and lettings. In 1871 Mr. C. Clarke, of Scopwick, sold thirty-three yearlings at an average of 19l. 18s. 6d., twenty-four two-shears at an average at 24l. 2s. (one making 157l. 10s.), and eleven three-shears at an average of 32l. 13s. 9d.; Messrs. Dudding, Panton, sold seventy-three at 14l. 6s. 4d. each; Mr. C. Clarke, Ashby, let a hundred at 13l. 13s.; Mr. T. Kirkham, Biscathorpe, let forty shearlings at 13l. 14s., thirty-seven two-shears at 16l. 3s. 9d., and twenty-three three-shears at 15l. 10s. 5d.; Mr. E. Paddeson, Ingilby, twenty at 13l. 12s.; Mr. Turner, Ulceby, sixty at 10l. 14s. 6d.; Mr. J. W. Kirkham, Hagnaby, thirty-six at 13l. 10s.; Mr. Havercroft, Wootton, sixty at 12l. 10s. 9d.; Mr. Caswell, Pointon, forty-eight at 17l. 1s.; Mr. Caswell, Laughton, fifty at 17l. 4s. 6d.; Mr. J. R. Kirkham, Audleby, forty at 15l. 2s.; Mr. T. Cartwright, Dunstan, thirty at 13l. 14s.; Mr. Pears, Mere, twenty-one at 12l. 12s.; Mr. Davy, Owersby, 120 at 13l.; Mr. Wright, Nocton, fifty at about 15l., &c. These averages, considering the extent of the lots, far exceed those obtained for any other breed, if we except Lord Polworth's lot of thirty Border

Leicesters, which averaged 30*l*. 10*s*. (less than Mr. Clarke's three-shears), and Miss Starkie's Border Leicesters, thirty-six, at 28*l*. 15*s*. The reader may perhaps wonder why the name of Mr. Marshall, of Branston—who has been the most successful exhibitor—is not included in the above list. The reason is, that he devoted his attention almost exclusively to the foreign trade, all the sheep he had for disposal being forwarded to distant parts of the globe. In 1861 he sold one to an Australian customer for 100 guineas, and in the following year he despatched 134 rams, principally to Australia and South America.

In order to show the relative value of Lincolns, as compared with other breeds, a brief report may be given of some practical experiments carried out by the Parlington Tenants' Club in 1861-2, as published in the "Year Book of Agricultural Facts for 1861": On the 4th of October, 1861, six sheep of each of the undermentioned breeds were turned upon rape, so that nature might have its course with natural food, and without tint, until the 11th of November, during which time the cross from the Teeswater gained 2st. 2lb., the Border Leicesters 5st. 1lb., the Lincolns 3st. 6lb., the Shropshire Downs 6lb., the Leicesters 1st. 13lb., and the Cotswolds 5st. 8lb.; while the Southdowns lost 11lb. On the 11th of November the sheep were again folded, the several crosses being then in a pretty equal state for taking on condition. If a lead could be supplied, t might be in favour of the Cotswolds, from the start this class had made in the latter part of the summer grazing, and whilst on the rape. The whole of the sheep had as many swede turnips as they could consume, and half a pound of linseed cake per day, with the exception of the Lincolns, and these, for forty-two days, had a quarter of a pound extra; but this extra cake was placed to their debit in like manner as the various weights of turnips consumed were to that of the several classes. For the comprehensive tables giving the result of the experiments reference must be made to the "Year Book" in question, but it may be stated briefly that the relative value which one class of sheep bore to the other (after deducting the value of the food from the value of the mutton and wool) when the sheep were slaughtered in February, 1862, was shown to be: The Teeswater cross, 11s. 7½d.; the North sheep, 12s. 5¼d.; the Lincolns, 1l. 10s. 5d.; the Southdowns, 17s. 3d.; the Shropshire Downs, 1l. 5s. 10¾d.; the Leicesters, 1l. 2s. 6d.; and the Cotswolds, 14s. 9½d. The whole of the sheep were sold on the same day in the Leeds market. It will be seen from the above that the Lincoln sheep showed a marked superiority over the other breeds.

CHAPTER XI.

THE DEVON LONGWOOLS.

By JOSEPH DARBY.

No variety of sheep is more valued or has become subject to higher breeding in central and eastern Devon, West Somerset, and certain districts of Cornwall, than the Devon Longwool. Peculiarly adapted to the warm, fertile vales and rich low-lying plains of this interesting part of the kingdom, it has held its own against all new comers, purely because none have ever been found more profitable. The late Mr. Andrew Hosegood, an extensive and old-established tenant farmer of the Williton vale, about thirty years since made use of the following practical remark founded on experience: "In this district I have known the Exmoor, Dorset, Sussex, Shropshire, Cotswold, and pure Leicesters tried, but experimenters always go back to the old sort—the Devon Longwool."

Probably no breed of equally high claims has been less written about, although derived from the ancient Bampton stock, universally acknowledged to have been an exceedingly useful and extensively propagated variety in the last century. Short descriptions are, of course, to be found of this parent race in Arthur Young's writings, and in the papers and surveys of the Board of Agriculture. But after a new variety had been created by judicious crossings, improved Bamptons received scant justice at the hands of journalists and book compilers on sheep-husbandry—attributable in some measure, perhaps, to the wide latitude of variety the intermixture of foreign blood from different sources was made to assume at first. While one flockmaster was judiciously content to rely solely on Dishley blood to repair defects, another decided on effecting a Lincoln cross; and a third conceived he might derive superior results by an alliance between the Bampton and Cotswold. There is no evidence of the Southdown having ever been resorted to, although the grey faces to be found in some of the older flocks naturally lead to the supposition;—bearing the fact in remembrance that the original Bampton possessed a white face, as well as the Leicester and every other long-woolled kind with which it has been known to have been connected.

The best flocks of Devon Longwools are, however, derived solely from Leicester and Bampton—a most valuable cross in every respect, as a more valuable sheep has been created than either; larger and more productive both in meat and wool than the former, and better shaped, smaller boned, and of more early maturity than the latter. From fifty to sixty years ago the improved breed was known by the appellation of the "Devon-Nott;" but by reason of many of the flockmasters adhering very closely to the Leicester type, the name "Leicester Longwools" began to be applied to these sheep afterwards; and authorities for a time seemed disposed to ignore their peculiar characteristics, and to consider them as only forming a sub-variety of the Leicester. From this another reason may be gathered why the distinctive claims of the breed have been so thoroughly overlooked in our agricultural literature, and why the Royal Agricultural Society, except in visits to the West, has never recognised it as worthy of forming a separate class. During the past thirty years these sheep have been designated "Devon Longwools"—an appellation remarkably suitable, and to which they are now generally deemed fairly entitled, as the most profitable flocks are not those which have a near affinity to the Leicester type, and many of the leading breeders have ceased to make new infusions of that blood for a great many years past, preferring to use Devon Longwool rams, for which there is a large and rapidly-increasing demand in the West of England.

The history of the ancient stock from which the breed is derived is lost in obscurity. In a letter by Mr. R. Proctor Anderson, written January, 1772, published in Arthur Young's

"Annals of Agriculture," the Bampton is designated "the best breed in Devonshire," and it is asserted that it had "existed in the neighbourhood of Bampton from time immemorial." If a portrait furnished by Vancouver as an illustration to his "Survey of Devon in 1808" may be taken as a faithful delineation of the true original Bampton, the fact of being the best in the county ought not to advance the animal too highly in modern estimation—great coarseness and want of symmetry in form being apparent. The contrast is so striking between the well-shaped, high-quality sheep here depicted by Mr. Harrison Weir and the old portrait of the Bampton referred to, that it is difficult to fancy the one animal descended from the other. But this contrast is probably not greater than that which existed between the old breed of Leicestershire and the improved Leicester or Dishley sheep after Bakewell had exerted thereon his magical powers of transmutation.

But Mr. Andernon, in the letter referred to, gives a description of the Bamptons of the last century. He says: "They are generally whitefaced; the best bred more like the Leicestershire than any other, but larger boned and longer in the legs and the body, yet not so long as the Wiltshires, with which they have been crossed, nor so broad-backed as the Leicesters. A fat ewe rises to 20lb. a quarter on an average, and wethers to 30lb. or 35lb. a quarter at two years old. 18lb. of wool have been shorn from a ram of this breed that was supposed to be 40lb. the quarter. The carcass is coarser than that of the Dorset, and the wool about 2*d*. per pound cheaper."

Billingsley, in his "Agricultural Survey of Somerset," published in 1798, affords a similar description, alluding to the Bampton Breed as "a valuable sort, not much unlike the Leicester, well made, and covered with a thick fleece of wool weighing in general 7lb. or 8lb., and they sometimes reach even the weight of 12lb., and sell at about 10*d*. per pound." He also adds: "The sale ewes are put to the ram at about the latter end of July, and the flock ewes about a month after. Young rams are preferred, as it is supposed that old ones degenerate in the quality and weight of their wool. The wethers of this breed when two years old and fattened on turnips attain the weight of about 25lb. per quarter, and, being driven to Bristol market, a distance of nearly sixty miles, are sold without their fleeces in the months of May and June." Billingsley alludes to the Taunton district in this statement, but another description is to be found in Vancouver's "General View of the Agriculture of Devon," published by the Board of Agriculture in 1808, which contains the following: "The sheep most approved in the division of Tiverton are the Bampton-Notts, the wethers of which breed, at twenty months old, will weigh 22lb. per quarter, and shear 6½lb. of wool to the fleece. The same sheep well wintered and kept on for another twelve months will average 28lb. per quarter, and yield 8lb. of unwashed wool to the fleece. The present price of this wool is about 1s. per pound."

Mr. Andernon concluded his letter with the very practical reflection, "This breed, I should conceive, may be greatly improved by crossing with the new Leicester"—an opinion with which the breeders appear to have fully concurred some years later, and which eventually was fully acted on. Towards the latter part of the last and at the commencement of the present century some considerable infusion of Dishley blood had already been effected. Vancouver, in the Survey before quoted, says: "The first cross of this breed with the new Leicester is growing greatly in esteem, from its improving the form and bringing the animal three months earlier to market; but however desirable this cross so far may be, more of that blood is generally objected to on account of the extraordinary nursing and care required to be paid to the young couples; the lambs being represented as very tender, and much oftener perishing through the severity of the season than the genuine offspring of the native sheep."

Making allusion to the neighbourhood of Crediton, Vancouver also observes: "A cross of Leicesters with about half Bampton is conceived to agree with the soil and circumstances of the country. This animal is unquestionably hardier than the new Leicester; but by the introduction of the latter blood the Bampton comes sooner to market, and at twenty months will weigh 24lb. per quarter and 7lb. of wool to the fleece." But the circumstances of different localities it is presumed unfolded variable experiences, then as now, in respect to the extent the alliance could be most profitably conducted; for, in another place, Vancouver remarks that the cross "is much approved through Buckland, Filleigh, and Peterockstow, particularly if carried to the fourth degree, or, in other words, four parts Leicester to one part native Nott."

Mr. Andrew Hosegood some years since bore testimony that when a boy he recollected the Bampton sheep as having grey faces, "they were very hardy and excellent for weight of mutton and wool," adding: "We have since crossed these very considerably with the Leicesters; by which means we obtain sheep that fatten quicker, but do not obtain so much weight." Another old-established breeder has left it on record that about half a century ago the blue tint in the face got to be preferred, and became in consequence rather general; until experience proved that animals bearing this characteristic were thinner in both flesh and wool, weaker in constitution, and less hardy. During the past ten or fifteen years white faces in consequence have become not only less objectionable, but generally sought after.

Tanner's prize report on the farming of Devon, published in 1848, refers only very cursorily to sheep; but it is distinctly stated therein that the kind most generally kept in the county was the Bampton and its crosses, which he found to be more or less partaking of the Leicester type as the farms were well-sheltered and of high fertility. Wilson, treating on the various breeds of sheep in the "Royal Agricultural Society's Journal," for 1855, says of the Bampton variety, "Like most of the old indigenous breeds of the county, it has gradually been displaced by the improved breeds; and now it is very difficult to find the pure Bampton unmixed with other blood, a few only remaining in Devonshire and West Somerset. They are usually met with crossed with the Leicester breed, and very much resembling them in shape, though somewhat larger in size, and hardly so fine in general character. They are without horns, and with clean faces and legs; they are hardy, but require good pasture. At two years old, if well kept, they average 120lb to 150lb. each. The meat is juicy, but like that of all large sheep, inferior in quality to the smaller breeds. The wool produce is good; the fleece, averaging 7lb., is rather coarse in quality, They are now so intermixed with Leicester blood as to partake more of the character of that breed than of the old stock; crosses with the Lincolshire and with the Exmoor breed are also met with."

This description scarcely did justice to the improved breed as it existed then, and certainly does not adequately represent the marked features and valuable characteristics of the Devon Longwool to be found in the best flocks at the present period. A well-bred animal of this variety differs from a pure Leicester, in having a longer and larger face, with greater width at the forehead and nose, the ears longer. The frame is more bulky, and of far greater length, although not quite so round or compact, but the girth is equal to that of the Leicester. The Devon Longwool also appears higher than a pure-bred Leicester. In good constitution and hardihood the former surpasses the latter; it will attain much greater weight of carcass, and more flesh in a given time, and is likewise reputed to come earlier to maturity.

Wilson's account does not at all serve to indicate the present realisations in meat and wool. The wether sheep are never kept until two years old, being fattened as hogs the first winter on turnips, and come out in the months of March, April, and May weighing from 22lb. to 24lb., and in some cases 25lb. the quarter; and when shorn they cut from 9lb. to 11lb. of clean washed wool each, although shorn as lambs the preceding year. This much is usually obtained by natural food, with very little, if any, assistance from oilcake or corn. In some instances high feeding is resorted to, but in those cases the hoggets ripen for the shambles at these weights at much earlier periods. The ewes are also affluent wool bearers, the fleeces of the best flocks averaging from 8lb. to 9lb. each. The lambs cut from $2\frac{3}{4}$ to $3\frac{1}{4}$lb. of wool each.

Mr. A. C. Skinner, Pound Farm, Bishop's Lydiard, in a communication bearing date November 30th, 1886, has given me useful information on this point. After stating that when he went into the management of his farm in 1862, he took from his predecessor a very good flock of Devon Longwools, he adds, "We now bring out the wethers from three to six months earlier than we did at that time, at about the same weights—80lb. to 88lb.—they being from eleven to twelve months old, and having had the aid of only a small proportion of cake or corn. We find in these days it is no use keeping them after they get about 80lb., as what they gain in weight we lose in price per pound, and their gain in flesh is very rapid. Our fleeces from two-teeth ewes average from 10lb. to 11lb. each, and the general flock, not including wethers or rams, average nearly 9lb. per fleece." To this statement may be added the fact that Mr. Robert Farthing, one of the oldest of existing Longwool breeders, realised 70s. each for the whole of his wether hoggets in the spring of 1886 after being shorn of their wool.

A leading breeder has remarked that "while good pastures are desirable for most breeds of sheep, the Devon Longwools will live hard, and on high land; they have often been tried with other breeds, and have been found to do well in dry seasons when grass is very short, keeping their own and improving, while other sorts living with them have wasted a good deal, and lost much flesh. I never house them—they stand all weathers out of doors."

Lambing commences about the second week in January, but is not general until towards the end of that month or the beginning of February. The period of weaning varies much. Some flockmasters take the lambs from their dams as early as the middle of April, but the generality wean them in May, and a few postpone the matter until June. The ewes are very fair nurses, and prolific, yielding many twins, although not equal in these respects to Dorset horns, or cross-bred sheep. It has also been remarked that the coarser-bred Devon Longwools are more prolific and better nurses than some of the better flocks of higher quality.

In the first edition of this work it was stated that prior to 1877 several West Somerset breeders, and especially Mr. R. Corner, Mr. Bird, and Mr. R. Farthing, had been in the habit of holding periodical ram sales for many previous years; and that in Devon, some of those who had been reputed as amongst the best breeders were Mr. John Wippell, of Brenton, near Exeter; Mr. Drew, of Exeter; Mr. Wm. Wippell, of Thorverton; Mr. Pentridge, of Bow; Mr. G. Radmore, of Court Hayes, Thorverton; Sir J. H. Heathcoat-Amory; and Mrs. Elizabeth Gibbings.

Some of these are now deceased, or retired from business; but many others have since taken up the running. Among the younger flockmasters of repute are Messrs. C. Norris, of Motion; J. Franklin, of Huxham; and N. Cook, of Chevithorne, Barton, all in Devon; and Mr. Alfred Bowerman, of Capton; and Mr. A. C. Skinner, of Bishop's Lidiard, in West Somerset.

Mr. R. Corner, of Torweston, Williton, did much to improve the breed, and he inherited a very good flock from his father, who, it is said, made 70 guineas of a ram more than sixty years ago. On another occasion, Mr. Corner, senior, made 100 guineas of a ewe and two of her offspring—a four-months old lamb and a ram sixteen months old. Rams were annually sold from the Torweston flock for over sixty years, and in 1872 Mr. R. Corner let his best sheep for 70 guineas, and in the following year others were let at 60, 40, 38, and 31 guineas; the respectable average of 19l. 8s. 1¾d. being obtained for thirty.

Mr. Robert Farthing, of Farrington, has held annual auctions for his rams in Taunton market for over thirty years, and the thirty-four shearlings offered in July, 1886, were described at the time in the *Field* as a superior lot, "among which there were magnificent sheep with grandly-formed, rotund barrels, stylish heads, depth and fulness at their breasts, thick even packing at their shoulders, and their legs of mutton generally well developed." The flock of Sir J. H. Heathcoat-Amory was started in 1868, and was soon brought to very high perfection—indeed, so much so, that since 1875 twenty-one first prizes, together with a considerable number of second prizes, have been won at the shows of the Bath and West of England Society alone. Mr. Greenslade, who has had the sole management of the flock from the commencement, recently stated that he has won not less than 21s. per week for it in premiums since he commenced showing. The annual ram lettings and sales of the Knightshayes Court rams excite great interest, and realise tolerably good prices even in the present depressed times, the averages varying from 10 to 15 guineas per sheep. The most magnificent of the rams bred at Knightshayes Court was probably the champion prize-winner Comet, which combined so much grand character and style with high quality that he was kept in service until five or six years old, and many of his progeny have been prize-winners.

Most of the other leading breeders have annual ram sales either at home or at neighbouring markets and fairs, and have had little difficulty during the past ten years of realising at least 10*l*. each for their best. Indeed, Mr. A. C. Skinner's average in 1882 was 12*l*. per ram for all he had to dispose of; but times have not been so good since then, and few breeders have gone to so much expense as Mr. Skinner in the selection of sires for his flock. He bought the first-prize sheep of Messrs. Bird which triumphed at the Royal International at Kilburn, and successively employed after him Mr. Franklin's first winner at the Devon County Torquay Show, and Mr. Norris's champion ram at the Plymouth Show of 1885. Latterly he has been using a son of Comet, before-mentioned, also a first winner bred by Sir J. H. Heathcoat-Amory, and hired from him.

Fixity of type is everything when attempts are made to advance such a breed as the Devon Longwool, which up to a comparatively recent period were exceedingly diverse in the varying characteristics of different flocks. That this evil is rapidly being removed must be evident to frequenters of the showyard, there being much more uniformity of character in the exhibits than there was a dozen years since.

Some flocks that furnish some of the best rams produced in the county of Devon have been so long bred to pure Leicester rams that their owners claim for them the more aristocratic designation, although these sheep, from great size and bulk of frame and heavy productions of both fleece and carcass, are not the kind that Bakewell probably would have deemed Leicesters. Still these originally were the ones resorted to for rams by many possessing ewes of a coarser type. Ten years ago the Court Hayes flock of Mr. Radmore, although he considered it to be Leicester and not Devon Longwools, furnished the stuff out of which the latter were made, as the largest sheep breeders about Bampton and Tiverton were accustomed to hire his rams, and had been in the habit of doing so for many years. He had selected his tups carefully during the last half century from the pure-bred Leicester flocks of Messrs. Smith of Dishley, Burgess, Farrow, Spencer, Sandy, Pawlett, Col. Ings, Sir Tatton Sykes,

&c., always aiming to get the biggest and heaviest-fleeced rams procurable, if well shaped. By such means, Mr. Radmore was able to produce sheep of high quality, combined with great bulk of frame and profitable flesh and wool production. Heavy fleeces formed a special characteristic, as their average, taking the entire flock, amounted to 12lb. each.

Whatever limits may hereafter be drawn as to the true characteristics of the Devon Longwools, and what sheep ought to be included in that denomination, the generality of breeders have long ceased to draw anew from pure Leicester blood for the improvement of their flocks, which they deem to be far better effected by resorting to the best tups of the breed itself, now so highly perfected, of which truly splendid specimens are brought out at the autumnal fairs and markets, more handsome in appearance and of greater intrinsic excellence every succeeding year. Mr. R. Corner declared that he was never able to do better than by working the best of his own rams, which he had been in the habit of doing for many years prior to the disposal of his flock. There is now much less difficulty for the leading flockmasters to get change of blood from Devon Longwool flocks, which have been closely assimilated to the type of their own.

A great demand is made for these sheep by graziers who occupy the Somerset marshes, on which they have been found to thrive far better than any other kind. Col. Luttrell, of Bagworth Court, in 1863 published the following statement: "When I first came to reside amongst the rich marsh lands of Somersetshire, I was surprised to see so few black-faced sheep; and on making inquiries, was told that the Down sheep did not pay for grazing on rich pasture lands, and that it was difficult to get them fit for the market without the aid of corn. Being rather sceptical on this point, in the spring of 1861 I bought 300 two-toothed sheep, of which 100 were Hampshire Down and 200 Devon-Notts. I picked out thirty of the best Downs to take the first run of the pasture, with sixty of the best Devons. All the Devons became fit for market during the summer, and were sold; the Downs showed but little improvement. On October 28th I put thirty Downs with thirty of the next best Devons on a piece of after grass. At this time I had both lots of sheep weighed, and I again weighed them November 28th, when I found that the thirty Downs had increased in weight 243lb.; whereas the Devons had increased 446lb.

Another fact may be added which either proves a remarkable propensity in Mr. Robert Farthing's Devon Longwools to lay on flesh rapidly, or the marvellous changes these Somerset marshes are calculated to infuse into this variety of sheep in brief time. His old ewes, to the number of about fifty, after weaning their lambs last year, were turned on the marshes, and in the week subsequent to the Welsh Show, held on the 11th of May, they were ripe for slaughter, with carcass weights averaging over 24lb. per quarter.

When the Royal Agricultural Society visited Plymouth in 1865 there were no separate classes for Devon Longwools, nor were the claims of the breed recognised at the earlier shows of the Bath and West of England Society. Separate classes for the breed were, however, part of the programme of the Taunton Meeting in 1870, with the result of Mr. Richard Corner, of Torweston, carrying off first and second prizes for yearling rams; but in the older ram class he only won second prize, Mrs. Elizabeth Gibbings, of Higher Brenton, near Exeter, receiving the first. At Dorchester, in 1872, and Plymouth the following year, Mr. R. Corner made a clean swoop of the whole of the prizes for this breed of sheep; but in 1874, at Bristol, there was a more even distribution of the prizes, for Sir J. H. Heathcoat-Amory won a first and second, and Mr. G. Radmore a first prize, leaving Mr. Corner one first and two second prizes. The success of the Knightshayes flock was quite in the ascendant at Croydon the following year, where they won the whole of the three first prizes, Mr. Corner only being able to get the seconds. At Hereford the following year, Sir J. H. Heathcoat-Amory was also chief winner by receiving two firsts and two seconds, Mr. Corner only carrying off one second prize, and Mr. C. Norris, of Motion, Exeter, gained first prize for ewes. At the Bath Centenary Meeting in 1877, however, Mr. Corner fully regained his supremacy, for he won two first and two second prizes, his chief opponent, the Knightshayes flock, getting one first and one second prize.

The retirement of Mr. Corner from the showyard afterwards placed Sir J. Heathcoat-Amory altogether to the front, still not to the extent of having been able to get a monopoly of the prizes, for Mr. C. Norris, Mr. J. N. Franklin, and latterly Mr. A. C. Skinner, have proved serious competitors, and have succeeded in winning a fair proportion of prizes. For some reason or other the Bath and West of England Society has withdrawn during the past five years the privilege of giving special prizes to this breed of sheep; but in the classes for "any other Longwools" they have usually few, if any competitors. When the Royal Agricultural Society visited Bristol in 1878, prizes were offered specially for Devon Longwools, as well as at the Kilburn International. At Bristol, Mr. Corner, not having as yet retired, shared the prizes very evenly with the Knightshayes flock; and at Kilburn, Messrs. W. and G. Bird and Mr. A. Bowerman won two of the first prizes, Sir J. H. Heathcoat-Amory carrying off the other. At Reading in 1882 the classes were not exclusive, but the latter was quite at the top with the exception of the ewe class, where Mr. J. N. Franklin took first prize. In the 1886 show season, in the classes open to other breeds besides Devon Longwools at the Royal, Sir J. H. Heathcoat-Amory won first and second prizes with two-shear rams, and the same premiums for shearling rams. This was a reversal of the decisions previously made at the Bristol and Wells Show, at the former of which Mr. H. T. Radmore and Mr. C. Norris won the prizes for rams; while at the Somerset Counties Show Mr. A. C. Skinner won both prizes for shearling rams, the best of the Knightshayes flock only having third premium. At Axminster Devon County Show, at which Sir J. H. Heathcoat-Amory did not compete, Mr. Joseph Ham, Budlake Broadclyst, had shearling rams which were preferred to those of Mr. Norris and Mr. H. T. Radmore; but Mr. Skinner carried off first prize, Mr. C. Norris winning first and second prizes, both in the older ram and in the ewe classes.

CHAPTER XII.

ROMNEY MARSH SHEEP.

THE sheep peculiar to the Romney Marsh District are longwools. The land is generally fertile, being rich alluvium, and the marshes carry a heavy stock of sheep. The climate, owing to the absence of trees, and to the bleak flat character of the country, open to the east, is very severe in winter and spring; none but hardy animals could thrive; and this is the principal feature that renders the Romney Marsh sheep so valuable, and which has hitherto prevented the introduction of sorts that surpass them in fattening properties. The extremes of heat and cold exceed any other part of Kent, and of late years it has been considered that shade in summer is quite as important as shelter in winter; consequently trees are being encouraged rather than destroyed, as formerly. Indeed, so thoroughly are the sheep of different districts adapted to their respective localities, that it is often wiser to improve than supplant; and this, if true in the general, is peculiarly applicable to a country which presents such remarkable features as the large tract known as Romney Marsh. Hence, although attempts have not been wanting to introduce other sorts, these sheep maintain their position, and are in as much favour now as they were when described by Youatt—or rather by Mr. Price—some sixty years ago; and, although the feeding properties and the form are considerably altered, those familiar with the breed will recognise it by the following description: "The pure Romney Marsh bred sheep are distinguished by thickness and length of head, a broad forehead, with a tuft of wool upon it, a long and thick neck and carcase; they are flat-sided, have a sharp chine, are tolerably wide on the loin, have the breast narrow and not deep, and the forequarter not heavy nor full; the thigh full and broad, the belly large and tabby, the tail thick, long and coarse, the legs thick with large feet, the muscle coarse and the bone large; the wool long and not fine, and coarsest on the thighs; they have much internal fat, and are great favourites with the butcher. They have much hardihood, they bear their cold and exposed situation well, and they require no artificial food during the hardest winter except a little hay. The wethers seldom reach the market until they are three years old; then they weigh from 10st. to 15st. (8lb.), and the ewes from 9st. to 11st."

Formerly the wethers were brought to market at from two to two and a half years old; in more ancient days many were wintered three times, and sold early after shearing. Now, however, owing to more generous treatment, and also in many cases to the influence of a strain of other blood, the shearlings are brought to market during their second summer; and, where arable land prevails, many are fatted on turnips and sold when shorn. We find them shortly described by Professor Wilson in an article on various breeds of sheep, especially with reference to wool-bearing qualities, published in 1865 in the sixteenth volume of the "Royal Agricultural Society's Journal." He speaks of them as having white legs and head, the latter long and broad, with a tuft of wool on the forehead; long thin neck, narrow fore quarters, the body long, with flattish sides and sharp chine, loins wide and strong, belly coarse, thighs broad and thick, with coarse bone and muscle. The wool is an important feature, specially valuable for the length of staple, fineness of quality, and bright glossy character, which makes it in demand for Flanders and France, being principally used in the manufacture of a fabric known as "cloth of gold." The usual weight of the fleece is from 7lb. to 10lb.

The Romney Marsh sheep have undoubtedly considerably improved since the days of Youatt, partly owing to more scientific management, and partly to the influence of foreign blood. The Kentish farmers would naturally try experiments with the fast-feeding Leicesters, seeing that in the days when Bakewell first made those famous it required from two to three years to bring the Marsh sheep to market. We believe the Leicesters were first tried on their own merits, and could stand neither the climate nor the close stocking so essential to profit. A dash of the blood would, however, increase propensity to feed, without seriously affecting the constitutional vigour which is so marked and essential a feature of the marsh-born sheep. On the arable lands adjoining the marshes, where there would be more shelter and artificial treatment, a more decided cross would be found beneficial. The best treated sheep at the present day are capable of reaching very creditable weights as shearlings; the heaviest pen at the Smithfield show of 1872, for example, were 260lb. each (live weight). Greater aptitude to feed, more symmetrical frames, and a diminution of bone and offal have been gained by the alliance, whilst the characteristic features remain intact.

We have never knowingly tasted the mutton, but we believe it stands far before any other long-wool breed for closeness of texture and good flavour.

An impression prevails that formerly the Romney Marsh breed were larger, as they were undoubtedly coarser sheep; it is quite possible that the influence we have referred to, whilst giving quality, also reduced the legginess of the breed without diminishing the weight. Though much more symmetrical than of old, with well-covered backs and with springy touch, the want of scale is still apparent; the sides are flat, and there is not that depth of forequarter which is so marked a feature of the Leicester. It is probable that the peculiarity of form is connected with the hardy character which gives these sheep such peculiar value.

In his essays on "The Farming of Kent," Mr. Buckland alludes to the pasture land of the Marsh as differing greatly in quality and productiveness. He describes it under two heads—breeding and fattening land; the former keeping from two to three ewes per acre during winter, and about twice that number (that is, we conclude, the dams and their produce) in summer. A fair average of the fattening land would be four to five sheep per acre. Cattle are purchased to keep the grass under, but occupy a very secondary place to the sheep, which are the chief source of profit.

We remember passing through a portion of the Marsh during winter, and being struck with two things—the exceeding greenness of the grass, and the number of big, robust-looking ewes, which rather crowded than dotted the plain. A keen east wind penetrated our bones, despite our Ulster, yet the sheep minded it not a bit; and we learnt, to our great surprise, that, save in deep snow or prolonged frost, they fended for themselves, and then only got a mouthful of hay. No wonder that they held their own; they were on their native ground, and not to be disturbed. There are still large graziers who object to using hay, even when the ground is covered with snow, preferring the sheep to scratch down to the grass. In former times it was but too common for the ewes to be left entirely unprotected during the lambing season, and great were the losses in times of severe frost. Now the ewes are placed in a sheltered inclosure near the homestead, or a temporary ewe-pen is erected, wherein the ewes lie at nights, supplied with hay and a few turnips, if they can be spared, and where the lamb is sheltered for the first few days of its existence, care being taken not to render it delicate by too much protection. The present value of breeding stock will encourage greater care than ever, and any reasonable expense is justified in order to save the lambs. Great benefit has been found from closely thatched hurdles, set up crossways with strong stakes, so as to be storm-proof. Thus, from whatever quarter the "stormy winds do blow," the lambs can gain shelter, of which they are sure to avail themselves. More care is bestowed upon the supply of proper food. In some cases, in addition to hay, bruised oats, or other nourishing food, is given to the ewes; and no doubt such care, by increasing the supply of milk, tends to bigger and healthier lambs, and so repays the outlay. The grass, whether from the fertile nature of the alluvial or the presence of salt, is peculiarly nutritious and healthy; and, with occasional assistance in the way we have indicated, the ewes thrive wonderfully.

In a district where the spring months are generally so cold, early lambing is not desirable, and consequently we find the Marsh farmers seldom turn out the ram before the beginning of October. The ewes are tolerably prolific; from twenty-five to thirty per cent. of doubles is not uncommon. They nurse well, and the lambs soon learn to eat the sweet grass, and grow away famously. They are not generally weaned until July, and it is a good plan to dip them at the time they are separated from the ewes. In order to bring out the hogs after the clip they should be put upon turnips in good time. Those who have little or no plough land send their lambs to the upland farms for wintering. It is a question whether it would not answer as well to sell the lambs out and out, as it is difficult to get them properly attended to. We believe it would pay well to allow them half a pound a day of cotton cake whilst on turnips, as well as a certain quantity of hay. When really well done, it is quite possible to bring the hogs into market from turnips, weighing from 17lb. to 20lb. a quarter; but, according to ordinary management, they have to be finished off on the feeding grass in summer, where they thrive rapidly and are much esteemed. By the second autumn they kill from 25lb. to 30lb. a quarter.

The principal ram breeders whose blood is in repute are Messrs. Rigdens, F. Murton, Henry Page, and Lord Hothfield, who has only recently become an exhibitor at Smithfield. The value of shearling rams sold by auction varies from 5*l.* to 12*l.* each.

CHAPTER XIII.

SOUTHDOWN SHEEP.

THE breed of sheep known as the Southdowns, from the fact that they originated on the line of hills so called, which extend from Beachy Head on the east, to the river Adur, that falls into Shoreham Harbour, on the west, occupy, as regards short-woolled sheep, a position similar to that which the Leicesters hold among the longwools. Both were the starting points of improvements that are still going on; and not a type of short wool but has been benefited by a cross or two of Southdown blood. Like the Leicesters, the present value of the blood is rather in its crosses than its purity, albeit we have nought but eulogy for these sheep as seen on their native hills. Nothing can be more dissimilar than the active, restless Southdown, and the sober, staid Leicester; yet both have a common origin, and their peculiar features and characteristics are doubtless due to local influences. In early days, when the majority of the district, and certainly all the high land, was downs, sheep that could travel well and were hardy would be invaluable: the very nature of the life, out all day roaming the downs, driven at night into folds on fallows, would tend to activity; and an animal that could bear fatigue would be of the highest value.

Mr. Thomas Ellmann, of Glynde, Sussex, occupies a similar position in reference to the Southdown as is awarded to Mr. Bakewell in Leicester history. Hear the description of the originals given by the indefatigable writer, Arthur Young, in 1788: "The true Southdown, when very well bred, have the following points: no horns, a long speckled face, clean and thin jaw, a long but not a thin neck, no tuft of wool on the forehead, which they call owl-headed, nor any frize of wool on the cheeks; thick in the shoulder, open-breasted, and deep; both fore and hind legs stand wide; round and straight in the barrel; wide upon the loin and hips; shut well in the twist, which is a projection of flesh on the inner part of the thigh, that gives a fulness when viewed behind, and makes a Southdown leg of mutton remarkably round and short, more so than in most other breeds; thin speckled legs, and free from wool; the belly full of wool; the wool close and hard to the feel, curdled to the eye, and free from projecting or strong fibres. Those flocks not bred with particular care and attention are apt to be coarse-woolled in the back, but some are fine all over; weigh fat from 12lb. to 15lb. a quarter." Contrast the above with the best specimens, as shown by the Prince of Wales, Lord Walsingham, the Duke of Richmond, Messrs. Gorringe, Penfold, Chapman, Ellis, and others, and remark the attributes of the improved sheep as described by Mr. Ellmann himself, and quoted by us from Youatt: "The head small and hornless; the face speckled or grey, and neither too long nor too short; the lips thin, and the space between the nose and the eyes narrow; the under jaw or chap fine and thin; the ears tolerably wide and well covered with wool, and the forehead also, and the whole space between the ears well protected by it as a defence against the fly; the eye full and bright, but not prominent; the neck of medium length, thin towards the head, but enlarging towards the shoulders, where it should be broad and high, and straight in its whole course above and below. The breast should be wide, deep, and projecting forwards between the fore legs, indicating a good constitution, and a disposition to thrive. Corresponding with this the shoulders should be level with the back, and not too wide above; they should bow outward from the top to the breast, indicating a springing rib beneath, and leaving room for it, the ribs coming out horizontally from the spine, and extending far backward, and the last rib projecting more than the others; the back flat from the shoulders to the setting on of the tail; the loin broad and flat; the rump long and broad, and the tail set on high, and nearly on a level with the spine. The hips wide, and the space between them and the last rib on either side as narrow as possible, and the ribs, generally speaking, presenting a circular form like a barrel. The belly as straight as the back; the legs neither too long nor too short. The fore legs straight from the breast to the foot, not bending inward at the knee, and standing far apart both before and behind; the hocks having a direction rather outward, and the twist, or the meeting of the thighs behind, being particularly full. The bones fine, yet having no appearance of weakness, and of a speckled or dark colour. The belly well defended with wool, and the wool coming down before and behind to the knee and to the hock, the wool short, close, curled, and fine, and free from spiry projecting fibres."

In most respects a good typical flock of Sussex Downs at the present day would answer to the above description; in one respect, however, a marked alteration has occurred—viz., in colour. Speckled faces and legs would now be looked upon with horror, as indicative of bad blood. A uniform tint now prevails, varying from brown to fawn, or almost grey. The Sussex-bred sheep are, as a rule, lighter of feature than the sheep cultivated on richer and flatter soils; the forehead is covered with short wool, and the cheeks are moderately woolled. Owing to better resources for feeding, the sheep are larger, reach maturity sooner, and are more cultivated. The hill farms, however, cannot produce great weights; activity rather than size is here the desideratum.

To Ellmann is due the credit of first improving the Southdowns; but it is equally true that his attention would have been in vain but for the facilities afforded in the growth of root crops for a regular and nutritious diet. So long as the sheep were expected to work hard for subsistence during the day, travelling long distances to and from their feeding ground, and often having to cover much ground in order to obtain a bellyful, and were folded at night on bare fallow, neither early maturity nor perfection of form would be possible. We have but little direct evidence as to the method pursued by Ellmann. Experiments were tried no doubt; and it is possible that either he or others may have introduced a dash of Leicester blood, which would give quality; but it is not necessary to imagine this. If he selected the best specimens from his own and neighbours' flocks, he would, by care and judgment, gradually produce a more symmetrical sheep. Arthur Young, whose description has been given of the breed before it had been undertaken by Ellmann, thus speaks of his flock when seen in one of his later tours: "Mr. Ellmann's flock of sheep is unquestionably the first in the country, the wool the finest, and the carcasses the best proportioned. Both these valuable properties are united in the flock at Glynde. He has raised the merit of the breed by his unremitting attention, and it now stands unrivalled." Mr. Ellmann retired in 1829, when his flock was brought to the hammer, and realised prices which in those days were considered extraordinary. 770 ewes of all ages averaged 3*l*. 1*s*. 6*d*. each, 320 lambs 36*s*. each, thirty-six rams 25*l*. each, thirty-two ram lambs 10*l*. each, and 241 wether lambs made 21*s*. each.

Mr. Ellmann's most noticeable successor was Mr. Jonas Webb, of Babraham, Cambridge, who commenced operations about the year 1823, and, although we have no records, he doubtless visited Glynde, as well as other noted Sussex farms, for the foundation of his flock, which eventually became, and was, for many years, by far the most valuable collection in the country, and the source from whence all the highest flocks were invigorated. No man before or since made such prices for Southdown rams, and the Babraham lettings were meets that drew from far and near, and have been happily alluded to by the late H. H. Dixon and others. It was never our privilege to witness this celebrated farmer at home, though we well remember his portly form at Shorthorn sales, when he bid with pluck and judgment; and had life been spared, he would probably have attained celebrity also in this direction, as he was a good all-round judge. Mr. Robert Smith, in his report on the exhibition of live stock at Chester, alludes to Mr. Webb's practice and success. At the outset, he purchased the best ewes he could get from the leading breeders in Sussex, and then, like Bakewell, set to work to remodel them into his own class of "Southdowns," and, having obtained his type, maintained it by close breeding. We are not aware with what degree of authority this statement is made; probably, like Bakewell, Mr. Webb observed reticence as to his practice, and if he experimented the world would be none the wiser. His sheep were noticeable for greater size than the denizens of the downs—a natural consequence of physical influences. He farmed largely, and his flock was sufficiently numerous, and sufficiently diverse in its origin, to allow of selection without close affinity. We advance this remark, because formerly some believed stock could not be bred from too close affinities—an error which closer investigation as to the practices of eminent authorities would remove. We used to hear the admirers of certain families of Shorthorns, for example, deprecating the introduction of a strain of foreign blood, which probably has saved the line from wearing out; and, judging from all that we know, we venture to say that, even in the case of a large flock like the Babraham, which would long yield sires of remote consanguinity, the time would come when but for fresh blood vigour would cease. Mr. Webb first exhibited at the meeting of the Royal at Cambridge, taking both prizes for ewes. Afterwards he showed only rams, finding that the forcing requisite to prepare the females for successful exhibition seriously injured breeding properties. On one occasion, we do not remember at which meeting, every animal drawn out by the judges from which the final selections were to be made belonged to Mr. Webb, so that he carried everything before him. He was in the habit of letting his rams by auction, the hirer having the privilege of purchasing, save under special conditions, by doubling the hiring figure; very high prices were realised, especially at the dispersion of his flock.

The Dukes of Richmond have been supporters of Southdowns for more than a century. In 1825 the late duke turned his attention to their improvement, and purchased valuable sheep. Many prizes, especially at the Smithfield Shows, were gained; but of late years the success of the flock has not been so great, and for many years Lord Walsingham's name was more frequently associated with the prize lists than any other breeder. On the death of the late lord the flock was brought to the hammer, and a fresh start made, when large prices were realised. The Merton flock had a large infusion of Jonas Webb's blood; and on such a good foundation the intelligent energy of the agent, Mr. Woods, and the great practical knowledge of the shepherd were brought successfully to bear. The Merton sheep, like those of Mr. Webb, were of a different type to the Sussex Downs, from which they originally sprang—so potent are the influences of physical conditions. The great characteristic was length and spread, the fore quarter resembling Bakewell's Leicester rather than the active Down sheep. Great weight was thus acquired, although not so valuable for the butcher; the quality and ripeness evinced the tendency to feed and the care bestowed. It would be tedious to recount the successes of this flock, which for some years had it very much their own way, notwithstanding which, and much to their credit, the Sussex breeders struggled on, and gamely came to the shows with undiminished ardour, rendering their class one of the most attractive features of the London fat show. The Sussex breeders disapproved of the type of the Merton sheep, and on one occasion it will be remembered that Mr. Ellmann, the son of the great flockmaster, protested against the decision of his colleagues. Be this as it may, the Merton sheep were splendid examples of the perfection to which animals may be brought by care and judgment; and the rams were eagerly sought for by foreigners.[*]

Our notice of distinguished breeders, brief as it necessarily is, would be unpardonably deficient were we to omit the name of the late Mr. Rigdon, of Hove, Brighton, who had cultivated and brought to great perfection the original Sussex Down, and had, in his late years especially, achieved well-deserved honours. His rams, both at Cardiff and Wolverhampton, were splendid specimens, models in form, and of rare type and quality. He was also first for shearlings at Islington both in 1871 and 1872, and took first rank as a breeder. His annual ram sale attracted many customers, and the sheep were let at high prices.

The Southdowns took such a decided lead among shortwool sheep, that their influence was sought as the first step in the improvement of other breeds. The Wilts and Hampshire Downs, however little indebted originally, owe much of their present quality and truth of form to Southdown influence. The improved Hampshires, which now predominate, have Southdown blood in their veins. The Shropshires, which are probably more cultivated than any other description of sheep, have

[*] A few years since the flock was dispersed, large prices being obtained, and since no sheep have been exhibited.

benefited by the infusion of Southdown, whatever may have been the effect of the other crosses that have been tried. As a cross with the Leicester or Lincoln, the produce is most valuable for feeding purposes. Mr. Overman has repeatedly taken first honours at Smithfield, with animals so bred. Like the Leicesters, the chief value of the Southdown has been in its crosses. As a pure breed we lack weight for food consumed, although what we have is of the best. They, doubtless, are well adapted to the spare sweet herbage of the downs; their light weights and active habits enable them to pick up a living where the longwools would starve. On rich lowlands they are not much cultivated, being superseded by their crosses and animals like the Oxfords, which combine weight with quality. The present Mr. Ellmann gives his opinion that the improved Lincoln, derived from a combination of Leicester, affords the most valuable cross with the Southdown.

Although we have quoted from early writers on the subject, we have not yet attempted to frame for ourselves a description of the Southdown, and we approach the subject with dread. The Sussex Down is noticeable for the light shade of feature, profusion of wool on forehead and on sides of face, short head, flat forehead, large full projecting eye, fine nose and muzzle, short neck, level contour, great leg of mutton, barrel-shaped carcass, level underlines, fine bone, and fine close wool. The Southdown as cultivated in Norfolk or Berkshire is a larger animal, with darker and larger features, and more substance of fore quarter; the shoulders are generally well laid, and the width of bosom and thickness through is very noticeable.

We collect some particulars as to Sussex management from Mr. Ellmann's lecture before the Royal Agricultural Society, delivered March 22, 1865. One great point to be kept prominently in view by the flockmaster is that the Downs must be close fed, otherwise the herbage becomes coarse, and the quality of the grass depends entirely upon its being kept short. In summer the sheep are kept upon the downs with the aid of a fold of tares or rape, the rape being sown at intervals, beginning from the 1st of May. For weaning purposes vetches and rape are preferred, seeds being considered "doubtful" food. The wether lambs are often sold at Lewes fair, being prepared for market by sainfoin, with a moderate allowance of oilcake or oats. The lambs are protected in the ewe-pen for three or four days after birth, but are early brought into the fields, as the shepherds consider too much attention tends to weaken and render them tender. As regards autumn management, Mr. Ellmann speaks highly of rape, but cautions the public against its indiscriminate use in a wet season without the addition of dry food; this, we think, applies to any description of sheep. The food is raw, and contains a large percentage of nitrogen, consequently it causes inflammation of the uterus; the affection commences in the liver, followed by general debility. In such cases opium is recommended as the best remedy. The Southdown, though naturally active and restless, bears close stocking, especially when not accustomed to the range of down land; the ewes are fairly prolific, but the proportion of doubles depends materially upon the condition of the ewes when the ram is introduced, and the nature of the food during the time the ram is with them. Except the dam and lamb are supplied with extra diet, which is done when fat lambs are produced—for which we do not think the Down so suitable as many other breeds—the maturity of the Southdown is slow as compared with Longwools. They can be brought out at about 1 year to 16 months, but frequently come out of grass in the autumn. The size and quality command top prices. With the establishment of lamb classes at Islington, great progress has been made as regards early maturity, Mr. Upton's lambs, which took the breed cup at the show of 1886, being extremely perfect, and the class so good that it was highly commended.

CHAPTER XIV.

THE HAMPSHIRE OR WEST COUNTRY DOWN SHEEP.

By E. P. SQUAREY.

HAMPSHIRE or West Country Down sheep are bred on the chalk formations of Berkshire, Hants, Wilts, and Dorset, and occasionally are found in Sussex and Surrey. This breed undoubtedly dates its origin from the crossing of the old Wiltshire Horned sheep and the old Berkshire Knot with the Southdowns which were introduced into Wiltshire and Hampshire early in the present century.

For a long time the high quality and charming character of the Southdowns, with their fine form, satisfied the most advanced of the Wilts, Hants, and Dorset farmers; and it was not until they realised how much they had lost in the size, early maturity, and hardihood of constitution which existed in the old Wiltshire type, that they bethought them of recurring to those animals to give additional substance and development to the Southdowns. These crosses were made with varying success, and depended simply on the instinctive capacity of the farmer to properly select the animals for this purpose. Whilst one aimed at the production of a large-framed, longwool-producing, hardy animal, another devoted his attention to the maintenance of the high quality and beauty of the Southdown, with earlier maturity and greater size.

From about 1815 to 1835, the Hampshire Downs of the north of Hants and the south of Wilts were totally dissimilar in character. It was evident that the leading ram breeders of each district had aimed at and had secured a different type. The North and East Hampshire sheep were large, muscular, early-matured animals, growing a fair quantity of wool of moderate fineness; the head large and well set on, of dark brown colour, verging towards black, covered with coarsish hair, with Roman nose, the neck with greatly developed muscles; the ears thick, of the same colour as the face, and an occasional tendency to recur to the original type, by producing "snig horns;" the legs with large bones, and in the most strongly marked type, the wool growing below the hocks and knees. An occasional white spot was exhibited on the face, ears, or legs; but the efforts of the ram breeders were uniformly directed to avoid this, and to procure perfect uniformity of colour. On the other hand, the Wiltshire breeders had adopted a more largely framed and probably less handsome animal than their Hampshire brethren. They were less careful as to the uniformity of colour, and an ewe with speckled face or ears was not dismissed from their flocks, provided she had size and other good qualities.

Amongst the earliest and most distinguished breeders in Hampshire were the Messrs. Dale, Budd, Pain, Digweed, Dowden, &c.; and in Wiltshire Messrs. Dear, Cusse, Smith, Blake, and others; the late Mr. Waters, of Stratford, near Salisbury, kept a smaller but more handsome and truer type of sheep.

About 1845, when the Hampshire Downs were gradually asserting their superiority over the Southdowns in the counties of Wilts, Hants, and Dorset, the necessity for an improvement in the general quality and tendency to make flesh was apparent. Foremost amongst the improvers in this direction was the late Mr. Humphrey, of Oak Ash, near Newbury, who may be said to have achieved for the Hampshire Downs that which Mr. Jonas Webb achieved for the Southdowns. To him was given the instinct to perceive the results that would accrue to the Hampshire Downs by a strong dash of the largest and best fleshed of Mr. Jonas Webb's Southdowns. Carefully tested, and the issue of the various crosses watched and applied with marvellous ability, and at a great cost of time and money, the result is realised in the present almost perfect animal known as the Improved Hampshire Down.

Since 1863 Mr. Rawlence, of Bulbridge, near Wilton, has maintained and increased the reputation of this breed of sheep, having obtained a large number of prizes at the Royal Agricultural Society, the Smithfield Club, the Bath and West of England, and local shows. To Mr. Morrison, of Fonthill, whose sheep are almost entirely descended from the Bulbridge flock, has been accorded the unique honour, as a breeder of West Country Downs, of receiving the silver cup at the Smithfield Show of 1872 for his shearling wethers, estimated to weigh 280lb. each, as the best pen of short-woolled sheep in the yard.* The other breeders of very excellent animals in Wiltshire are Messrs. Waters, Dibben, Bennett, Moore, Coles, and Read. Whilst in Dorset, Messrs. Saunders, of Watercombe; Fookes, of Cerne and Homer, have obtained celebrity with animals more closely allied with the Southdown. The old Hampshire flocks are admirably represented by Messrs. F. Budd, Parker, Arnold, Barton, Lewis, Fitt, and others.

It may be of some interest now to inquire into the circumstances under which the Southdown flocks in Wilts and Dorset have gradually merged into the improved Hampshire Downs.

Almost concurrent with the inclosure of the common lands, about seventy years since, large areas of the Down pasture lands, which had served as admirable feeding grounds for the Southdowns, were broken up. A little later artificial manures were introduced. These conditions induced the farmers to largely increase the growth of artificial crops for sheep feed, such as turnips, rape, vetches, trifolium, rye, and Italian ryegrass. The consumption of these artificial crops by sheep led the breeders generally to turn their attention to the system adopted by the Hampshire men, of selling their wether lambs in the late summer or early autumn, instead of keeping them, as was then the custom, until they became two-teeth or four-teeth sheep, when they were sold at a smaller price than the lambs now realise. Under these conditions it was important to secure early maturity and greater size, and the flockmasters, with very

* A still higher honour was reached in 1886, when Mr. Chapman's shearlings took the Champion Cup of 50l. as best sheep in the Hall.

few exceptions, at once crossed with the Hampshire Downs, and now successfully compete at all the early fairs with their Hampshire brethren.

In Wilts, Berks, and Dorset more attention has been paid to the quality of flesh and wool, and thereby has been produced an animal generally superior to that of the Hampshire breeders, who until lately sacrificed everything for size, colour, and character of face.

Since the conversion of so much down into arable land, the modes of keeping sheep in both counties have been nearly similar. The ewes are put to the tup at somewhat different periods, varying according to locality, from the latter end of July to the middle of September. They are generally kept on turnips and hay during the winter, except when the crop is short; the hay is cut into chaff with a portion of straw, with the addition, in some localities, of malt-dust, bran, corn, or cake. Some years ago the ewes were taken to a lambing yard near the homestead, but it is now become general to make temporary yards in some convenient place on the farm, near the green food, and where the manure is most required. If the weather is fine the lambs go out on turnips in a few days after they fall, but some farmers prefer keeping them in the yard or pen for two or three weeks. The couples are generally kept on turnips and hay until about the first week in April, when, the water meadows being ready, they go there by day and fold out on Italian rye-grass, rye, winter barley, or trifolium at night; the tup and wether lambs getting a little cake or corn before the hurdles. On farms where there are no water meadows there is usually a larger quantity of late swedes provided, after consuming which the couples are kept on rye, winter oats, barley, trifolium, and mangold-wurtzel, until the vetches are in flower.

Farmers who keep their stock in the highest condition wean their lambs about the first or second week in May, when they are generally kept on sainfoin or clover by day, and are folded on vetches by night. When the vetches are consumed they are put on rape or cabbages, with the aftermath clover or sainfoin; the sale lambs have large folds of the above, the ewe lambs, or the stock ewes, clearing up any food which they leave. By this management the wether or sale lambs arrive at great size, and realise high prices at early fairs.

Under the present high prices of mutton, the best wether lambs, which had heretofore been sold in July, August, and September, to be fattened in Kent, Surrey, and Middlesex, Hertfordshire and Essex, for the London markets in February, March, and April following, are now for the most part taken for immediate slaughter, and the second-class animals have taken their places in the districts named for winter feeding. This fact indicates how greatly during the last few years the stock of mutton existing in the country has been decreased by larger consumption or other causes.

In the years 1848, 1849, 1850, the statistics of flocks in the neighbourhood of Salisbury, numbering 10,000 sheep, gave the following results: Yield of lambs, 91 per cent.; mortality of ewes, 5·5 per cent.; mortality of tegs or hogs, 3 per cent. per annum. This breed of sheep is scarcely so prolific as the Dorset horned ewes, but under good management and more even and liberal feeding, the yield of lambs has certainly been increased.

The wool is of fine quality but short staple; the average growth may be taken at $4\frac{1}{2}$lb. per sheep. If the existing higher price of the long staple wools over these finer descriptions should be likely to be permanently maintained, we have little doubt but that the length and character of the Hampshire Down wool might be modified to meet the requirements of the manufacturer, without sacrificing their admirable qualities as producers of mutton.

The ewes are usually bred from for three years, and as full-mouthed ewes are sold at the autumn fairs to breeders of early lambs, who generally put them to a Lincoln or Cotswold tup, and fatten the ewe and lamb together. This first cross produces an animal with great aptitude to fatten, and if kept till they become shearlings they carry a large quantity of mutton and wool.

The altered system of farming on the chalk formation of Wilts, Hants, Berks, and Dorset, and the introduction of the improved Hampshire Down sheep, has led to a very considerable increase of mutton and wool per acre during the last fifty years. In the absence of statistics it is difficult to give a reliable estimate of such increase, but inquiries amongst those most competent to form an opinion lead to the belief that the increased production in weight of mutton and wool exceeds that at the end of the past century by at least 50 per cent.

To those who desire to see the Hampshire or West Country Down sheep in perfection it may be interesting to attend the Overton and Weyhill fairs, in Hampshire, and the Britford and Wilton fairs, in Wiltshire, where the best specimens of these varieties of sheep may be met with.

CHAPTER XV.

SHROPSHIRE SHEEP.

ALTHOUGH a comparatively recent breed, these valuable sheep are probably more widely distributed than any others, and merit increasing patronage, as they possess many sterling qualities, which it is the object of the present paper to describe. Although moderns in their improved character, the original stocks were the Longmynds in Shropshire, and the denizens of Cannock Chase in Staffordshire. Plymley, who is quoted by Tanner in his prize essay on Shropshire published in the "Journal of the Royal Agricultural Society," vol. 19, p. 42, thus describes the sheep: "There is a breed of sheep on the Longmynd, with horns and black faces, that seem an indigenous sort; they are nimble, hardy, and weigh near 10lb. per quarter when fatted. The fleeces upon the average may yield $2\frac{1}{4}$lb., of which $\frac{1}{2}$lb. will be the breechens or coarse wool, and is sold distinct from the rest. The farmers of the hill country seem to think the greatest advantage they derive from the access of foreign stock is from the cross of the Southdown with the Longmynd sheep; the produce they state to be as hardy and to bite as close as the Longmynd sheep; and the weight of the carcass is increased." Plymley's work was published in 1803; and, after such evidence, it is surprising that anyone could contend for the purity of the Shropshires.

Mr. H. Evershed, in his essay on Staffordshire, described the dry surface of Cannock Chase, and its good climate, as favouring a heavier heath-sheep than occurs elsewhere. The original sheep had a short light fleece of about 3lb., and a carcass which might be fattened at three years old to eight or nine stone. Their descendants, whilst retaining the same hardy character, are much larger, mature earlier, yield a heavy fleece, and a frame weighing ten stone at thirteen months without extraordinary treatment. We have quoted these authorities in order to show that it is to the Southdown chiefly, though not entirely, that the present form and character of the Shropshire are due; indeed, about the only objection that could be urged against the breed when the first edition of this work was published in 1877 was that, although for the past twenty years it had received much attention, there was still a lack of uniformity; but this defect has now almost entirely disappeared, as breeders are at last tolerably agreed as to the particular type that is most desirable. The variety could only be accounted for by the supposition that different crosses in different proportions had been tried; and we think there is no doubt this had been the case.

A great impetus was given to breeders when the Royal Agricultural Society recognised the importance of the breed by giving it a separate class, which was first done at the Canterbury show in 1860. The wisdom of the step has been abundantly illustrated by the numbers and quality of the entries at all subsequent shows—which have for many years past far outnumbered any other breed. One reason for the difference of character which so long prevailed may be found in the fact that, while many breeders achieved from time to time prominent positions, there was no one in early days who took such a decided lead as to impress his type permanently, as was the case with the Leicesters and Southdowns.

Of the earlier breeders, we must single out for special notice Mr. Samuel Meire and Mr. George Adney as men who, pursuing a different practice, laid the foundation of the present breed. Mr. Meire carried on his operations at Berrington, until he gave up that farm and retired to a small estate of his own at Harley, the same parish in which Mr. Adney had farmed. Mr. Meire was a good judge of stock, and set to work upon the coarse Shropshire, going chiefly for three points—straight spine, with well sprung ribs, oblique shoulders, and good rumps. These points could not be obtained by cultivation or selection alone, and Mr. Meire introduced the Southdowns, buying or hiring rams from the late Mr. J. Ellmann, of Glynde. Aptitude to feed, with the short back and chine, were derived from a cross of Leicester blood introduced with great judgment. Having thus obtained what he desired, Mr. Meire endeavoured to fix the same by close breeding. That his sheep possessed much constitutional vigour is proved by the history of his celebrated ram, Magnum Bonum, who served for eleven seasons, his dam living to be twenty years old. He was the sire of Perfection, used by Mr. Foster, of Kinver Hill, which got the first prize shearling at Chester. At the same show Mr. Foster secured both prizes for ewes; the first-prize pen bred by Mr. Meire, and described in catalogue " as two 11 years 3 months and 2 weeks old, two 9 years 3 months and 2 weeks old, and one 7 years 3 months and 2 weeks old; pen of five Shropshire Down ewes, dark brown face and legs." The fact that such aged ewes could be brought up in such condition as to beat blooming shearlings is a proof of constitutional vigour. In 1853, at Gloucester, Mr. Foster and Mr. Meire secured all the prizes, and every sheep was descended from Mr. Meire's stock. When Mr. Meire gave up the Berrington farm he brought a few choice ewes to Harley; the number, we believe, never exceeded forty. The first year's sale of rams averaged 12 guineas for fourteen sheep, the second 26 guineas. Ill-health soon compelled him to give up ram breeding; he preferred rather light uniform-tinted features, and went for close wool and quality rather than size. Mr. Henry Smith's flock, of Sutton Maddock, which was so well known at one time, was at first principally descended from Mr. Meire's stock, the great characteristic of all his sheep being quality. No man took more pride than Mr. Smith in his flock, so long as his health permitted. We visited the farm in the autumn of 1864, during very dry weather, and found everything burnt up; notwithstanding, the ewes were in excellent condition. They were on some dried-up seeds, without any water, yet looking uncommonly healthy. Small in appearance as compared with some flocks, because closer to the ground, they were thick proofy sheep, with straight backs, oblique shoulders, and big rumps—quality, aptitude to feed, and true form, were unmistakable. The colour of face was dark grey, with flat foreheads; the legs black. Later on, both the Kinver Hill and Sutton Maddock flocks were altered in character by the influence of Oxfordshire blood. The size was increased thereby, but we very much question the policy of the cross. It may be here mentioned that the late Lord Chesham, whose flock for many years and until its dispersal a year or two since stood A 1, purchased a good many sheep from Mr. Smith, and thus a considerable percentage of Mr. Meire's blood must have existed in his sheep.

Enough has been said to show that Mr. Meire was a great improver in his day, and that his sheep made a considerable impression wherever they went; it is to us a matter of great regret that his operations were arrested at so early a period. Mr. Adney pursued quite a different plan to Mr. Meire; he stuck to the Shropshire as he found them, making his improvements by selection; his judgment was undoubtedly good. His most fortunate produce was the ram Buckskin, who was descended from a Southdown cross, and if the picture representing Mr. Adney and his sheep is a likeness, this was evident in his fine rather flat head and grey character. Old Patentee was by Buckskin, being a twin out of a ewe bred by Mr. Adney; he did not follow his sire, having a large, plain, and dark head; he was an extraordinary getter, and his blood exists in every Shropshire flock of any note. Mr. Adney's enemies declared that he sold many more rams than he bred, picking them up all over the country. His judgment was far in advance of his contemporaries, and if this rather questionable practice be a fact, he would doubtless realise a considerable amount. Mr. Thomas Horton, of Harnage Grange, Mr. Adney's nephew, possessed a good deal of his blood, and at one time, when his Duke of Kent, a grandson of Buckskin, was in his prime, his reputation was considerable; but his flock lacked uniformity, and latterly his animals wanted quality. This was not his only prize winner—Lord of the Isles and Lord Salop were both prize winners. Up to and for many years before his death, no flock was more highly valued by breeders than that of the late Mr. Masfen, of Pendleford, who, although he was not an exhibitor, had a very high reputation for judgment, and his rams made a higher average than most of his contemporaries.

The Messrs. J. and E. Crane, of Shrawardine, Shrewsbury, followed closely in the steps of Mr. Adney, and for a series of years, dating from the Chester meeting in 1858 to that of Worcester in 1863, they were never out of the prize list, securing their chief honours with ewes, for the quality and character of which the flock was then justly celebrated; and several first and second prizes, besides commendations, have been obtained since that date. One of the brothers died several years since, and the survivor has reached the average span. We visited the farm in 1864. Mr. James Crane then resided at Shrawardine, on a fine farm of 410 acres, with a considerable portion of good meadow on the banks of the Severn. At that time it was evident that, for some reason or other, the flock was not quite in its right place, and judging from appearances, we thought the error might have been in too close breeding. The ewes, though uniform in colour and character, and of much substance, wanted symmetry, and neither showed well walking or standing; the legs were too much under, the loins rather high and the rumps low. The means that were adopted to correct these defects appeared to us judicious—viz., by the use of Mr. Coxon's Nobleman, a grand old sheep, noticeably deep forward and thick through the heart. He was by a grandson of Patentee by Patent. The lambs by Nobleman were very promising. In 1864 they secured a fine three-shear, Lot 1 of Mr. Thornton's, sire of the shearling which took the first prize at Newcastle, noticeable for length, beautiful head, and general symmetry; he was bred by Mr. Claridge, and was got by one of Messrs. Crane's rams. The averages made by rams and ewes at the August and September Shrewsbury sales have always been high, and sufficiently attest the reputation in which these sheep are held.

The flock is now carried on by Messrs. Crane and Tanner, and maintains its high reputation.

We may here pause for a moment to notice as a fact that the Shopshire breeders were the first to inaugurate the convenient practice of selling their draft ewes by auction—a practice which gave opportunities to outsiders, and helped to extend the influence of the breed, at the same time that it insured much larger prices than had previously been made by private sale.

In 1872, at the dispersion of a Staffordshire flock, one pen of ewes made an average of 23 guineas; a few years since such figures would have appeared fabulous. In the year 1864, and, we have no doubt, on subsequent occasions, Messrs. Crane's ewes were sold as high as 8 and 9 guineas each. The Shrewsbury and Birmingham sales have become of late years largely attended, and great numbers of sheep change hands.

It would be impossible for us in the limits of a single article to attempt even an enumeration of the breeders who have assisted to make these sheep famous. Confining our attention to the county, we have at the present time two men who deservedly stand somewhat in advance of the ruck—viz., Mr. Evans, of Uffington, and Mr. Mansell, of Adcott. Both have been careful to establish in their flocks uniform character, and the type is now recognised as much what should be aimed at. We are not sufficiently acquainted with the origin of either flock to describe the steps by which they attained their present high position. At the Wolverhampton show in 1871 certain extra prizes were offered, and the class was made peculiarly attractive—not in vain, as no less than 542 sheep were shown, and, notwithstanding some necessary roughness, the quality, on the whole, has never been surpassed. Probably the first prize shearling of Mr. T. Byrd's was rather a lucky one, as the character of the flock from which it sprang does not rank high, and it was far from a perfect animal. Mr. Mansell, who grew its sire, took the second place with a very superior sheep, which many thought the best animal. In the class for old sheep—one of the best we have seen—Mr. Mansell's two-shear was first, notwithstanding serious defects as to wool and colour. Mr. Evans here came second. His sheep was a magnificent specimen, and rightly placed, if only he would work, but his legs were gummy. At Cardiff, in a much smaller class, Lord Chesham's shearling, which took the first prize, was by Mansell's No. 8. Mr. Evans only here succeeded to third honours. In the better class for old sheep Mr. Mansell achieved a decided success, securing first and second prizes; his sheep, though somewhat dissimilar in type, were equally admirable. The first-prize sheep (a two-shear) was a remarkable animal, almost perfect, barring a rather short quarter. The second animal was much smaller, but more compact, wonderfully level and neat. We rejoice to see the efforts of such men as Mr. Mansell and Mr. Evans crowned with well-deserved success. Both have steadily worked for a good type, and we think have succeeded in their object; none make higher figures at their annual sales. Another noticeable and probably the best entry yet seen was at the Shrewsbury meeting of R. A. S. E., in 1884, when great efforts were made by individuals and the Shropshire Breeders Society to place their animals in a leading position. The extent of the exhibition may be gathered from the fact, that, out of a total sheep exhibits of 485 entries, the Shropshires number 246, and never before had such uniform excellence been seen. Lord Chesham took first and fifth and reserved for single shearlings, and was first with a splendid lot of five shearling rams.

Our notice would be incomplete if we omitted some of the more remarkable of the Staffordshire flocks. In 1872 one of the most celebrated was dispersed at extreme prices even for those sensational times—we refer to Mr. C. Keeling's, of the Ewetree Farm, Penkridge. We visited him in 1864, and found the land rather weak naturally, but kept up in high condition by the enterprise of the occupier. The flock at the time of the sale had been in existence nearly thirty years. Although this flock, like most of those bred in Staffordshire, originated from the Cannock Chase sheep, the quality it attained was due to judicious crosses from Masfen, Horley, and especially Patentee blood. Many of the ewes which we admired for their deep flesh and symmetry were by Gratitude, the highly commended two-shear at Canterbury. His origin is curious. When Mr. Byrd hired old Patentee, Mr. Keeling was allowed to send one ewe to him. He chose a fine specimen by old Norton, from Masfen's sale. Gratitude was a twin. Mr. Keeling's sheep were uniform in character, the features rather dark, not well-woolled on the forehead; big, thick sheep, showing much constitution. We believe the reason for the sale was that Mr. Keeling was leaving his farm. The late Mr. John Coxon, of Freeford Farm, Lichfield, ranked high as a Staffordshire breeder. His flock originated in 1825, descending from the Whittington Heath sheep—a breed of hardy sheep very similar in type to those of Cannock Chase. The flock comprised at the date of our visit

129 ewes. Thirty rams were annually reared, of which twenty were brought to the hammer. Grand-looking sheep, and much appreciated by breeders. The average of 1864 was 17*l*. 3*s*. 4*d*.; forty yearling ewes made 67*s*. each. Since that time higher prices have been realised. Nobleman, the third-prize old sheep at Worcester, and which has been already alluded to, was the sire of the yearlings, and very promising they looked; indeed, the stock throughout were highly creditable. Mr. Coxon was a good feeder, and occasionally illustrated the feeding qualities of the Shropshire: in 1862 one of his old wethers weighed 59lb. a quarter. Mr. Coxon superintended the late Col. Dyott's flock, which has also achieved considerable success. Of late years Mrs. Beach, of the Hattons, near Wolverhampton, has made her mark, carrying off at the Wolverhampton show a special prize as the winner of the greatest number of prizes. The land is particularly favourable for early development, and few can bring out lambs or even shearlings in such perfection. Her sheep are not large, but possess great symmetry and particularly matchy heads.

Sufficient has been advanced to show that the Shropshire sheep, though of comparatively recent origin, are at the present widely spread and much valued. Let us now consider how far these favourable opinions are justified. We know of no breed so prolific: the increase in all cases is to a certain extent, and often materially, influenced by the nature of the land, as yielding nourishing or inferior food. On an average, if the ewes are well cared for before and during the time the ram is with them, at least 50 per cent. of doubles may be looked for; and when Shropshire rams are put upon Longwool ewes, the increase is much greater, as the following facts sufficiently prove. On a strong, poor farm we purchased forty Banffshire ewes every autumn—*i.e.*, a description of Border Leicester with a slight Cheviot cross—and served them with a Shropshire ram, either a shearling or a ram lamb. In 1872 thirty-six ewes produced seventy-eight lambs, all sold fat. The next season the forty ewes produced eighty-two lambs; but owing to unfavourable causes, we lost ten lambs, and seventy-two fine lambs were sold fat during the summer. This prolific tendency is a point of great importance, for it is not with the Shropshires as with some of the larger breeds, that a fine single lamb is more esteemed than a double. The ewes are good mothers, and can do justice to their offspring; moreover it is always possible to assist nature by nutritious diet. Next, the Shropshire is a hardy sheep, suitable for a large range of soils, standing moisture better than severe cold, and capable of close folding without sensible loss of size. The yield both of mutton and wool is far greater than from the Southdown or other shortwool. Hampshires may arrive at greater weight, but they require more time, the proportion of bone and offal is greater, and the wool much less; it is no uncommon event to find a flock of Shropshire sheep on good land yielding an average of $6\frac{1}{2}$lb. to 7lb. a fleece. The quality of the meat, both from the fineness of texture, the presence of fat in the tissues, and the rich dark colour, is fully equal to the best Southdown. And for all these reasons the Shropshire sheep are rent-payers, and deservedly and increasingly popular. They suit the moist climate of Ireland remarkably well, and not a few buyers from the Emerald Isle may be seen round Mr. Preece's Ring at Shrewsbury or at Bingley Hall. With generous treatment from their earliest days—and such treatment is surely profitable as well—Shropshire hogs can be brought out in May to weigh from 18lb. to 20lb. a quarter. Cases could be cited where much more has been done, but we speak of what is possible where the land is heavily stocked. Altogether, they take rank as the most important shortwool breed of the present day. It was to be desired that greater uniformity of character should prevail, and with such breeders as Lord Chesham, Messrs. Mansell, Evans, Coxon, and Mrs. Beach in the van, satisfactory progress has been made.

The Shropshire sheep of the present day exhibit much of the quality of the Down, with considerably more size; the features are rather longer, of a uniform dark but not black tint, the eye full and large, the forehead moderately flat and well woolled, the ears rather large and thin, standing well out from the head. Much improvement in symmetry has taken place of late years. Formerly the shoulder was frequently upright, the spine not straight, the top far from level, and the fore quarter generally light; now the best bred sheep are as true-grown as the Downs.

The character of the wool is of great importance, especially where the climate is moist. An open condition of wool is to be deprecated; the staple should be fine and close, with which a good weight is quite possible. Although capable of making considerable weight upon good keep, we do not consider the Shropshires can mature so rapidly as such breeds as Leicester or Cotswold; the closer texture of meat requires a longer time to deposit. With ordinary management the shearlings are brought to market during the summer off grass, when their quality and moderately small weight render them very suitable to the season.

CHAPTER XVI.

OXFORDSHIRE DOWN SHEEP.

By Messrs. A. F. M. DRUCE and C. HOBBS.

NO BREED of sheep has grown more into public favour, or has more rapidy extended in numbers, within the last fifteen or twenty years, than the Oxfordshire Down. It is now about sixty years since Messrs. Druce of Eynsham, Gillett of Southleigh, Blake of Stanton Harcourt, and Twynham of Hampshire, undertook the construction of a new breed of sheep, that should in great measure possess the weight of the Longwool with the quality of the Down. Probably the advantage of such a breed was first apparent in the offspring of a cross occasionally resorted to in the case of draft ewes, but from pursuing which farmers had hitherto been deterred by the tendency of the offspring to breed back to either side; and for many years after the breed had become recognised as distinct, the want of uniform character was a source of criticism. Some slight admixture of Sussex Down may have been introduced by those early breeders; but we are of opinion that the Cotswold grey-faced ram and the Hampshire Down ewe were the chief, if not the only, materials which by judicious blending and careful selection have resulted in a class of sheep which, under suitable conditions, are probably as profitable as any that can be mentioned, both on account of size, weight of wool, aptitude to fatten, hardy character, and valuable meat. The success of the early promoters of the project led many others into the field. It was not until 1850 that they were styled the Oxfordshire Down, the county of Oxford being their stronghold; previous to this date they were properly regarded as crossbreds, and known as Down Cotswolds, under which designation they achieved sucesses at the Smithfield shows. At the Royal Meeting at Warwick there were thirty-seven entries: the first prize in the old class (comprising Oxfords, Shropshire, and Hampshire Downs) was taken by the late Mr. Samuel Druce for a sheep of this breed. In 1852 Mr. Blake's flock was distributed, and those who remember attending the sale at Stanton Harcourt must have been struck with their general uniformity of character, size, quality, heavy fleeces, and aptitude to fatten.

Mr. Philip Pusey, seeing the advantage these sheep were likely to confer on the public generally, as also on the flockmaster, became himself a warm supporter of the breed. At his death they lost a great patron, who, had he been spared, would have materially aided their rapid spread and improvement. His flock was brought to the hammer in 1855; also that of Mr. Gillett, of Brizenorton; and soon after of Mr. Gillett, of Southleigh, both ram breeders standing high in public estimation. The distribution of these flocks laid a good foundation for many others; not only did many tenant farmers give up the old breeds to make room for the improved sort, but the Oxfordshire sheep found favour with landed proprietors; and amongst others his Grace the Duke of Marlborough had a flock on a farm in his own occupation at Blenheim.

As soon as the breed became established, some of the most successful breeders began to exhibit their sheep at the Royal Agricultural Society's show, and, as they had no special class, their animals were shown with short-woolled sheep and crossbreds. This unsatisfactory state of things continued until 1861, when the stewards in their report stated "That the judges are of opinion that Oxfordshire Downs should not be excluded from competition at these annual shows, as they believe them to be animals possessing great merit, and worthy of having a class to themselves." The Royal Agricultural Society decided on a separate class, and the Oxfordshire Downs made their first appearance as a recognised breed by the great society in the Exhibition year of 1862, at Battersea, where they numbered sixty-two entries, and were highly spoken of by the judges, who, however, objected to their want of uniformity—a deficiency again referred to by the judges at the Royal in 1865 and 1868.

The Reports in the "Royal Agricultural Society's Journal" of 1870 and 1872 speak in high praise of their general excellence and great improvement in uniform character. We still see difference in type in the rams offered to the public; but, knowing that a heavy fleece can be obtained, with wool thickly set on the skin, and holding the opinion that a fine quality of mutton is not to be found under an open coat, we think that a great advance will be made when a lashy Cotswold fleece is no longer to be found among flocks bearing the name of Oxfordshire Downs. For rent-payers they are not to be excelled, and with their robust constitutions and early maturity, bearing as they do such an abundant supply of mutton and wool, they have made their way into most counties; and many hundreds of rams are yearly sold by the different breeders.

It was a proud day for this flourishing breed of sheep when at the Smithfield Club show in 1872 his Grace the Duke of Marlborough took the champion prize with his splendid wethers, as the best pen of sheep in any of the classes; Mr. John Treadwell's ewes being amongst the last three pens left by the judges for selection.

We now shortly describe the characteristics of a good type of the Oxfordshire Downs. It should have a nice dark colour, the poll well covered with wool, adorned with a top-knot on the forehead; a good fleece of wool, thick on the skin, not too curly; a well-formed barrel on short, dark legs (not grey or spotted); with good, firm mutton. The tegs are usually sold fat, from eleven to thirteen months old, at an average of ten stone, and are much sought after in the London and other markets.

The following figures, taken from the Smithfield Club show catalogue, will give some idea of the live weight of a pen of three shearlings when about twenty-two months old:

	Cwt.	qrs.	lb.	
Shearling wethers	7	3	27	1870
Ditto ditto	7	1	20	1871
Ewes	8	1	26	1870
Ditto	7	3	9	1871

The average age of the pen of three ewes in 1870 was 57½ months.
Ditto ditto ditto 1871 61 ditto.

The weight of wool for a whole flock will average about 7lb. per sheep; rams have been known to cut as much as 20lb. when shearlings. Great numbers of shearling and ram lambs are now sold by public as well as private sale. The principal breeders who of late years have distinguished themselves at the Royal and other meetings are the Duke of Marlborough, the Earl of Jersey, Countess of Camperdown, Sir H. W. Dashwood, Bart., Sir John Shelley, Bart., Messrs Wallis, Treadwell, Howard, Druce, Rogers, Stilgoe, Hobbs, Longland, Street, Parker, G. Adams, G. H. Morrell, H. Gale, Pratt, W. Arkell, H. Barnett, R. Dickens, W. Cooper, W. H. Fox, J. Worley, E. Gillett, R. W. Hobbs, C. Hobbs, jun., J. Miles, C. Chappell, H. Overman, and J. Case. There are other breeders who, although regardless of showyard honours, meet with a good sale for their sheep. At the ram fair at Oxford, held annually (the second Wednesday in August), there are large numbers of Oxfordshire Down rams disposed of at remunerative prices, and comprise specimens from the flocks of Messrs. Roberts, Gillett, Bryan, Franklin, Gale, Chillingworth, Button, and others. They find a ready sale, not only in the home counties, but in Scotland, Ireland, and several countries on the Continent. Most satisfactory prices have been realised recently, rams having changed owners at from forty to sixty guineas each. The cross with the Hampshire ewe for early fat lamb for the London market has been for some years, and is still, in favour with several flockmasters.

The Oxfordshire sheep are adapted more particularly for mixed soils, and stand close stocking and confinement; that is, they can be kept entirely in hurdles, and will probably do better so than if allowed a range. The different sorts of food usually grown on the "mixed soils," so as to keep them as much as possible on the arable land, are as follows: "January—kohl rabi, swedes, and turnips; February—the same; March the same, and mangold wurzel; April—mangold wurzel, winter oats, rye, and trefoil; May—trifolium and vetches; June—vetches; July—the same, and clover; August—the same, and cabbage; September—cabbage and rape; October—rape, early turnips, and mustard; November—turnips and kohl rabi; December—kohl rabi and swedes.

The stock ewes are generally divided in August, and rams selected to suit each lot; they run over the stubbles, and are penned on rape or cabbage at night. They then clean up the pastures till Christmas, having bean or pea straw at night. It is considered unwise to give them many turnips before yeaning. They are then brought into the foldyard for lambing, and are fed on hay, cotton cake, and a few roots, and remain so till the lambs are sufficiently strong to go into the turnip field. They will be found very good mothers, being strong and prolific, producing a considerable proportion of twins; this, however, depends much upon the nature of the land. The lambs when taken into the field have a fold in front of their mothers, where they are supplied with hay, corn, and, as the case may be, cut swedes, or crop off the greens; the ewes with twins are also given corn. The lambs are usually weaned when about twenty-two weeks old. The plan now adopted is to have the fold thoroughly well set, and allow them to remain in front of the ewes, and after a few days they will become quite reconciled.

They are a healthy class of sheep, and cases of giddiness are seldom known in any of the flocks. The management closely approaches that practised in Hampshire and Wiltshire, where the attention to ewes and lambs has become proverbial—early maturity (*i.e.*, 20lb. a quarter at a year old) requiring great attention during the young stage; and we are satisfied from experience that an early acquaintance with suitable artificial food and a frequent change of the natural produce are points of the gravest importance. The master's eye is required daily to note progress. A check to the young system is often bad to recover from, and it is a great argument for the folding system that the sheep are so frequently under the eye that any marked change must be noticed at once.

Mr. Druce adds: Just a cursory reference to a ram named Freeland, whose impress upon the most fashionable flocks is still marked, and whose name may certainly be found at the head of the pedigree of many of the sheep which command the highest prices now, bred by A. F. M. Druce in 1874, took all the chief prizes wherever shown in 1875, and was afterwards let to Mr. Treadwell for the season. His career the next year was equally successful, and was let at Mr. Druce's annual sale the last Wednesday in July for 85 guineas to Mr. T. S. Cooper to go to America. He was there exhibited at the Philadelphia International Exhibition, and gained every honour possible against *all breeds*, being thus commented upon by the judges: "For excellence in quality, uniformity of symmetry, great constitutional development, and for being a very superior specimen of the breed to which it belongs." Also in 1877 at the centennial meeting of the Bath and West of England Society, a two-shear ram of this breed named Campsfield was pronounced by the judges as the best ram of any breed in the yard.

Mr. C. Hobbs adds: As time moves on this breed of sheep continues to gain in popularity, and are still making their way into new districts. They have become much more uniform in character, and as producers of good quality and heavy weight of mutton and wool at an early age are now difficult to equal. Many are the enterprising men throughout the country who, in addition to producing large numbers of these sheep for the butcher, sell annually vastly increased numbers of shearling rams and ram lambs, which are distributed not only in the United Kingdom, but to France, Portugal, Russia, Belgium, Holland, Germany, Denmark, Sweden, Poland, Canada, United States of America, South America, Australia, South Africa, &c.

CHAPTER XVII.

THE ROSCOMMON SHEEP.

By the late R. O. PRINGLE,
Editor of Irish Farmer's Gazette, Author of "A Review of Irish Agriculture," &c.

SHEEP farming in Ireland is not carried on to the same extent as it is in England and Scotland. Thus, in the report which was issued by the Board of Trade, showing the agricultural returns for the United Kingdom, in 1872 England had 75·2 per cent. of sheep to every 100 acres under crops, fallow, and grass; Scotland, 157·4 per cent.; and Ireland, only 27·1 per cent.; the total average for the United Kingdom being 68·8 per cent. On the other hand, Ireland had 25·8 head of cattle for every 100 acres as above, being 5·1 per cent. over the average for the United Kingdom. The difference in the proportionate number of sheep and cattle reared in Ireland, as compared with Great Britain, arises in a great measure from the large number of small holdings, say under fifty acres, which exist in Ireland; and for such holdings sheep are not a suitable description of live stock. In illustration of the influence exercised by the prevalence of small holdings on the breeding and rearing of sheep, we have only to compare the number of sheep in Ulster, a province noted for its large proportion of small farms, with Connaught, where wide ranges of pasture prevail. Ulster contains 5,478,867 acres, and had 550,938 sheep, while Connaught contains 4,392,043 acres, and had 1,307,799 sheep; the proportion in Ulster being one sheep to about 10 acres, and in Connaught one sheep to 3½ acres.

Connaught has been for many years the chief sheep-breeding province in Ireland, and graziers from other parts of Ireland have long been in the habit of resorting to the great fair held at Ballinasloe, in the month of October of each year, for supplies of ewes and wethers. These Connaught-bred sheep have always been noted for their large size, but they were coarse and ungainly in their shapes. Culley, who visited the fair of Ballinasloe about the close of last century,* describes their defects in very strong language. He says: "I am sorry to say I never saw such ill-formed, ugly sheep as these; the worst breeds we have in Great Britain are by far superior. One would almost imagine that the sheep-breeders in Ireland have taken as much pains to breed plain, awkward sheep as many of the people in England have to breed handsome ones. I know nothing to recommend them except their size, which might please some old-fashioned breeders, who can get no kind of stock large enough. These sheep are supported by very long, thick, crooked, grey legs; their heads long and ugly, with large flagging ears, grey faces, and eyes sunk; necks long, and set on below the shoulders, breasts narrow and short, hollow before and behind the shoulders, flat-sided, with high narrow herring backs, hind quarters drooping, and tail set low; in short, they are almost in every respect contrary to what I apprehend a well-formed sheep should be; and it is to be lamented that more attention has not been paid to the breeding of useful stock in an island so fruitful in pasturage as Ireland." Culley, however, found some traces of improvement in the general character of the Connaught sheep, owing to the partial introduction of English rams, and he remarked, when referring to the descendants of those imported sheep, that "it is both extraordinary and pleasing to see how much they exceed the native breed." It had not, however, been an easy matter to improve any class of live stock in Ireland previous to that period, as the exportation of breeding stock from England to Ireland was strictly prohibited. Some enterprising persons had from time to time contrived to evade the law and to smuggle into Ireland rams and ewes of English breeds, which were sold

* Culley reports that 95,000 sheep were shown in the fair at which he was present, and that there had often been more.

at large prices to Irish breeders. This smuggling trade does not, however, appear to have been profitable to those who engaged in it, owing probably to the risk which attended it; and a curious illustration of this is supplied by a rare pamphlet in my possession. That pamphlet was published at Dublin in 1777, and was written for the purpose of supporting a petition to the Irish Parliament by Mr. Thomas Johnston, of Derry, in the county of Tipperary. Mr. Johnston, it appears, had speculated largely in smuggling over "stallions and mares, boars and sows, rams, ewes, and bulls" from England; but, owing to the hazard and expense of the undertaking, it had proved unremunerative, and Mr. Johnston petitioned the Irish Parliament to consider the losses he had sustained while serving the nation. Amongst the items set forth in his petition was "thirty rams and ewes, which were lost at Carlingford, on account of the ship's being seized for want of a cocket"—for smuggling, in fact—"and perished for want of food." It was a somewhat curious plea to use in support of his petition; but it is clear that the sympathies of the Irish Parliament were with Mr. Johnston, as the House "was pleased to grant him the sum of £200, not having it in their power, from the number of other Parliamentary grants, to extend their bounty farther at that time." Mr. Johnston's petition was supported by a memorial signed by forty-eight individuals. The memorial ran as follows: "We, the Gentlemen Breeders of the Kingdom of Ireland, certify that Mr. Thomas Johnston, by his great skill in the choice and unwearied assiduity in the importation of live stock into this kingdom for several years past, and at a vast expense, hath improved the stock in general of Ireland, in a most extraordinary manner, to such a degree, that we are well convinced of the great benefit we have received, and sincerely wish he may meet with Parliamentary encouragement."

When the restrictions placed upon the importation of live stock into Ireland were removed, the improvement of the Connaught breed of sheep was vigorously conducted, and the Roscommon breeders having taken the lead in improvement, aided by the dry and wholesome nature of their sheep walks, the improved breed became recognised as the Roscommon sheep. It is, however, within the last twenty-five years that a systematic course of improvement has been chiefly carried on, resulting in what is now known as the "improved Roscommon breed."

It is admitted that the first improvement effected in the old Connaught breed of sheep was caused by crossing it with Leicester rams. By this cross not only was the form of the animal improved, but the wool lost much of that coarseness which it originally possessed. The pure Leicester was tried in Connaught, but it did not succeed. It was said to want constitution to suit the wet climate of the west, and this was shown more particularly in the case of lambs of the pure Leicester breed, which were subject to scour, and could not be brought through the winter without considerable loss by death. The cross, however, with the native breed succeeded admirably, and when the effects of that cross were established, the Roscommon breeders evinced great skill and intelligence in so selecting their breeding flocks as to perpetuate certain desirable characteristics in the style of the sheep and the quality of both their flesh and wool. Selection has been, in fact, the chief means latterly employed in producing the "improved Roscommon sheep."

The old Connaught breed of sheep were never fattened until they were three or rather four years old, when they made great weights, but the mutton was coarse. In consequence of the improvement which has been made in the breed, shearling wedders are now often sold fat to the butcher, making from 25lb. to over 30lb. per quarter; but as a general rule the Roscommon graziers hold them over until they are thirty months old, at which age they are generally sold in Ballinasloe fair, at prices varying from three to four guineas each, to Leinster graziers, by whom the sheep are kept until they are about three years old, when they make from 36lb. and upwards per quarter. Draft ewes, fed after being cast for breeding, weigh from 34lb. to 40lb. per quarter, and the quality of the mutton is unexceptionable. It must be understood that the Roscommon sheep are, in general, reared entirely upon grass, with the help of some hay during winter. Turnip-feeding does not, as in Great Britain, form a material point in sheep-farming as conducted in Roscommon, there being only one acre of turnips grown in that county to each 109 acres of area. These sheep, from first to last, are for the most part reared and fattened without seeing a turnip. In all cases where turnip-feeding is pursued, the Roscommon sheep prove that early maturity, along with heavy weights, has become one of their characteristics; so that if turnip growing were extended in the west of Ireland, it is only reasonable to believe that Connaught would produce much larger supplies of sheep than is done at present. With the pressure on the meat market which now exists, this is therefore a point which deserves to be seriously considered.

The wool of the Roscommon breed is a soft, deep-grown, rich wool, and the quality of the fleece is invariably looked upon as a material point in the selection of rams and ewes by the leading breeders. The fleece of grass-fed wedders and hoggetts usually weighs about 10lb., but about 8lb. may be taken as the average; at the same time much heavier weights have been obtained. Thus, the fleece of Capt. J. Blood Smyth's prize ram Prince Arthur, whose portrait is given in the accompanying engraving, weighed 24lb.; and Capt. Smyth's shearling rams generally average about 14lb. of wool. Some of his ewes, shorn while rearing lambs, have clipped 14lb. and even 16lb. of wool.

Notwithstanding the well-known and very evident improvement which has been effected in the Roscommon sheep, and the highly favourable opinion entertained of their merits by the extensive graziers of the midland and southern counties of Ireland, until very recently Roscommon sheep were classed at the shows of the Royal Agricultural Society of Ireland, and also at those of the Royal Dublin Society, in a mixed class of "Long-woolled sheep, other than Leicesters." A class so arranged frequently contained Cotswolds and Lincolns, as well as Roscommon sheep, and the decisions generally rested more on the particular inclinations of the judges, than on the relative merits of the animals. Of late, however, both societies have recognised the Roscommon sheep as a distinct breed, and offer special prizes in the different sections, as in the case of the Leicesters and other established breeds. This recognition of the breed by the Irish Royal Societies has already been productive of great good effect, as it has induced several breeders of Roscommon sheep to bestow greater attention on the improvement of their flocks.

Messrs. Roberts, Taaffe, Cotton, Flanagan, and Flynn may be mentioned as leading breeders of improved Roscommon sheep in the county which gives its name to the breed. Mr. Richard Coffey has also a very superior flock in Westmeath, and has been very successful as an exhibitor; while Capt. J. Blood Smyth, Fedamore, county of Limerick, has done much to establish the character of the improved Roscommon breed. Capt. Smyth has been for several years a leading exhibitor, and it is to his flock we are indebted for our illustration of the breed.

The ram Prince Arthur was the first-prize ram in his class at the Ball's Bridge Show of the Royal Agricultural Society in 1871, and the ewe in the engraving was one of Capt. Smyth's first-prize pen at the same show.

CHAPTER XVIII.

NEGRETTE MERINO SHEEP.

THE accompanying engraving, from a drawing by Mr. Harrison Weir, represents most faithfully three specimens (two rams and a ewe) of the famed fine-wool-producing Negrette merino. Their skin hanging in folds as if too large for their frames, with wool close as it can grow, down to their very hoofs, and their faces well covered, stamp them as animals intended by nature to grow wool.

The merino does not pay for feeding artificially like other breeds; he will produce wool on the worst of grass, and when well fed will do little more. The quality of the wool deteriorates in proportion as he is overfed, and very little is gained in mutton, the nature of the merino being to secrete on the kidneys the surplus fat; so that, although producing very good and juicy meat, the merino will not yield fat mutton, nature having evidently intended him for a wool-producing animal.

The Negrette merino is a native of Spain, and his fine wool was at one time so great a source of revenue to the Spaniards, that no merino was allowed to be exported from Spain without the licence of the king, under the extremest penalties—at one time even that of death. The following account of how they were introduced into this country may be of interest, as also may some particulars of sales of merinos, taken from old agricultural magazines.

King George the Third, as is well known, was an ardent agriculturist, and, being determined to try the celebrated Negrette merinos on his own farm, he in the year 1787 took measures for the collection and importation of a few. It was a kind of smuggling transaction: and as they could not be shipped from any Spanish port without a licence from the King of Spain, they were driven through Portugal, embarked at Lisbon, and landed at Portsmouth; but, being hastily selected from various proprietors, they were not considered good enough to make any experiment with. It was accordingly determined in 1791 to apply direct to the Spanish monarch for permission to select some from the best flocks. This was liberally and promptly granted, and a little flock was draughted of the Negrette breed, the most valuable of all. They were transferred to Kew, and the experiment of the naturalisation of the merino, and the crossing them with British sheep, was commenced. In 1804, thirteen years after the importation, the first sale took place, and what are even now thought fair prices were obtained. One of these Negrette rams sold for forty-two guineas, and two ewes for eleven guineas each, the average of the rams being 19*l*. 14*s*., of the ewes 8*l*. 15*s*. 6*d*. At the second sale in August, 1805, seventeen rams and twenty-one ewes were sold for 1148*l*. 14*s*., being an average of about 30*l*. each. In 1810 thirty-three merino rams sold for 1920*l*. 9*s*., being an average of more than 58*l*. each. One full-mouthed Negrette ram was sold to Col. Searle for 173 guineas, one for 134 guineas to Sir Holme Popham, one for 116 guineas to Mr. Gale, one for 105 guineas to Mr. Sumner, and one for 101 guineas to Mr. Woods; and of the ewes, a full-mouthed Negrette was sold to the last-named gentleman for seventy guineas, two others for sixty-one and sixty-five guineas each, one to Mr. Down for ninety-two guineas, and another to Mr. Willis for sixty guineas; seventy averaging 37*l*. 10*s*. per head.

The first advocates of the merino thus saw their predictions most fully realised, and in the following year, says an old historian, a merino society was formed, with a great number of members of station and influence. Sir Joseph Banks was the president; fifty-four vice-presidents were appointed, and local committees were established in every county of England, Scotland, Wales, and Ireland. From this very period, however, is to be dated the decay of the merino in English public opinion. These sheep could not be adopted generally in this meat-eating country of England, being deficient in the principle of early maturity and general propensity to fatten, which in the colonies is not so much considered as their valuable fine-wool-producing qualities. Possibly, also, mismanagement may have had something to do with this failure, as English flockmasters did not understand the character of these sheep, and by grazing them on wet and rich pastures caused a delicacy and tendency to disease; but they have proved themselves hardy whenever properly treated, and have much improved the fleece of all sheep with which they have been crossed, by giving it increased weight, and adapting it for finer fabrics.

In our colonies the merino cannot easily be over-estimated; they are kept there in flocks of many thousands, and consequently receive but limited attention. By nature they are undoubtedly intended to live a migratory life; they produce more horn on the wall of the hoof than any other breed of sheep, and are able to travel much longer distances in search of food without being footsore. Their short strong legs and deep ribs do them good service; they can live in a dry season on a scanty pasture, and thus are enabled to produce good fleeces where other animals would starve. In short, the great value of the merino consists in their being able to produce the finest fleeces of wool of good weight on the scantiest pasture. The merino, with his very thick fine wool, and covered as he is from the tip of his nose to the hoofs, suffers more from wet than heat. The heads of the males are usually ornamented with a pair of handsome spiral horns; the females are mostly polled.

The remnant of the pure royal flock fell into the hands of the late T. B. Sturgeon, of South Ockendon Hall, Essex, shortly after the death of King George III., and they have since been bred pure by Mr. Sturgeon and his sons, who have spared neither trouble nor expense to maintain their high character. The effect of food, climate, and fresh blood imported from the best flocks to be found, is seen in the increased size of the sheep, their improved form, and fine heavy even fleeces. Messrs. Sturgeon and Sons, of Grays Hall, Grays, Essex, are glad to show these sheep to anyone interested in them; they are within five minutes' walk of the Grays Station on the London and Tilbury line.

The management is not very different from that ordinarily pursued. The ewes are put to the ram when 20 months old, in

September, after having been carefully examined to see that the wool is even and good, and that each animal is in all respects fit for breeding purposes. Care is taken to select a ram with points calculated to correct any defect in the ewes, and only distantly related, so as to avoid too close breeding, which here, as elsewhere, tends to reduce size and weaken the constitution. A studbook is kept, in which the number of each lamb is entered, when born, also its sire and dam. After the ewes are tupped they are all put together, and kept in one flock on grass. About a fortnight before lambing time they are yarded at night, both for protection and to enable the shepherd to attend to them. Lambing usually commences early in February. As soon as they are recovered from lambing, the ewes and lambs are put on turnips, upon which, with hay chaff, and a little bran and cake, they do very well until the end of March, when the turnips are usually finished; they are then put on rye, rye grass, and then on to the permanent pastures. About the end of May, or the beginning of June, the lambs are separated from the ewes, so as to allow the ewes time to recover from the effects of suckling their lambs before they are shorn, which usually takes place in the latter end of June, or the beginning of July. The ewes then live on the grass till tupping time. The ram lambs, when separated from the ewes, are shut up in sheds, and fed on hay chaff with a few oats or cake, so as to have them always prepared by dry feeding for shipment to the colonies, they being only bred for exportation.

The value of the wool from the flock at Grays is quite equal to the best colonial when clean washed. The general practice, however, is not to wash the sheep clean, as it is found preferable to sell the wool in grease, rather than subject the sheep to that process. It is very gratifying to know that the enterprise of our countrymen at Gray's is bearing good fruit, and that the value of Sturgeon merino rams has recently been duly acknowledged in the Cape newspapers, from which we make the following brief extracts. In the *Empire* of February 26, 1886, in a leading article, after proving by the opinion of competent judges that the merino sheep are best adapted to South Africa, the importance of three points is insisted on: (1) wool, large quantity and quality; (2) ability to thrive on poor pastures; (3) adaptability to a hot climate. The Rambouillet has his claims, but is not the sheep for South Africa. The Negrette Merino, or Sturgeon Merino, as he is called in the colony, has long been introduced into the Australian flocks, and is not unknown in the Cape and Natal. One of these rams from the Sturgeon flock was recently purchased by Messrs. Allport and Co., of Beaufort West, which is described as a splendid animal, with a heavy fleece of very fine close wool, is two years and four months old, and his live weight when he left England was 189lb. net. Mr. F. J. Allport, in a letter in the same paper, which bears date August 20, 1886, entitled "Our Wool, and Breed of Merino Sheep," after doing full justice to the advantages which have followed the use of Rambouillet rams in the increase of size and weight of wool, contends they have been used too freely, and the result has been a tendency to coarseness in the wool. After pointing out the advantage that must follow the introduction of first-class Australian rams, such as those bred by Sir S. Wilson, at Erceldoun, but which would cost 50 guineas each delivered in the colony, he says: "Then we have the *English Sturgeon Ram*, bred by Messrs. Sturgeon and Sons, of Grays, Essex, one of the most valuable breeds of long-woolled sheep in the world, and great favourites of mine. They say you may find finer fibre than these, you may possibly find more wool, but on no other sheep will you find so great a quantity and so fine a quality combined. When fat they attain a weight, if first class, of 160 to 170lb. live weight when three years old, are very powerfully built, with good constitutions, short on the leg, and have a very fine elastic staple of wool, long, soft, and strong, and shear a fleece of about 20lb. in the grease of a twelve months' growth. I lost a splendid imported ram of this breed, which died suddenly on the 4th instant from internal inflammation. He arrived in the colony too late to be of any service last season. His live weight was 174lb., and his skin when dried, after his death, weighed 26½lb.; the wool being a fourteen months' growth, three inches long, with a stretch up to 4½ inches of beautiful fine wool, worth 1s. per lb. in England. It was a great loss to myself, as well as to the district of Beaufort, but I have such confidence in the purity and value of this blood that I at once ordered another of the same high standard, costing 60 guineas in England, which

I hope will arrive in the colony about the 1st October next, and, if I am more fortunate, he will prove of great value in improving the quality of our wool in greater fineness, length and strength, so as to enable us to compete successfully with the best wools offered in the London market."

CHAPTER XIX.

EXMOOR SHEEP.

IN the far corner of West Somerset, sea-girt on the one side at Minehead and Porlock, and bordering North Devon on the other, is a wild tract of mountainous country, the greater portion of which still consists of open, uncultivated commons, bearing little else but heather, the home of the blackcock and wild red deer. But a race of mountain sheep has always been native to the region as far back as either history or tradition will carry us, no one being able to tell whence it came. Probably, however, the Exmoor, as the breed is most popularly called, had origin from the same ancient stock as the Portland, and even the Dorset, for the sheep that once existed on the Mendip Hills seemed a connecting link between them. Youatt, in comparing the Mendip sheep with the Dorset, states that the former would lamb quite as early, and bear two crops a year if required, but that they had smaller horns, white countenances, and were more diminutive in size; and he adds, "these sheep appear to be an intermediate race between the Exmoor and the Dorset." Those desirous of tracing the family resemblance further may be informed that Exmoor ewes are frequently applied to the production of fat lambs; while, if fed well, they are bountiful in doublets, and sometimes triplets, which they are usually allowed to rear, being the very best of nurses.

As to points, a leading breeder, in communication to ourselves, says: "I take it the chief merit of Exmoors lies in their round barrels and good constitutions, with fine-flavoured mutton and fair quantity of wool." The late H. H. Dixon enumerated them as follows: "A very strong constitution, which will bear being buried in a snowdrift for several days; a fine, curly horn; a broad, square loin; round ribs; a drum-like and not a square carcase, on short legs; and close-set fleece with wool well up to the cheeks." To the above description it may be added that they possess white faces, legs, and fleeces, and have been termed "the little white ivories;" but, although they stand tolerably well on their fore legs, there is a failing point behind the shoulder, and none, even of the improved sheep we have yet seen, girth particularly well there. They are also rather indifferent about the neck.

The elevated mountain ranges, which are from 1000 to 1800 feet above the sea level, are intersected by narrow, circuituous valleys, where the farms and hamlets appear. These valleys are tolerably fertile in their nature, and cultivation often extends far up the hill-sides, by the agency of the turn-wrist plough. Catch-water meadows are also formed wherever a trickling stream can be utilised, and luxuriant verdure always follows in the train of irrigation here. These facts should be borne in mind in perusing the appended accounts of sheep management, which varies materially on different farms, being dependent on the extent to which the flockmaster is provided with fertile inclosures and water meadows.

Arthur Young, in "Annals of Agriculture" for 1794, makes a passing allusion to Exmoor sheep, which had been brought under his notice in a visit to Monksilver. He states that they were at that period sold at South Molton market as hoggets, at from 9s. to 16s. each, and, after being kept on the hills two or three years for the annual profit of their fleeces, they were fattened on turnips, and sold without their fleeces, the average weight of which was 3lb. to 4lb., and that of their carcases 16lb. to 18lb. per quarter. The weight of the fleece is about one pound heavier now, with which exception this description will apply to the present period. Billingsley, in his Survey of Somerset furnished to the Board of Agriculture, gives a very similar account, adding thereto the following: "Though these sheep in appearance are vastly inferior to those before described (the old Bampton breed), being in their youth subject to a precarious subsistence on the forests and hills, it is the opinion of many sensible farmers that they are altogether as profitable stock."

The prize report on the farming of Somerset in the "Journal of the Royal Agricultural Society" for 1850, part 2, contains the subjoined description of the management of Exmoor flocks at that period, of additional interest from being written by the present Sir T. D. Acland, Bart., M.P., one of the principal breeders:

"The hill-country farmer generally keeps a breeding flock of horned ewes and a flock of wethers, which run on the hill summer and winter. The number of his ewes will be limited by the extent of his water meadows, on which he relies in great measure for the keep of the couples after the lambs are dropped. The number of hill wethers depends on the extent of the common right attached to the farm. About the 20th of June all the sheep are gathered for sorting and shearing. The mouths of the sheep are examined, and those whose teeth are broken are drafted and kept back from the hill to be sold or fatted off. The ewe hoggets replace the draft ewes, and the wether hogs of the former season are shorn with the hill wethers, and turned off to the hill after being signed with some large mark which can be known at a distance. They cost nothing but the trouble of an occasional gathering until next year, and the only profit they yield is about 5lb. of wool. In their fourth or fifth year they may be brought on to grass. They are also used as labourers on the farm, to eat the grass down close in the fall of the year, and are sometimes marched in close phalanx up and down a ploughed field to tread in the wheat. The ordinary sheep of the country, when fat, do not weigh above 11lb. or 12lb. a quarter. Where pains have been taken to improve a flock, they may reach, on the average, 16lb. or 18lb. a quarter, and some are brought up to 24lb. a quarter when fed on the Bridgwater marshes. There is also great difference in the quality of the wool of a common and of a well-bred sheep. It is the practice of farmers who have good land as well as common, to put their draft ewes with a small-headed and high-proof Leicester ram, to sell the lambs fat in May, and the ewes as soon as they get fat. There are great objections to horned sheep. It is almost impossible to prevent them from being infected with the scab while they are on the open hill; they also acquire such restless habits that they are always breaking the fences when brought into the inclosed ground. In fatting them much judgment and practical knowledge is required, for they do not get on well in hot weather; and it frequently happens that when they are first put on turnips they lose ground, or 'pitch,' as it is called, for two months in the autumn, and are slow in regaining it afterwards. For these and other reasons, farmers who occupy good land in the vale with their hill farms, are getting tired of their horned sheep, and use their hill farms only as summering ground for knot sheep and bullocks."

The period that has elapsed since this statement was penned has been quite a revolutionary one in effects on sheep management. Advanced prices of meat have held out premiums for quickening mutton production, and even mountain breeds have been improved in their feeding capabilities to bring them to earlier maturity. Great pains have been taken to effect this object with some of the largest of the Exmoor flocks, and a large amount of success has attended such efforts. Mr. James Quartly, of Mollond, Mr. R. Stranger, of North Molton, Sir T. D. Acland, Bart., M.P., Lord Poltimore, Messrs. Tapp, of Twitching, and Mr. Robert Paramour were the principal breeders who were instrumental in the work of improvement; some of Sir T. D. Acland's tenants on Winsford-hill have good flocks likewise, viz., Mr. Harding, Mr. Lovelace, Mr. Darby, &c.; and Mr. Rock's sheep are also greatly improved at High Bray. As may be supposed, these highly-cultivated flocks are treated very differently from some of the common kinds of Exmoor; for instance, the wethers are not allowed to run so long on the hills. "Since mutton has been such a high price," says Mr. Birmingham, Sir T. D. Acland's farm manager, "the wether sheep have not been allowed to run so long as three or four years. If at a moderate price, the wethers pay well for running because of their wool."

Mr. R. Stranger, it may be presumed, follows high-breeding with his young sheep from the onset, or he could not make them out so quickly. In a courteous reply to our request for information on this point he says: "I sell my wethers when about 18 months old, and they weigh from 15lb. to 20lb. per quarter. A few, however, are retained for Christmas markets. My pen of three which took the first prize and silver medal at the Smithfield Club Show last year were 3 years 9 months old, and their living weight 6cwt. 0qrs. 9lb. The other three pens in the same class weighed respectively 4cwt. 1qr. 25lb., 4cwt. 1qr. 5lb., and 4cwt. 0qr. 24lb.

Wool has advanced in price, however, no less than mutton, which may be taken as the true reason why the practice of running Exmoor wethers three or four years on the hill commons for the value of their fleeces is still adopted very extensively. The education they receive in this semi-wild state over so long a period adds prejudicially to their feeding profitably subsequently; for an intelligent farm bailiff of the district has been known to observe "that you may as well try to fatten a hurdle as one of the old running wethers of the common Exmoor breed." But the mutton is of first-class quality after the work is effected, and the annual yield of wool at prices which prevailed a few years since was a consideration, being held to be the easiest and cheapest mode of converting the hill commons to profitable account.

Mr. Birmingham, who has been previously quoted, answered some of our queries on this subject as follows: "In many places the Exmoor sheep which run on the commons are inferior and difficult to fatten. In such cases the wool is the first consideration. They are hardy, and will stand any amount of bad weather, and last to a great age. In the other class of sheep that have been improved of late years, flesh has been studied as well as wool. On the Winsford-hills, and at Anstey, Withpool, and Molland, may be seen fine specimens of improved Exmoor sheep, that are killed when $2\frac{1}{2}$ years of age; their mutton is splendid, and always commands a high price. The lambs are taken care of until hoggets, when they are turned to common with the old sheep until taken in to be fed. The ewes are not fattened until broken-mouthed. The common kind of Exmoors fatten to about 15lb. per quarter, but the highly-bred to 19lb. per quarter, and some wether sheep for Christmas I have known to reach 28lb. per quarter."

Mr. J. M. King, jun., of Steart, Dunster, has kindly afforded some valuable information as to the present management of Exmoor flocks. He says: "The usual period of lambing is from March to the middle of April, and the weaning takes place about Midsummer. They are turned on to the hills early in the spring, and many of them remain there the whole year round. the only time they are got in being for shearing and weaning, the latter of which is effected by keeping the lambs for some weeks in the inclosures. The Exmoor is a larger, higher quality, and in every respect better animal than the Welsh, and has been much improved in the last fifteen years. Stores command very high prices, and have advanced from 25 to 30 per cent. in value during the last two years. Winsford Fair, held about the middle of August, would be about the likeliest place to obtain them. The demand for the rams is limited, as the breed can scarcely be considered one for general use; but still much care and attention have of late years been bestowed upon it, and good rams often fetch from ten to fifteen guineas."

Exmoor ewes will lamb earlier than the time mentioned by Mr. King if required so to do. Mr. Stranger says: "The lambing season begins in January, and continues through February and March; early lambs are generally the best for rearing." Some other points of management are particularised in Dixon's "Prize Essay on Mountain Sheep," published in 1866. He says: "The original Exmoors milk better than the 'improved,' and old ewes especially. There are instances of ewes rearing three lambs well after the first fortnight. The ewes are always brought down to the lower ground to lamb, and get a few turnips and oats; and then come in again from the hills in November to the poorest inclosed lands. They are put to the tup at two years old, and are generally drafted after three crops of lambs, though some old favourites go on far longer." There is a demand for the draft ewes of the Exmoor just as for those of the Dorset breed, to bear lambs for fattening in other parts of the country; to answer which purpose more effectually they are tupped with Leicester rams. The same cross is also freely resorted to for rearing purposes, as, whenever the land is a little better, the west-country cross-breds are kept, generally obtained from Exmoor ewes and Leicester rams; and a great many small farmers, even of the hill districts, not having a large run of common, find it more profitable to keep this kind instead of the pure Exmoor. Dixon, alluding to this matter, observes: "They (the Exmoors) hold their own on the purely hill districts; but since the Commons Inclosure Act many farmers have crossed them with the Leicester. A larger sheep has been secured, but at the expense of stamina and numbers. These 'knots,' as they are styled, are generally without horns. Ewes of the sort have been partially adopted by some of the Cornwall farmers; and Mr. Anstey uses the tups as well."

The Exmoor breed has the reputation of being better adapted to the requirements of its native district than any other mountain variety. Blackfaces, Cheviots, and Welsh sheep have all in turn been tried, it is said, and found wanting, and the experimenters, after giving them a fair trial, have been glad to return to the old sort. The only exception appears to be Mr. F. A. Knight, the owner of Exmoor Forest, who has adopted Cheviots, and is believed to be well satisfied that they are more profitable in both wool and flesh than the native breed. But around Simondsbath reclamation from the heather has been effected to a far greater extent than at any other spot on the same high level in the western hill country, turnip and rape crops being in consequence more plentiful than elsewhere.

We may repeat the statement made by H. H. Dixon in 1866 respecting well-bred Exmoors, that, "under very high pressure they have done wonders of late years." The improvement effected, moreover, has been brought about by a careful and judicious selection of rams, comprehensive weeding of ewes, and skilful matchings; or, in other words, by breeding on the in-and-in system. Whatever may be thought of the Exmoor, sufficient merit is exemplified under cultivation to invest the sheep with high claims for perpetuity of existence. The improvement of the breed has recently progressed quite as fast as the reclamation of the hill commons; and if sufficient quality and capability to put on flesh rapidly can be imparted to this hardy and prolific stock, for the future requirements of agriculturists on this elevated tract of country, we may hope that it will be one of the few mountain species that the hand of civilisation will spare.

CHAPTER XX.

THE BLACK-FACED OR SCOTCH MOUNTAIN SHEEP.

THERE can be no doubt this is the oldest variety of the fleecy tribe extant in Scotland, though its origin is somewhat obscure. For several centuries prior to the eighteenth, the character of the Scotch sheep were more uniform than afterwards. On almost every holding—and there were many in those days—a small flock was kept, herded and folded near the homestead, the mountains being then overrun by wolves and foxes. These ancient sheep were designated the "dun faces," from the brown or tawny colour of the hair on their face and legs. The fore-quarters were singularly light, the neck long and often low set, the tail very long, the face "dossy," wool short but fine, horns by no means common, feeding and developing properties deficient, symmetry wanting, and maximum weight far below that of the Scotch sheep of after years. From these animals which are not yet extinct in Scotland, we are inclined to think that the black-faced sheep mainly sprung. True, some tradition says that the black-faced sheep were introduced into Scotland to the Ettrick Forest from a foreign land, several centuries ago, by one of the Scottish kings; the inference being that from this royal flock the breed spread gradually over the country, eclipsing the native type. While this hypothesis gains credence in certain quarters, others assert—and we confess to have a slight leaning with them—that the Scotch mountain or black-faced sheep of the present day owe their origin to the ancient dun faces, and that they have been improved by breeding from carefully and judiciously selected specimens to an extent which renders the features of the parents scarcely recognisable in the progeny. If the early history of the black-faces has a foreign association, it must be admitted they have crossed very successfully with the native sheep. It is quite possible that the sheep of the royal farm were not of foreign extraction at all, but only greatly improved by cultivation and attention. In the present century we have seen improved rams of the pure black-faced breed harmonise astonishingly with ewes of the old variety. Whether foreign blood was introduced or not, the South-west of Scotland unquestionably has the credit of raising the black-faces to the degree of perfection they have for a considerable number of years displayed both north and south.

In the course of the eighteenth century agriculture in Scotland rallied from the grasp of the darker ages. The mountain lands became safer for sheep, and in not a few instances these animals exchanged attitudes with the cattle and horses, the fleecy tribes generally taking the higher ground. The Weirs of Priesthill, Muirkirk, Lanarkshire; the Gillespies of Douglas Water, and others, well nigh a century ago, did much to make the black-faces what they have long been, and still are. Lanarkshire is regarded as the headquarters of the black-faces, not so much because that county produces the largest number of the finer specimens of the breed, as from the fact that in it were effected the earlier and more valuable improvements. The principal market for black-faced stock is held at Lanark; but Ayrshire, Dumfriesshire, Peeblesshire, and Roxburghshire produce large numbers of high-bred animals, the first-named two counties in particular. By tups from the best southern flocks the northern mongrel sort of stocks have been immensely improved. Those in the northern counties of Scotland who succeeded in growing the better class of animals brought ewes as well as tups from the south. Slowly during the present century the original native breed in the Highlands was supplanted by the better class of stock from the south; and, excepting Orkney and Shetland, there are not now many specimens entire of the time-honoured dun-faces in Scotland.

It is interesting and instructive to note the effects of climate and geological formation on the distribution of our domesticated animals; the more closely we follow the teaching of nature the more successful will be the result. The Scotch black-face is more widely distributed, and occupies a much greater area than any other race of sheep in these islands. They can be traced in unbroken succession from the Hebrides on the north far into the interior of the north midland counties of England—the mountain chain geologically known as the carboniferous limestone and coal measures, which take their rise in North Staffordshire on the south, passing through Derbyshire and Yorkshire to the borders of Scotland on the north, forming the great watershed of an extensive district. The range of hills is separated by the vales of Kendal and Edon from the yet higher Silurian mountains of Cumberland and Westmoreland; passing to Scotland we find the Silurian formation stretching from south-west to north-east of the island, forming what are known as the Carrick, Moorfoot, Pentland, and Lammermoor Hills. The same geological formation extends through Cantyre, the Grampians, and part of the county of Sutherland. The whole of this extensive range is tenanted principally by the black-faces, whilst at a lower elevation they share with the muirfowl and plover the wild heaths and rushy dells of Yorkshire, Durham, and Northumberland in England, and the peaty and barren moors of Galloway and Ayrshire in Scotland.

The northern sheep are not nearly so well bred, nor are the pastures quite so good as those in the south of Scotland, yet the infusion of new blood has been so effectual that the features of the "duns" are almost obliterated, with the exception already made. The finer breeding flocks have long occupied the higher grounds in Lanarkshire, Ayrshire, Dumfriesshire, and Stirlingshire; but the number of black-faced or mountain sheep is not so great as it was about the beginning of the present century. The Cheviot sheep had encroached extensively on the black-faced provinces throughout Scotland, notably along the west coast, within the last fifty years. Some people say that the Cheviot is as hardy as the black-faced, and that it will arrive earlier at maturity, and carry a greater quantity of mutton, possessed of quite as fine a flavour. The writer's experience leads him to believe that the white-faces are not nearly so hardy as the black, but the former will take on earlier a little more flesh, though the flavour of the mutton is not so fine as a rule. The result of the very bad winters which have marked the decade from 1866 to 1876, has been in many instances to reinstate the black-faces in the homes from whence they had for a time been expelled to the Cheviots.

Not the least formidable obstacles in the way of improvement among the mountain sheep were the promiscuous pasturing of the smaller flocks on common runs, the too frequent neglect of proper castration, and of preserving the best specimens only to breed from on the male side. "Commonties" gradually disappeared, more careful treatment was extended to the pastures as well as to the sheep, and for the last quarter of a century no department of Scotch agriculture has been better attended to than the selection of first-class tups by breeders from the best stocks. Even the owner of a mixed stock of two or three score has for a considerable number of years endeavoured, with success generally, to secure a well-bred tup. The Scotch mountain sheep are the most migratory of the race. After grazing in comparative solitude in the corries for some time they become wild, especially those specimens displaying closer affinity to the ancient native breed. Their pedestrian powers are great; in fact, we have frequently seen wethers so wi and agile that a common shepherd's dog could not get abreast of them in a race—could not even give them a "turn."

The black-faces are very lively, instinctive animals, and become much attached to certain parts of the pasture. The breeding stocks, or individual animals composing them, have choice portions of the runs or glens, where some of them graze night and day during summer, and others, after passing the day in the lower grounds, hie towards their favourite quarters as evening approaches. Partly owing to the often scanty nature of their pastures, the mountain sheep occupy more time feeding than almost any other animal. It is astonishing how punctual to time these sheep will start from the lodging quarters for the lower pastures, and *vice versâ*, day after day. Shepherds have observed that any deviation from the ordinary hour of departure for or from the sleeping ground is ominous of the character of the weather for the next few days. As a rule, if the animals are late in "drawing towards" the higher ground or unusually early to leave it, bad weather may be expected (unless, of course, there is a marked and increasing scarcity of food); while on the other hand, if the sheep leave the better pastures early and return late, beautiful weather is often betokened, and not unfrequently follows.

After the system of wintering the breeding flocks in the low country was fairly inaugurated, several of the older animals set out for the wintering ground as soon as they had a scent of snow in the summering regions: and in like manner, if fine weather set in before the removal from the lowlands (the 1st of April), shepherds often found it difficult to prevent some of the animals from straying in the direction of the summer grazings. A few days before lambing many of the ewes wander to distant spots of the walks, and, unless carefully watched at this season, considerable loss is occasioned by lambing unobserved. What is very remarkable, ewes were known to have gone year after year to the same locality, several miles distant from the rest of the flock, to lamb. So uniformly was this system practised by the older ewes in some of the less strictly watched flocks twenty or thirty years ago in the hilly district, that when a certain ewe happened to be missing in April or May, search was at once made in the particular part of the hill where she was known to have lambed before, and in almost every case with success. For several years the ewes in the larger flocks have been carefully herded on the lower parts of the walks until the lambing season is over.

We have known of lambs only five or six months old leaving flocks or pastures which they had lately joined, and going back to their native runs, a distance of several miles; and in the case of ewes newly removed from well-known to strange ground, we have experienced their clandestine return to the former, a distance of fifty or sixty miles, in a marvellously short time, with one of Scotland's noblest rivers to swim. When these animals are fairly bent on returning to their favourite haunt, they travel night and day, and overcome barriers which in any other circumstances would be insurmountable.

No other variety of sheep shift so much for the means of subsistence. It is consistent with our experience that the higher-bred animals will do less to provide their own dietary than those flocks considerably improved from the ancient type, but by no means perfect in breeding, which abound in the uplands of Aberdeen, Kincardine, Perth, Banff, and Inverness. During heavy snowstorms it is wonderful how many of the black-faced flocks survive. Neither in respect of quantity nor quality of provender are they difficult to satisfy, particularly in winter. If they have some shelter they will scrape among deep snow for heath, fogage, or whins, and tide over the storm with a very scanty supply. The proportion of the breeding sheep indulged with turnips in the course of winter is comparatively small. About thirty years ago, when more of the mountain sheep were wintered on the summer grazings in the uplands than was afterwards deemed advisable, it was not uncommon—and heavy snowstorms were more prevalent then than now—to have large numbers buried under immense accumulations of snow. It is surprising the length of time the sheep have been known to survive under such circumstances. We have seen several animals of the black-faced breed not very far removed from the ancient "dun-faces," taken out from below a huge wreath of snow alive, after being buried there five weeks. Some of the animals died immediately after exposure, while others, though weak for some time, lived several years. All they had to subsist on during those five weeks was what they could get at lying; they were of course unable to rise under such a pressure. Heather roots, grass, and even the soil were eaten to a mournful extent so far as the creature's head could reach; in fact, nothing grew for many years on those bitten specks. In the midst of a heavy snowstorm great difficulty is often encountered in removing the sheep from one place to another. Unless one of their number lead the way it is invariably hopeless to attempt driving a few abreast. If a taste is got of turnips or fine grass in the neighbourhood of the pastures, close attention is required to prevent future inroads on these crops sometimes by night and often by day. On all hill farms to which uninclosed grazings are attached, the sheep stock are by stringent clauses in the agreement passed on by valuation to the incomer. The importance of the animals being "hefted," *i.e.*, accustomed to their grazing grounds, cannot be overrated—and in some districts part of the flock belongs to the landlord, and is rented with the farms.

Rams are required at the rate of one to forty ewes. During the rutting season the tups often fight terrifically, not unfrequently killing each other; and that their heads can stand such fearful crashing is a marvel. A black-faced ram is extremely powerful, especially with his head. Going backwards for twenty or thirty yards each, a pair of Highland rams then start, and meet each other with a fearful violence, their heavy arched horns cracking loudly, and their hind quarters rising considerably in response to the collision of heads. The lambing season is usually from the middle of April till a similar period in May. Twin lambs are the exception. The lamb is remarkable for the amount of cold and hunger, especially the former, it will endure, even in the first few days of its existence, and also for the early use of its feet. Two or three minutes after birth it will be on its feet, though rather tremulously, and a few minutes more it walks easily. The defensive-like attitude with which it weathers the biting blasts of the spring snow and other storms affords early evidence of the natural hardiness of the breed. The ewe's care over her young is great. With moderate treatment as to food before lambing the ewe is invariably possessed of a large quantity of milk, and it is only in cases of desperation as to nourishment, as a rule, that they will leave their lambs. Such is the attachment generally of a Highland ewe to her offspring, that we have seen many of them remain within a few yards of their dead lambs for several days, even though the latter had not survived birth, and though the former were on the brink of starvation.

Attended, as the conversion of smaller holdings into large ones was, by much heart-burning to the evicted, and pointed to as it still is by not a few as one of the worst steps ever taken for the country, it must be admitted that it conduced to the introduction of a much improved class of sheep. It was the larger flock-owners, as a rule, who were the first to bestir themselves in producing finer stock. The mountain sheep farming may be said to have assumed three different characters in the course of the last thirty years, viz., the purely breeding stocks in the southern counties of Scotland, the wether stocks in the northern counties, and the mixed ewe and wether flocks in the central and northern counties.

Among the more celebrated breeding flocks are those of Mr. Archibald, Overshiels, Stow; Mr. Aitken, Listonshiels, Mid-Lothian; Mr. Malcolm, of Pottalloch, M.P., Argyllshire; Mr. Moffat, Gateside, Dumfriesshire; Mr. Hope, Cadenhead; Mr. Craig, Polquheys, Ayrshire; Mr. Craig, Craigdarroch, Ayrshire; Mr. Greenshiels, Westown; Mr. Paterson, Glenlaggart; Mr. Paterson, Birthwood; Mr. Tweedie, Castle Crawford; Mr. Denholm, Betlaw; Mr. Wilson, Kennix; and the Messrs. Watson, Coulterwater—in Lanarkshire; Mr. McDonald, Strathmashie, Inverness-shire; Mr. Pagan, Invergeldie, Perthshire; Mr. Foyer, Knowhead, Stirlingshire; Mr. Stewart Ballid, Inverness-shire; Mr. Kennedy, Sherramore, Inverness-shire; Mr. Grant, Inverlaidnau, Inverness-shire; Mr. Robertson, Achilty, Ross-shire; Mr. Willison, Parisholm, Biggar; Mr. Murray, Eastside, Pennicuik; Mr. Welsh, Erickstane, Moffat.

In the case of the high-class breeding stocks in the southern counties the ewe hogs are often wintered separately from the old sheep on the lower grounds, sometimes getting turnips. Meadow hay is an important element in the dietary of the black-faced sheep all over the country in a heavy, protracted snowstorm. To some hill farms are attached a portion of the low-lying grass-land called a hog fence, where the sheep of the first year are wintered; but, as this is the exception rather than the rule, the hogs are more generally sent to the arable and dairy farms of the lowlands, where they have the run of the seed-layers, stubbles, and old pastures. They generally arrive about the end of October, and remain until about the 5th of April. The price per head is from 5s. to 6s. for their six months' keep. They return to their native pastures strong and in good condition; and to this system of wintering may be traced much of the improvement which has been effected in hill stock within the last twenty years.

The principal breeders in the south invariably sell the wether lambs in the Lanark and Sanquhar markets in the autumn months, and retain through the winter only the purely breeding animals. Those who confine themselves to wether stocks in Perthshire, Aberdeenshire, and Inverness-shire, and the jobbers, buy the wether lambs in these markets at very high prices, ranging from 12s. to 24s. a head, according to the quality of the lambs and the nature of the sheep market. After about three years' keep on the non-breeding farms these animals turn to the southern markets in the capacity of three-year-old wethers, realising from 35s. to 40s. each. By this time some of the wethers are ready for the butchers, and the majority about half fat—in which latter case they are bought by fleshers and dealers, and put on turnips in England and elsewhere a few months. It should be mentioned that large numbers of black-faces occupy the higher grazings in the north of England. As a rule, in black-faced stocks, the ewes or gimmers do not have lambs until two years old; earlier breeding, being detrimental to their development, is as far as possible avoided. The ewes are drafted off at the age of five or six years, and the blanks in the breeding ranks filled by the gimmers. The crook or cull ewes are bought by English farmers and others, mostly for the purpose of crossing for one cross only with white-faced rams, and are afterwards fed for the slaughter-house.

Several of the wether hirsels in Perthshire, Aberdeenshire, and Inverness-shire number from 6000 to 10,000 animals. The hogs are sent to the coast side for the winter at an expense of from 3s. 6d. to 5s. a head, and in most cases the two-year-old and three-year-old wethers are wintered on the lower portions of the summer grazings. Smearing is extensively resorted to in the Highlands, and contributes much to the animals' comfort, especially where shelter is deficient.

Though most of the breeders named have reared longer than Mr. Archibald, Overshiels, has done, we believe that his stock is probably the finest extant, and the Highland Society's prize list for many years corroborated the opinion. Assuming that Mr. Archibald's stock is thoroughly representative of the higher class animals at the present day, let us glance briefly at the treatment received by those animals we have so frequently had occasion to admire. In the early days of the flock rams were introduced from most of the best-bred stocks in Scotland, but latterly the changing of blood has been chiefly from the different strains in Mr. Archibald's own possession. In addition to Overshiels, he rents the hill farm of Midcrosswood, on the west end of the Pentlands, and an arable farm at Duddingston, near Edinburgh. The young stock are lambed on the two hill farms, and kept there with their dams until weaning time, which is early in August. The tup lambs are then taken to the arable farm at Duddingston, and put upon good sound grass, where they remain till about Martinmas. Thereafter turnips and hay constitute the dietary until March, when those intended for the show yard are clipped and fed on oilcake, beans, and grass till the date of competition. The tups intended for sale get similar treatment, excepting the cake and beans. The ewes and gimmers for exhibition are fed on grass and turnips, and the bulk of the breeding stock grazed on good pastures during summer, and wintered on low grounds.

The improvement in the quality of the wool has by no means been commensurate with the progress in the breeding and the enlargement of carcass. In fact, one of the aims of modern breeders has been to increase the length and weight of the fleece, comparatively regardless of its quality. On an average the weight of the black-faced fleece would be about 3lb. As wool growers these animals take a minor place compared with their rank as mutton producers. The price per stone is not much more than half that realised for the finer sorts of wool from white-faced sheep. On the other hand, the mutton of black-faces commands deservedly the highest market rates. Some fanciers of the breed prefer a speckled face, others with white predominating, and not a few with black the ruling tinge. The more richly speckled face is supposed by some to denote a delicacy of constitution. Be this as it may, we confess an admiration of the face with black the strongest element. The horn should be hard and free from a bloody aspect, otherwise the animals may be regarded as soft and unhealthy. Neither in the case of ram nor ewe should the horn turn close to the head. We have observed, as a rule, that those beasts whose horns threatened to pinch their heads, even though cutting relieved the animal of any pain therefrom, were less healthy and throve less satisfactorily than those whose horns sprung wider.

A model specimen of the black-faced breed should have long arched hard horns turned quite clear of the side of the head, bare face and legs, an entire absence of any doss on the face, or of a brownness on the legs. Either black or white should be the ruling hue in the face, not an equal portion of both. A Roman nose supported by well-filled profiles is a desirable feature. The neck should not be very long nor low. The shoulder being high, the neck should be well filled. The back not very long but well up, the ribs fairly sprung, the quarters deep, and the mutton full below. A great breadth of back, and monster arching of rib are evidently not characteristic of the breed, but they stand well for depth of frame and squareness of quarter. A well-woolled, highly-bred tup walks out very gaily, indicating only two or three inches between the flow of the wool and the ground. He should stand erect on his legs, with well set up quarters and considerable breadth of brisket. About a good black-faced ram there is a great deal of symmetry and not a little style. Tups whose horns rise a good deal from the head at the root, before turning cannot be too carefully avoided, for they invariably inflict a heavy loss among the ewes in the lambing season by the size of the lambs' heads. Black-faced sheep are liable to "Braxy," and "Sturdy." On most hill farms a few acres of roots might be grown, which, with a liberal supply of hay, would be invaluable during protracted storms.

CHAPTER XXI.

CHEVIOT SHEEP.

By JOHN USHER, Stodrig, Kelso.

THE CHEVIOTS—a range of hills in the border counties of England and Scotland—are believed to have given to the early breed of Cheviot sheep " a local habitation and a name." They seem to have been a native breed, although a legend still gains credence, especially among shepherds, that the first of them were imported into the country by the Spanish Armada, having swum to land from some of the shipwrecked vessels of that ill-fated expedition that were drifted on the Western Isles. They are generally described as small sheep, very light in bone and wool, with brownish heads and legs, and hardy constitution; their scraggy frames bearing very little resemblance to the well-proportioned Cheviots of the present day. Nevertheless, from their adaptation to the soil and climate, they appear to have spread over a great part of the elevated lands in the south of Scotland long before an attempt was made to improve them.

The earliest recorded attempt at improvement was about a hundred years ago, and was eminently successful. The merit of this is universally accorded to Mr. Robson, of Belford, although Cheviot breeders of the present day differ materially regarding the cross he made use of. We have it from Mr. Robson Scott—a grandson of Mr. Robson—that he travelled over the greater part of England for the purpose of seeing various breeds of sheep in different districts, with the view of selecting rams to cross his flock of Cheviots. The sheep he considered most suitable were of a breed then existing in Lincolnshire, of which he purchased several rams to put to selected ewes The cross answered admirably, greatly improving the flock in every respect, without materially lessening its hardy character. Mr. Robson then occupied several high and stormy farms on the Border, and the crossed breed throve well upon them. Twenty years afterwards he made a second visit to Lincolnshire to obtain another infusion of the same blood, but found the breed had become so much larger and less hardy that he declined to venture on them. The theory of Mr. Aitchison, of Lynhope, a high authority in Cheviots, as well as other eminent breeders, is that the breed Mr. Robson imported were Bakewell's Leicesters, with which he crossed a few select Cheviot ewes, and that the offspring of this cross were sent to the hills to cover his extensive flocks. The great resemblance between the two breeds raises a strong presumption in favour of this hypothesis; but, on the other hand, the tenderness of the Leicesters makes it very improbable that such a cross could stand the winters of so stormy a climate. We have besides, in later times, been cognisant of instances where a slight dash of the Leicester blood was introduced, and proved detrimental to the hardihood of the breed, and experimenters were generally fain to retrace their steps. Of the two assertions, therefore, we incline to that of Mr. Robson Scott, more especially as it is not merely derived from tradition, but, as he solemnly affirms, from an oral statement he had from his grandfather. Under any circumstances, Mr. Robson stands confessed the great improver of the breed, although, like Bakewell in Leicesters, the means he used are involved in some obscurity. This early cross gave a correctness of form and symmetry that has never yet been surpassed; greater bone has no doubt been introduced in the present day, but, in the opinion of many Cheviot breeders, to an unprofitable extent, as greater bone often implies reduced numbers. During the years that have intervened since the above was written, we have had additional light on this disputed point. A grandson of Mr. Robson's shepherd, when the said cross was introduced, assured me on his grandfather's authority, that the sheep were really purchased from Bakewell. We have now come to the conclusion that the sheep were Bakewell's Leicesters, when in a state of transition, and before coming to the perfect type which they subsequently attained. This is borne out by the fact that twenty years afterwards, Mr. Robson found them too tender for his purpose—perhaps they were then also too expensive for his purse.

Mr. Robson's flock proved the nucleus from which Cheviot breeders drew their supply of rams for many years. His mode of selling is said to have been somewhat unique. A ticket was attached to each sheep with the price put on him, so that customers could choose according to their taste and means. The impetus given to the breeding of Cheviots was immense; they rapidly found their way into other districts of Scotland and the north of England, supplanting the black-faced breed, which, like the aborigines in India and America, may be said to retire before the advancing wave of civilisation. Let it not be supposed, however, that we disparage the black-faced breed of sheep. For hardihood and beauty they are unsurpassed, and still yield a profitable return in regions where Cheviots could not live. Our earliest associations in sheep farming are connected with them; and we well remember a severe snowstorm in Lammermoor, late in April of 1827, when the Cheviot ewes, losing the instinct of maternal affection, left their newly-dropped lambs to perish in scores, while the black-faced stuck closely to theirs, and the loss in them was trifling. Our memory still clings to their black and mottled faces, bright eyes, and beautifully arched horns, with all the freshness of a first love.

Early in the present century the Cheviot sheep were largely introduced into the northern counties of Scotland, chiefly by farmers of large capital on the Borders. Numbers of small crofters were turned out of their holdings, which were changed into extensive sheep walks. There can be no doubt that this movement, although unpopular at the time, was the means of increasing production, and proved in every case of judicious management a most profitable investment.

In later times the condition of Cheviot flocks has been greatly ameliorated by draining, shelter, providing a plentiful supply of food for use in stormy weather, and other modern improvements. Mr. Aitchison, of Lynhope, may be said to have been the pioneer both in the advocacy and practice of the system of cutting a considerable quantity of hay, not only on the open grounds, wherever the deepness of the soil afforded an extra covering, but by having several inclosures on each farm where hay could be produced sufficient for its requirements, thus making them self-sustaining. These inclosures are also useful as a run for the weaker ewes and lambs, and afford an early bite, so essential to ewes in the lambing season. To use Mr. Aitchison's own forcible language, "Hay is the sheet anchor of the stock farmer." We doubt not some of our readers will recognise in Mr. Aitchison a man not only intimately associated with the improvement of Cheviot stock, but of agriculture in general, and recall with a thrill of pleasure his deep-toned voice, clear enunciation, and fervid eloquence in returning thanks at the banquets of the Highland Society's of Scotland for the toast of "The Tenantry," or the halo of romance he threw over his subject when he proposed "The Peasantry of Scotland."

The practical management of a Cheviot flock is, on the whole, exceedingly simple. Generally speaking, they go at large over the farm during the whole season, individual sheep never taking a very wide range. The area required for each varies from about two to four acres, according to quality. In some cases the hogs are kept separate from the ewes, which gives an opportunity of supplying them with more generous treatment in stormy weather, but frequently they are allowed, shortly after weaning, to graze together. This gives them the advantage of a mother's care, for they generally recognise each other. In some cases they are allowed to go on without being weaned at all, but we think such a system must be injurious to the future progeny. Ewes have their first lambs in April at two years old, and are sold as drafts at five or six, being replaced by the best of the ewe lambs. They are invariably sold for producing a crop of lambs by Leicester tups. These, with the wedder lambs, the small ewe lambs, and wool, usually form the whole produce of the farm. This applies to Cheviots in the southern counties of Scotland—in the north the practice differs considerably. There the wedder lambs are not sold, but kept on till sold as wedders at three years old. The wedder hogs are never wintered at home, but sent into winter quarters in Ross-shire and neighbouring counties—some as far as Aberdeenshire—where they have the *outrun,* as it is called, on arable farms, viz., nearly the whole grass, on which they are kept till the weather becomes stormy, when they are folded on turnips. They are sent about the 10th of October, and remain till the beginning of April. The cost of wintering varies from about 7s. to 9s. each.

With the exception of the great Inverness market in July, where large sales are made by *character,* for delivery later in the season, markets for Cheviots are held in autumn, the most important being Lockerbie for lambs, and Falkirk September and October trysts for ewes and wedders. Besides these, auction marts have sprung up in various quarters, where large quantities are disposed of; and although we question the policy of selling stock in bulk in this way, thus superseding old-established markets, and paying 3d. or 4d. in the pound for doing what farmers ought to be able to do for themselves, there can be no doubt that for the sale of single sheep it is admirably fitted. Mr. Aitchison was the first to introduce the system of selling Cheviot tups by auction at his farm of Menzion, in Peebleshire, more than forty years ago; the practice is now universal. One of the most attractive sales of the season is held at Beattock. Mr. Bryden, of Kennelhead (late of Moodlaw), long known as a most successful breeder; the Messrs. Carruthers, of Kirkhill; and Mr. Johnstone, of Cappelgill—almost equally celebrated— have an annual sale there, and draw purchasers from all parts of Scotland. In 1872 about 120 tups averaged 10 guineas each. Similar sales are held in various localities, one of the most important being at Hawick in September, where, among others, the lots of Mr. Aitchison and Mr. Elliot, of Hindhope, always command a large attendance and a deal of spirited bidding. The latter gentleman has for many years been a most successful exhibitor of Cheviots at the Highland Society's and other shows, carrying everything before him. We cannot resist giving an anecdote, which shows that his fame as a breeder must even have reached the ear of royalty. Happening on a recent occasion to be an exhibitor at the Smithfield Show, Mr. Elliot took the opportunity of visiting the Home Farm at Windsor, when he had the honour of being commanded to wait upon the Queen. Her Majesty, with that graceful condescension for which she is remarkable, received him with a cordial shake of the hand, desired him to be seated, and entered freely into conversation with him. While Mr. Elliot may well be proud of such an honour, we doubt not but Her Majesty was also gratified by the interview, and thought him both in appearance and intelligence an admirable type of the Scottish Borderer. Market localities, and names of celebrated Cheviot breeders mentioned in the above paragraph, refer to some nine or ten years ago. Changes have taken place during these years, and some of the most eminent breeders have gone over to the "majority," leaving their fame behind them.

Harking back from this digression to our subject, there is perhaps no finer animal of the sheep species than the Cheviot tup. Possessing the general conformation of the Border Leicester, he is altogether a more stylish sheep, carrying his head higher, with greater fire in his eye and grace in his movement. Compared with the Leicester, he is as a cavalier to an alderman.

Besides reproducing their own kind, the Cheviots are valuable for crossing with the Border Leicesters; the former giving hardihood, the latter great tendency to fatten. By infusing the two breeds in different proportions, other breeding stocks are raised, suited to medium soils and temperatures. Thus, taking the Leicesters as the centre of agricultural improvement, the others may be said to radiate. First, we find three-parts bred in the intermediate; next, half-bred in the higher altitudes; then we come to Cheviot entire on their native mountains; and above and beyond them our old favourites the black-faced, among their fastnesses of rock and purple heather.

Cheviot sheep are seldom shorn before July, the weight and fineness of the fleece depending on the nature of the pasturage; the texture being finer on dry, sweet herbage than on coarse grass, and bringing a higher price. It has a steadier demand than almost any other, being extensively employed in the manufacture of tweeds, now so commonly used in clothing, from the prince to the peasant. Coming down from the poetry, so associated with the Cheviots in the lights and shadows of pastoral life, to the inevitable prose—for to mutton they must all come in the end—that of the Cheviot sheep may fairly be put down as one of the luxuries of life. It has always been a nice point whether this or the black-faced is the finer, and we recall an incident which occurred many years ago; in which the father of the present writer bore a part. He was a great enthusiast in black-faced sheep, and having the honour to be a special favourite with Sir Walter Scott, and an occasional guest at his table, begged his acceptance of a few wedders to convince him of the superiority of the black-faced mutton to the Cheviot, of which Sir Walter was in the habit of keeping, what is called in Scotland, a pot-flock. Sir Walter accepted them on condition that he would dine with him, along with a few friends, to test their respective merits, when a saddle of each should be presented, having received the same advantages of the culinary art. The verdict was in favour of the Cheviot, to the infinite delight of the great poet and novelist. Dissenting, however, from this judgment, we venture to remark that the quality of both depends very much on the feeding. For delicacy of flavour, we never tasted any mutton equal to that of a *yeld* young ewe or gimmer of either breed that happened to get fat on its native pasture.

In taking leave of the subject, it may be stated, without fear of contradiction that no animal has conduced so much to the prosperity of the Scottish farmer as the Cheviot sheep, and more especially to those who have engaged exclusively in hill farming. This may be partly attributable to the fact that stock farming is generally embarked in by men of capital, as it involves a considerable immediate outlay, and the farm being usually large, competition for them is necessarily limited; whereas arable farms are competed for by men who have made money in other walks of life, and the demand being greater than the supply, rents have in many cases become exorbitant. Stock farmers are, besides, not nearly so much influenced by the weather, and their expenses are nothing in comparison. The practical working of the stock farm is managed by a few shepherds, a class of men in the rural districts of Scotland distinguished for great moral worth and simplicity of character. They receive their wages in the grazing of one or more cows and a certain number of sheep. They are thus small capitalists, and their interests are identical with their masters. In arable farming, a very serious increase has of late years arisen in the expense of cultivation, not only by the rise in wages of agricultural labourers, but in implements, machinery, and, in fact, every department of skilled labour connected with the farm. During the last few years, severe winters, and lower prices of sheep, lambs, and wool, have put hill farmers much upon a par with the cultivators of the soil.

CHAPTER XXII.

DORSET HORNED SHEEP.

By JOSEPH DARBY.

THIS celebrated breed is, by common confession, one of the oldest and best of the upland horned races. Dorset horned ewes stand unrivalled for fecundity, for excellence as nurses, and still more for taking the ram at almost any desired period, and being pre-eminently adapted for the production of our earliest crops of lambs. Few other sheep were formerly to be met with in their native county until Southdowns expelled them from the chalk hills; but they were improved by crossing at a very early period, and the original breed had become rare well nigh a century ago. Mr. Claridge, in furnishing his report on Dorset to the Board of Agriculture in 1793, said "The original breed of Dorset sheep is very scarce to be met with, as most of the farmers have crossed their flocks with either Hants, Wilts, or Somerset sheep, which certainly improved them in size." This crossing must have been conducted with great care and judgment even at that period. Mr. Parkinson, who wrote very early in the present century, affords the following information respecting Dorset sheep: "I look upon the Dorset ewe," says he, "as the best horned ewe in the kingdom, those of Somerset excepted, and they are so near alike that few people, unless the natives of the two counties, know the difference. The best of the Dorset ewes are more correct in their shape than many of the improved breeds of sheep. Mr. Bridge says they have been much improved of late years; they used to be long-legged, which is by no means the case at present." He then describes them as straight in carcass, deep in body, the rump much larger than in most sheep, the horns thin and rather bending backwards, the eye quick and lively, the face thin, the mouth small, the head standing up well, the neck very proper, the scrag neither thin nor clumsy, the leg well let down towards the shank; and adds that they are "set full in the shoulder, which gives flesh on the back, and is an indication of flesh in every part," and also "the ribs are not so high or round in the upper part as in some improved sheep, which, when as high as promoted by Mr. Bakewell, proves a fault, and diminishes the weight." Youatt also has the following description of Dorset sheep: "Most of them, at least of the pure breed, are entirely white; the face is long and broad, and there is a tuft of wool on the forehead; the shoulders are low but broad, the back straight, the chest deep, the loins broad, the legs rather beyond a moderate length, and the bone small. They are, as their form would indicate, a hardy and useful breed; they are good folding sheep, and the mutton is well-flavoured; they average, when three years old, from 16lb. to 20lb. a quarter."

At Plush, near Cerne Abbas, on the farm of Mr. Miller, a flock of the original breed existed until about six years since, when it was dispersed, and none now remain. These sheep were much smaller in size than the improved Dorset, but neat, tolerably well shaped, and thickly woolled. They bore almost as strong a resemblance to the Portland breed as to the modern Dorset, and were regarded by some as the connecting link between the two, for the opinion was once very prevalent that the Portland in reality was the original ancestral type to which all Dorset horned sheep owe descent. If so, no greater metamorphosis has been effected in any race, for the superior animals turned out at the present day seem to possess few affinities or points of resemblance even to the Plush flock, except in fecundity and early lamb-bearing. If Dorset sheep had been greatly improved at the period when Claridge and Parkinson praised them, how vast has been the progress towards their perfect development during the past half-century. They were sufficiently good, when Southdowns came into competition with them for the occupancy of the chalk soils of Dorset, to cause the struggle between the two breeds to be a long and arduous one. Both Stevenson, in his Agricultural Report, and Spooner, in his book on sheep, allude prominently to this subject, and show that the Downs supplanted their rivals because better fitted to crop the close herbage of the chalk hills, and being smaller, many more could be supported on a given acreage. The wool question tended most to prolong the struggle, for the Dorsets would always shear from 1lb. to $1\frac{1}{2}$lb. more per head than their opponents. But the fleece of the Southdown in those days, although light, was held in high estimation, and the advanced rate per tod it then commanded in the market helped eventually to turn the balance. The Dorsets were ultimately expelled from the chalk region, and driven back to better land in the western part of the county. Contention between the breeds has, however, in isolated cases, been waged more or less ever since; thus Mr. Ruegg, in his prize essay, published in the Royal Agricultural Society's Journal for 1855, observes: " Mr. Pope, having a flock of pure Downs at Toller, sent some of the best of them to his rich land at Mapperton, a horn country, and found out that the poorest Downs on the thin land at Toller did better than the best Downs on the rich land at Mapperton." The horns therefore seem now to be in possession of their own country, from whence the Downs are not likely to eject them. In addition to early lambing, they twin oftener than the Downs. As lambs they fatten well, but as hogs they do not progress with the Downs. In the second year they regain their position. Mr. Damar, of Winfrith, put up 300 horn and 300 Down lambs, and, after eighteen months' run, found that the horns had paid 7s. a head more than the Downs, reckoning them at 2s. less cost, and at 5s. a head more in sale.

But more has perhaps been done for the perfection of the breed during the past thirty or forty years than during any previous period of progress, and but for a third set of favourites having entered on the field, they might again stand a fair chance of regaining the territory from which they were expelled, but which is now occupied by the improved Hampshires. Mr. Thomas Danger found Dorsets wanting in two essential points; they were less disposed to fatten than some others, and in form were rather imperfect, consequently he made it his special study to effect a great improvement in these two points, and succeeded in bringing the Hunstile flock to a high standard of merit. For many years he was accustomed to make almost a clear sweep of the prizes in the showyards for Dorset sheep. Mr. Abraham Bond, who entered into his labours, said of him: "Mr. Danger, my predecessor, made rapidity in flesh formation almost his only study, and to this is to be attributed his success." Mr. Danger's superior tactics have, however, been so very generally imitated that the defect, once so noticeable, of the young wethers being slow in grazing for the butchers, has quite disappeared. Similar testimony by Mr. H. Mayo, of Coker's Farm, Dorchester, was given about the same time, then not only one of the best breeders, but one of the principal prize takers of Dorset sheep. About ten years since he gave this testimony: "The grazing qualities of the sheep are greatly improved, and it is a general thing all through the country. I think success has been attained to such an extent that some of the best bred Dorset lambs arrive at maturity and put on flesh as fast as the Downs. The breed has altogether received wonderful development of late years, and I think the ewes breed more lambs than ever they did."

That an improved form and greater aptitude to fatten have been imparted to Dorset sheep in recent years is unquestionable. Wether lambs and tegs used, within memory of the present race of farmers, to sell at low prices compared to those of other breeds, purely because of their indisposition to fatten until the second year. But the defect has now been sufficiently removed to advance very materially the current rates of this class in the markets and fairs. How this great change has been effected must perhaps remain a disputed matter. The union with the Somersets took place nearly if not quite a century ago, and all the Dorsets obtained from the alliance was increased size and whiter faces. The two varieties have been so long and so closely amalgamated that they are justly considered at the present period one and the same. The Somerset breeders are now even more particular than their Dorset brethren to develope short-legged, compact sheep, quite the antipodes of the lanky Somersets of a past period. How, then, has the modern improvement in form and materially enhanced aptitude to fatten been accomplished in such brief space of time? The breeders reply by careful and judicious selections, in the first instance by a few leading breeders, and subsequently by the distribution of their rams and the infusion of their better blood into flocks generally. But working on the in-and-in principle is well known to require a lengthy period for the accomplishment of grand results, and it has been alleged on pretty good authority that the shorter method of infusing the blood of other breeds has been resorted to.

One of the most respectable farmers in Western Dorset wrote as follows, about ten years since: "I have known many of the best horn flocks for more than forty years, and I am not at all prepared to say they are better now, if indeed so good as they were then. I admit they are more uniform in size, shape of horn, and general character, but certainly not so large or hardy in constitution. During that time most of the flocks have been crossed by the Somerset polled sheep (a breed now almost extinct) or by Leicesters, and in some cases by Devons, which has made them more round and perfect in symmetry. I have myself seen within the last seven years Leicester rams with some of the best horn flocks in the West of England, though I hardly think the owners would like to admit the fact."

Whatever agencies the breeders have adopted, they appear to have made the breed vastly better for grazing, without depriving it in any material degree of the inherent native good qualities for which it has been so long and happily distinguished. Their fecundity has not been detracted from, and they still yean more twins, and give less trouble in lambing than other kinds. The ewes are reported to take the ram as early as ever they did, when permitted to do so. At the time Claridge wrote, the season for putting the most forward ewes to the ram was the last week in April for such as were intended to be sold the following autumn, and for the stock ewes about midsummer; and the lambing period he describes as the middle of September for the sale ewes, and the beginning of December for the bulk of the flock. But neither for stock nor sale purposes is the lambing required to take place so early now, nor has it been for many years. Spooner in his book on sheep observes: "They take the ram as early as May and June, and their lambs are usually dropped in October and November, so that they are the principal sources of the supply of house and early lamb, which about

Christmas and the following month is esteemed a great luxury, and accordingly commands a high price. . . . . At Weyhill, one of the largest sheep fairs in the kingdom, they form a very large proportion of the sheep offered for sale. It is the ewes in lamb that are thus driven in the month of October a distance frequently of fifty or sixty miles, which journey, occupying upwards of a week, they generally bear remarkably well." Formerly it was not uncommon for lambs to be yeaned on the journey, and a horse and cart accompanied the flock to convey such lambs as happened to be dropped. The breed must indeed be hardy to stand such usage, to which it is not now required to be subjected, as the journey from Western Dorset to the neighbourhood of Weyhill, and the other Wilts and Hants autumn fairs, can be accomplished by rail, if they have to be sent at all, which is frequently not the case. The railway system has brought about a complete revolution in causing the great mart for early lambing Dorset ewes to be transferred from Weyhill to Dorchester Powndbury Fair, held on the 29th September, at which from twelve to sixteen thousand from the principal horn flocks of the county pass under the baton of Mr. T. Ensor. This enterprising auctioneer offers a number of prizes for the best ewes, the competition being in three classes, according to the size of the flocks. The ewes frequently have lambs at their sides even at this early period, and realise from 45s. to 75s. per head.

Mr. H. Mayo, writing as late as 1871, after observing that the ewes will take the ram two or three months before any other breed of sheep, adds: "When the lambs are yeaned in October and November, and both they and their mothers receive good feeding, the former will generally be found ready for the butcher in about ten or eleven weeks; nor does it take long to make the ewes ripe afterwards, and they will average from 20lb. to 25lb. per quarter. To obtain early lambs for fattening we generally make use of a Sussex ram, as the lambs are considered a little better quality with the cross. We lamb our usual flock about Christmas, and shear about the middle of June, when the lambs yield from $2\frac{1}{2}$lb. to 3lb. of wool, and the ewes from 5lb. to 6lb." Mr. Paull, a large farmer of Piddletown, in asserting about ten years since that the Dorsets were prolific wool-bearers, and that horn wool was then 1*d*. per lb. more valuable than Down, added: "The wool of the horn lamb is very much sought after for its peculiar whiteness and the fine point it has."

Beyond the chalk range we find in western and northern Dorset an undulating surface, with low, flat, or rounded elevations, formed of some of the lower members of the oolitic series, interspersed occasionally with patches of greensand. The soil is tolerably fertile, and is redundant of moderately good upland pastures. Further on are the rich, inferior oolite sands and loams of southern Somerset, where the pastures yield a mild succulent herbage, and are ever green. The Dorset horns occupy the whole of this region, and, so far from being ever expelled therefrom, are rather extending themselves further into Somerset, in competition with Devonshire longwools, cross-breds, and Down sheep. The subjoined testimony, rendered for the first edition of this work, refers to locality adaptation. Mr. Abraham Bond said: "I think them very well adapted to the light-land corn farms on either side of the Quantock Hills, taking in a district on the Bridgewater side from West Monkton to Stowey, and on the Taunton side from the same point to about Cothelstone. I think the climate would be too hot for them on the Somerset lowlands, as they are peculiarly liable to attacks of fly in the summer months." Mr. Henry Mayo wrote: "They will do, I may say, on nearly every land, but most profitably on good pasture land. They are considered to do much better on a high sour farm than Down sheep, will generally breed more lambs, and fewer losses are experienced in lambing than from Down flocks." Professor Buckman, who farmed on the borders of Dorset and Somerset, stated: "The soil of this part of Dorset is mostly an inferior oolite, the lighter on the inferior sands. Hampshires and Dorset horns prevail, but Southdowns are not unusual, and all do well. Both Hampshires and Dorsets are very early and usually prolific. I fancy that for the past two years Dorsets have paid best, but Down mutton is to be preferred."

The turn of the tide has also re-established Dorsets into favour on many farms in the chalk country. Larger sheep than Southdowns are now required for the profitable management of even the hill farms, the Downs having been so generally broken up and converted to tillage. Mr. J. Homer, of Martinstown, writing a dozen years since, said: "Very few real Southdown flocks are left in this district, the improved Hampshire or Dorset horns being so much better suited to its requirements." But it is chiefly on those farms having water meadows along the margins of the Frome and the Piddle, and more particularly in the neighbourhoods of Dorchester, Maiden-Newton, and Piddletown, that Dorset flocks prove of highest value in the chalk country. Under these circumstances it is considered more profitable to keep the forward draft ewes, and perform the fattening business as regards both lambs and their dams on the water meadows. Thus Mr. Paull affirmed: "Horn sheep are well adapted to farms that have some good water meadows, as they possess good quality and fatten readily, and their lambs come to early maturity for market." And, again: "Horn sheep are dropped about Christmas. As most of these hereabout are fattened, they get cake as soon as they will eat, and all they can be made to consume, the object being to get them off as soon as possible, which in a fair season would be about April 1. Since meat has been so dear many farmers fatten the off-going ewes as well as the lambs, and they also are allowed whatever cake in reason they will eat, the same object being desired as with the lambs."

Nor is this practice confined to the district named. Further west many flockmasters have long since found it more remunerative to do the fattening at home, rather than sell off their draft ewes in lamb as heretofore. Distant localities having become so intimately connected by the railway system, that Dorset and Somerset can as readily supply the London market with lamb as the home counties, which forty or fifty years ago afforded the exclusive supply. Graziers residing in the neighbourhood of the metropolis still go westward to obtain Dorset ewes forward in lamb at high prices. But of late years they have had formidable competitors in graziers from Hants, the Isle of Wight, and elsewhere, who turn their attention very much to the production of early lambs, and find Dorsets best adapted to serve the end in view.

The existing status and prospects of the breed in the county of Dorset have been fully described by Mr. T. Ensor, than whom there can be no better authority, as, in addition to the monster auction for ewes, before referred to, he holds an annual auction for Dorset horn rams at his Dorchester repository in the month of May. In a communication rendered expressly for this article as late as December, 1886, he says : " Owing to the enterprise of many eminent breeders, Dorset sheep have entirely supplanted the Downs during the past few years, especially in the locality of Dorchester and on those chalk land farms having good water-meadows and pastures. This is no doubt greatly due to the circumstances that they possess good quality, fatten readily, and incur but little risk in lambing, while the lambs come to early maturity for market." As to their more recent improvement, Mr. Ensor writes : " Owing to the larger areas of turnips grown and the use of corn and cake in their feeding, together with the very careful selection they have undergone by eminent breeders, they have during the last few years been doubled in size, their fleeces are twice as heavy as before, while fattening propensity has been increased to the extent that the best Dorset lambs now arrive at maturity quite as early as the Downs: Although they have been so much improved, they retain their hardihood and fecundity as much as ever. Prizes are now offered by the Dorchester Agricultural Society to the shepherds who rear the largest number of lambs with the least loss of ewes compared to the number placed with the rams. In 1884 the two flocks in which the shepherds won the chief prizes were those of Mr. W. Hull, Durce, Dorchester, and Mr. Gale, Broadway, Weymouth. The number of ewes of the former gentleman was 700, from which 867 lambs were reared with the loss of twelve ewes. Mr. Gale's ewes consisted of 360, from which 466 lambs were reared with the loss of only three ewes. In 1885 these flocks gave very similar returns. Mr. Hull's ewes, 716 in number, sustained thirteen losses, and reared 879 lambs, and Mr. Gale's 360 ewes only lost five, while the rearage of lambs was 473. Three other flocks also stood high in the competition of that year, which were Mr. W. Mayo's, of Friar, Waddon, consisting of 607 ewes, of which 19 suffered mortality, the rearage of lambs being 762; Mr. R. Smith's of 450, the number lost being seven, and the lambs reared 574, and 360 belonging to Mr. Flower, of Stafford, Dorchester, of which only a single one was lost, 404 lambs being reared. Mr. Ensor remarks that these statistics show that "out of 3547 ewes, only sixty were lost, and that they reared 4425 lambs, or 125 per cent. of lambs with a loss of only 1·6 per cent. of ewes.

According to information rendered by Mr. W. Bond, Buckland House, Durston, Taunton, the experience of Somerset flockmasters is very similar, as he states that losses in lambing and barrenness are so rare that from 150 to 160 lambs may usually be calculated on for every 100 ewes placed with rams. Mr. Bond is accustomed to lamb down much earlier than his neighbours. The customary time for the flock ewes to yean both in Dorset and Somerset ranges between the middle of December and the middle of January, but the whole of the Buckland House ewes yean in October and the first half of November. With moderately liberal feeding his wethers come out fat at from thirteen to fourteen months old, with carcase weights of from 70lb. to 80lb. each. Of course there is no difficulty to get them fat at a much earlier age with high feeding, and Mr. Herbert Farthing states that wethers always well fed are usually fit for slaughter at about twelve months old, and made to yield carcases weighing about sixty-four to eighty pounds. The ewes often yield carcases weighing from twelve to thirteen stone. Horn flocks have increased very much in western and southern Somerset during the past few years. There are six, if not seven, members of the Kidner family who propagate the breed, and Mr. S. Kidner, Milverton, and Mr. John Kidner, Minehead, have taken first prizes at the Smithfield Club Show. Within a radius of about five miles will be found not only the flocks of Mr. H. Farthing and Mr. W. Bond, but those of Mr. Dunning, Court Barton, Creech St. Michael; Mr. England, Quantock, West Monkton; Mr. Bond Huntstile, Goathurst, and several others. In south Somerset there is Mr. Charles Harding, of Montacute, who has made his mark in the showyard, especially at the Smithfield Club Shows.

Mr. Herbert Farthing, of Thurloxton, has quite taken the lead in the showyard during the past twenty years, and, as is tolerably well known, has frequently had no competitor for the prizes offered by the Bath and West of England Society. Occasionally one or two flockmasters from the Beaminster or

Dorchester district may send pens of ewes, but as a ram breeder he is, as a rule, allowed to carry all before him. It appears that Mr. Farthing commenced exhibiting at the shows of this society at the Salisbury Meeting in 1867, and at the Smithfield Club Show in the same year, and has not missed putting in an appearance there ever since. On those few occasions when the Royal Agricultural Society offered special prizes for the breed, they were usually carried off by him, and what with his winnings at the Devon and Somerset County Shows, and at the Sherborne, Taunton, and other local exhibitions, he has received over two hundred premiums for Somerset and Dorset horn sheep. According to general testimony, both in Somerset and Dorset, the ewes are accustomed to clip from 5lb. to 6lb., rams clip from 10lb. to 12lb., and the lambs, which are usually shorn, yield about half as much as the ewes. As it appears the wool of lambs of this breed is in great demand, it would be folly not to clip them. According to Mr. T. Ensor, there is also a great demand from butchers for the carcases of Dorset lambs, his language being: "There is as keen competition amongst butchers for horn lambs, as for Down lambs, and perhaps even more demand for the former than for the latter. When the old draft ewes are intended to fatten lambs for the London market, Mr. Ensor says they are made to yean in October and November. Both ewes and lambs have good feeding, so that the latter are generally fit for the butcher from ten to twelve weeks old, when they average from 10lb. to 14lb. per quarter, and go to the London market realising from 40s. to 50s. each. The ewes are then finished off for slaughter, which does not take long, they being usually nearly fat when the lambs go. Some of the lambs from the flock ewes are also fattened when there are good watered meadows for the ewes and lambs to be turned into in March or the beginning of April. The lambs fatten readily, especially when they receive as much cake or corn as they will eat, which is generally the case, the object being to get them fit for slaughter as early in April as possible.

At Mr. Ensor's annual auctions in May, held in his Dorchester Repository, a considerable number of shearling rams and lamb rams are not only sold, but compete for prizes. They are from the flocks of Mr. Henry Mayo, Mr. W. Hull, Mr. C. Harding,

Mr. Pitfield, Mr. Flower, and Mr. Raper, of Little Bredy, and Mr. J. F. Raper, of Mayeston Farm. The ram lambs fetch from five to twenty guineas each, and the best rams from fifteen to thirty guineas. Dorchester is now about the centre of the district in the county of Dorset where the breed is propagated, for although these sheep are most numerously kept in the west and south of the county, they extend throughout the Isle of Purbeck eastward. In south Dorset, besides the breeders already named there are Mr. W. Mayo, of Friar Wadden; Mr. C. Hawkins, of Wadden; Mr. J. Mayo, of Broadway; R. B. Sheridan, Esq., of Frampton; the Messrs. Kent, and several others. Westward, Mr. Legg, of Beaminster; Mr. Bradford, Mr. R. Watts, and Mr. L. Groves all keep large flocks. Around Dorchester there are Messrs. Symes, Mr. T. Sampson, Mr. John Chick, Mr. T. Chick, Mr. W. Dunning, Mr. Symes, Messrs. Paul, &c. In the Isle of Purbeck the Earl of Eldon and Messrs. Kent take the lead.

The fecundity of Dorset sheep is so great that the possibility of taking from them two crops of lambs in one year does not exist merely as rare and exceptional, but has often been effected. Such excessive demands are deemed too trying to the constitutions of ewes for flockmasters to make them as a rule, but with high feeding the Dorset ewes can be made to take the ram at almost any period. The details of ordinary management do not differ very materially from those required for Southdowns and Hampshires, beyond bringing the ewes to lamb earlier. Although from an early period, always bearing a good reputation as sheep well adapted to the folding system, they are certainly more impatient of continuous close breachings on turnips than Down sheep; nor are they subjected to this overmuch in their native districts. West Dorset and South Somerset possess a large proportion of good upland grass fields to arable land; the custom in consequence of turning the flock into these to ramble at large by day, and bringing it back to the ploughed fields at night for folding on some green or root produce, is very general. Although breeding flocks have hitherto been almost entirely restricted to the counties of Dorset and Somerset, there seems little reason why they should not be extended to other parts of the kingdom, particularly to those districts of Devon or Cornwall where grass production in midwinter progresses under such happy circumstances that the natural condition of things appears extremely favourable to the adoption of the breed of sheep that will yield earliest lamb. Probably now that the grazing and fattening capabilities of these sheep have been so much improved, they will be more eagerly sought for.

CHAPTER XXIII.

WELSH MOUNTAIN SHEEP.

By MORGAN EVANS.

T is quite unnecessary to prove the very ancient origin of the Welsh mountain sheep. With black cattle, they formed at one time the principal stock of the Celtic race in the mountainous districts of Wales; and they must have been the only sheep both on hill and plain, on heather-clad mountains and sunny vale, from Anglesea to the Bristol Channel, from the Severn to Cardigan Bay. As the more fertile lowlands became cultivated, and free and undisturbed communication took place between Wales and England, the small ancient breed in these places became crossed with the larger kinds introduced from the adjoining counties, or they have been entirely supplanted by them, and at last driven before advanced agriculture into the poor hilly soil and mountain ranges of the Principality where they still linger. Leicesters, Cotswolds, or Downs are now to be found on all fertile and well-cultivated farms in the country. On medium and poor soils in exposed places a cross of these with the Welsh mountain sheep is commonly seen; whilst in the realms of gorse and heather, stretches of barren common, and the cottier tenements on the hillside, the ancient breed still holds sway, living on scanty food, rearing hardy lambs, and producing the sweetest mutton known to the palate of the epicurean Englishman.

Although there is a slight difference in character in the mountain sheep of separate districts, they were doubtless originally the same breed, and have the same common origin. Attempts have been made to divide them into two distinct classes, but the variations appear to be those only natural to accidental selection or to the effect of soil and climate. A minute description of all the peculiarities of these and of the different modes of treatment to which they are subjected is not possible in the space allotted to this chapter; we must therefore at present be content with a general view, although the Radnors claim a special notice to themselves.

The Welsh mountain sheep are principally white faced, but some have rusty brown faces, some speckled, and others grey. The males are horned, the females generally hornless. Sometimes the ewes have very short horns, and occasionally have these appendages large—equal in size to those of the rams. The poll is generally clean, but it is not uncommon to find rams with a tuft on the forehead, and also very woolly on the scrotum. These latter characteristics are considered by some breeders valuable indications of vigour and hardihood. As no great care, however, is taken in breeding these sheep, specimens of all the above variations in horn, colour of face, and amount of wool on the forehead may be found on the same mountain range, and even in the same flock. The head is small, and carried well up; the neck long, and the poll high. The tail is long, the rump high, and the shoulders low; the chest is narrow, the girth small, and the ribs flat. They have all the character of a wild active breed of animals, suited to scanty herbage on rocky slopes and precipitous hillsides. The average weight of the store ewes is about 7lb. per quarter. They feed slowly, and the wethers, when three years old and fat, weigh from 9lb. to 10lb. per quarter. The ewes are not prolific, producing generally but a single lamb on mountain land, where one lamb is enough and two would be too many to nurse properly. Both improved keep and crossing with other sheep are found to increase the number of twins.

The average clip of wool is about 2lb. The quality is usually fine, but in some districts it is coarse and mixed with long hairs about the neck and along the back of the animal. It is well known that wool is greatly affected by soil and climate. Continued exposure to cold and to most severe winds tends to change wool into hair. The difference in quality of wool appears to be due to the position and locality in which the sheep have been bred for generations rather than to any separate origin, for in all other features the Welsh mountain sheep is alike in all localities. Even in the same county, Cardigan, as mentioned by Youatt, the wool in the northern parts differs from that of the more southern parts of the county. The wool on the Pembrokeshire range of mountains adjoining is particularly fine, and in much demand by the local weavers, who formerly were the only purchasers of wool that were known in Wales.

The manufacture of flannels and woollen cloths was until recently an important branch of the industry of this country. Formerly all the woollen goods used were what is called "home made." The ordinary rural farmer walked and slept in woollen goods grown on his own sheep; the coat on his back, the blanket on his bed, were the natural produce of the farm. The spinning wheel could be heard humming at one season of the year for weeks in his house preparing blankets for his bed, dresses for his wife, or petticoats for his daughters. He knew nothing of English broadcloth. Corduroy or fustian breeches might be indulged in, but all else the good man wore, except his boots and hat, were made from the wool of his own sheep. The cloak, gown, jacket, skirt of his wife and daughter were of like material, and there was no dealing in hosiery in the shops; for the stockings of the family—black or grey—were spun and knitted in his own house. A weaver with a hand loom lived in every village, or a small water mill in a glen close by converted the home-spun yarn into flannel or cloths. The dyes used were few—black, blue, or red. Red shawls, or whittles, as they were called, were formerly much in use; and a goodly array of the female peasantry clad in these is said to have dismayed the French when they landed at Strumble Head, in North Pembrokeshire, in 1797, who thought they were soldiers, and anticipating an overwhelming force of infantry, laid down their arms. The agricultural labourer and his family had a horror of English goods. The servant maids at the fairs even now bargain for one or two pounds of wool along with their fixed wages. The wool is converted into clothes or bedclothes, and is generally the only dowry they have on their marriage, and these not being made of shoddy, last with care a lifetime. The stockings worn by the female population being invariably black, and black being much used in making greys for other purposes in the household, a few black sheep were thought an acquisition to the herd. The Welsh mountain sheep occasionally beget black lambs, and a few black ewes are generally kept for the benefit derived from their wool for family purposes. But it does seem strange that, with all the uses to which wool was put, no effort appears to have been made to increase the weight of fleece, even at the expense of a somewhat coarser fibre. Wool, however, was the perquisite of the females—at least, they had the entire management of all used in making woollen fabrics for family use. The counsel of the gentler sex, who abhorred wool that did not exactly suit their taste, would of course prevail. It must also be considered that, until a comparatively recent period, local weavers were the only purchasers of wool known to the Welsh farmers, and they of course patronised that wool only which made goods of the quality best suited to their limited machinery, and which would make flannels of a kind most in demand at the country fairs and markets, where their stalls are invariably found.

The Welsh mountain sheep are good nurses, and rear their lambs well. They are often sold from the western counties of Wales to go into some of the English counties for breeding fat lambs, and they succeed well when crossed with larger breeds of sheep. On exposed farms of poor soil in Wales they are frequently used for this purpose, or a cross of these is kept, the mountain ewes forming the original basis of the stock. The sheep are crossed with Downs or Leicesters, or with any large mongrel strain, and again recrossed with the mountain sheep if necessary, all depending on the class of sheep the farm is best suited to carry. The real mountain sheep are sold as wethers at three years old. The cross-bred come to earlier maturity, and the produce of these are sold as lambs in the May, June, or July fairs. In making the first cross with mountain ewes, a cross-bred small ram a little bigger than themselves is used, always selecting males with small heads and hardy constitutions. The rams may be too large and of too good a quality, the consequence being much difficulty in lambing and tender lambs, unable to stand the wind, rain, and cold. A friend of mine, one of whose farms is on high land, writes and says: "I have used the cross-bred mountain ewes, the largest I could get, which when fat made 11lb. to 15lb. per quarter. I kept them for years, and will go back to them again, I believe, this winter for breeding lambs. My sheep are too good for my poorest land. These little ewes were by far the best nurses I ever had, and four-fifths of them brought twins by small-headed Southdown rams, the lambs weighing from 8lb. to 11lb. per quarter at the fair on July 10. I exclude almost every horned ewe from the flock. Many of the diminutive little ewes had the lambs by their sides much heavier and bigger than themselves."

In the winter time, just before lambing commences, the farmers on the mountain side bring the sheep down into their small inclosures, and, in addition to the grass the sheep consume, they are given small quantities of hay or oats. The oats are always given in sheaf; the mountain sheep would not know what to make of clean corn, and would not look at it. All the lambs kept in stock as ewes or wethers are shorn in July or August. When they are weaned the mothers are milked for a month or two, and butter is made of the milk, or it is mixed with skim milk to make cheese. Milking sheep, however, is becoming less common every day, and where the practice half a century ago was almost universal, it is hardly known at the present time. The young Welsh farmer, economical and of small means as he usually is, finds in mountain ewes a good basis for his future flock. The ewes are bought cheaply in the summer and autumn months, after their lambs have been weaned or sold. By continued crossing with larger animals, he at last establishes a paying if not a fashionable class of sheep —ewes and wethers that are thrifty in seeking food, and which when killed die well. The improved strain is almost invariably commenced and continued by the purchase of ram lambs. Aged rams seldom or never exchange hands. Lamb rams in Wales, just as yearling bulls in Switzerland, are supposed to get more vigorous offspring than older sires; and twins are said to follow the younger rams more frequently than those of a riper age.

When brought into the inclosures, these sheep are found difficult to keep within bounds. Fences such as are usually found, low stone walls, turf banks, or hawthorn and hazel fences, are as nought to these wild creatures. A purchase of ewes at a fair to-day spreads in the direction of the four winds to-morrow, unless extreme precaution be taken, and the secret of their whereabouts is sometimes found to be the house-tops of neighbouring cottages. To prevent their marauding proclivities — for no professional shepherds are kept — they are bound with fetters—"lonkers," as they are called in some parts—made of woven rush or hempen fillets. These extend from the fore to the hind leg, leaving the extremity of each limb from twelve to eighteen inches apart. Sometimes an occasional sheep—the ringleader of the flock—has a fetter on each side; and if putting them on in the usual way be not found sufficient to stay the wanderings of the wicked one, both fetters are crossed, from the fore foot on one side to the hind foot on the opposite side—and it is surprising to see how they go about even under these difficulties.

Attempts have been made to supersede these sheep in their native mountain homes by Cheviots and other breeds, but the change has not been found to answer. No sheep suit the mountain tops of the country so well as the indigenous breed, and the most profitable on the lower ranges of poor soil and waste lands are a cross with the native stock. Welsh mountain sheep are likely to hold pre-eminent sway in their strongholds at high altitudes for many generations to come, and as long as the geological structure and the climatic influences of the country remain unchanged.

CHAPTER XXIV.

THE RADNOR SHEEP.

By MORGAN EVANS.

SOME of the oldest remaining indigenous breeds of sheep in our island are characterised by black faces. A remnant of one of these is found in the county of Radnor, on the hills of Brecon, and scattered along the western parts of Montgomery and Merioneth. The original Shropshires on the Long Mynd and Morfe common were black or speckled-faced, horned sheep, and were doubtless allied to those of a like character in the adjoining counties of Wales, and probably had one common origin. The most important class of native dark-faced sheep in Wales at the present time are the Radnors—a hardy, active race, that under improved management have developed into a breed of fair size, carrying a good weight of fleece, whilst at the same time in outward form and hardihood of constitution retaining much of the primitive type of wild mountain sheep.

Having treated of Welsh mountain sheep elsewhere in this volume, and the way they are usually managed, there is little to be added in describing the more purely local breed of Radnors. The best kind of Radnors are those having black faces, but a large number are of a tan, grimy, or grey colour, and others, though of questionable purity, have faces partly white. The rams are horned, and the ewes should be hornless —a sexual variation thought by some high authorities to be almost always the result of breeding and domestication. Radnor and other Welsh mountain ewes frequently show a tendency to produce horns; but such excrescences are not cultivated in females of the flock. Several of the Radnor ewes exhibited at agricultural shows of late years have had " short stumps, scarcely amounting to horns." The Radnors are short-legged and active, hardy when exposed to severe weather, and thrifty in seeking for food on the scant herbage of the mountains. Like all sheep of their class they are light in the fore quarter, and, compared with modern improved breeds, slow-feeding animals; but the mutton they produce is of excellent flavour. When fed in the usual way, on the mountain side and the adjoining pastures, the wethers at three and four years old produce Welsh mutton of the true flavour—a flavour not unnaturally supposed to be in great measure due to the peculiar grasses mountain sheep feed on in such places. Much that is called Welsh mutton in London, even if it does come from Wales, is not of the true sort, but is often called Welsh mutton for the very reason that it is not. Just as all the so-called "Dorset butter" cannot possibly be produced in Dorsetshire, so neither can the hills of Wales maintain mountain sheep enough to be the source of all that goes by the name of Welsh mutton in England. But anyone who has more than once tasted Welsh mutton of the right kind, properly cooked, should not afterwards be mistaken, or be easily deluded by his butcher into thinking the joints supplied him, because small, are necessarily portions of Welsh mountain sheep fed on their native hills.

Mr. Darwin says, " Sheep are perhaps more readily affected by the direct action of the conditions of life to which they have been exposed than any other domestic animal." And again, "A slight difference of climate and pasture sometimes slightly affects the fleece." The improved Radnor is a superior animal to the white-faced Cardiganshire sheep, or to its remote ancestors, and requires a somewhat better climate and herbage. They have of late years greatly increased in size, and the weight of fleece and quality of wool have been much improved. The old breed was very small, and a great point with breeders of past times was a very large tail, heavily woolled, and a large quantity of coarse wool or hair about the breech. The Radnors have recently been developed into more useful, respectable-looking animals. The wethers are sold at three or four years old, and now weigh, when fat, from 14lb. to 15lb. per quarter. The ewes, sold at about four years old, weigh, when fat, 13lb. or 14lb. per quarter. The wool is of good quality, and the fleece weighs from 4lb. to 5lb., which must be considered a fair amount in proportion to the size of animal. Indeed, the old British sheep could not have been very short of wool, for Howel Dda in the tenth century, in appraising the qualities (*teithi*) of a cow, says: " The worth of her teat is fourpence every year that she lives, or a white sheep with a white lamb that can with her fleece protect her lamb between her four feet from a May shower."

The Radnors are found on the hills of Brecon, Montgomery, and some parts of Merioneth, as well as in the county from which they take their distinctive name. The fairs held at Kington, Knighton, and Builth, are perhaps the most celebrated places where they are offered for sale. The ewes are sold in large numbers to graziers. A great many go into the adjoining counties in England to breed fat lambs by crossing with Shropshires, Leicesters, or Cotswolds. The ewes are prolific, excellent nurses, and produce good lambs when crossed with larger improved breeds, and they feed quickly after their progeny has been disposed of. Having these good qualities, it is not to be wondered at that they are in considerable repute for the special purposes to which they are adapted when brought on the more fertile pastures of the English grazier.

A great part of Radnor is hilly uninclosed land and common. In 1849 it was stated that nearly two-thirds of the county was uninclosed. A considerable area of land has since that time been brought under cultivation, and the hill sides encroached upon, although not to any very considerable extent. The suitability of the Radnor sheep to the uninclosed districts in which they are reared may be readily assumed from the preceding remarks. They do well during the summer months on the mountains. The young sheep and ewes are usually brought into the inclosed pastures during severe weather in winter. The aged wethers, however, are generally left to take care of themselves in their natural habitat, though the winds blow and the storms rave ever so wildly.

The breed is only of local importance, and is not likely to extend beyond its present limits. Nor is there any probability of the Radnors being superseded on their native hills for some time to come. The gradual improvement in their character that has been effected of late years will tend still more to make them pre-eminent in the districts they now occupy. With increased attention to breeding and greater care taken of them in the winter months they may yet attain to a higher average merit, and become still more profitable. Both the easiest and most natural way of improving the Radnors, if crossing is resorted to, is by giving them a dash of Shropshire Down blood. The Shropshire sheep, now become so aristocratic in their new form, blend well with their neighbours and poor relations under the control and skill of the breeder. The Radnors cannot, however, be pushed far in the direction of weight or rapidity to fatten without losing the especial qualities for which they are so prized, and becoming unsuited to the hills. As lowland sheep on fertile pastures they can never hope to rival the popular improved breeds already in existence. As mountain sheep in their own district, there is no reason to wish their displacement by any other breed. It would be difficult to find any sheep so suited to the uplands on which they are kept, and to the treatment to which they are subjected. One thing also which tends considerably to perpetuate local breeds of cattle or sheep is that they meet with a ready sale in their own districts. They have markets of their own, where they are popular amongst the farmers of the country around, and purchasers from abroad visiting the fairs go there specially for an accustomed class of animal. English drovers, for instance, go to Pembrokeshire or Carnarvonshire for black cattle, and look with indifference on store stock of any other colour, which they can readily obtain nearer home. And the English grazier who goes to Kington or Knighton fairs for Radnors would not be likely to return with a flock of Cheviot ewes, should there be any offered for sale at those places. A local breed in store condition always commands the best price in its own locality. On this ground a change in the mountain sheep of the district is not advantageous, at least to the introducers of the new breed; and it will almost always be found better to improve the breed of sheep already acclimatised and in demand than to attempt to replace them by stock foreign to the country.

E E

The accompanying illustration by Mr. Harrison Weir represents the prize aged ram of the Radnor breed exhibited in 1872 at the Royal Agricultural Society's Show held at Cardiff, and also one of the prize ewes at the same exhibition. The former were shown by Mr. J. R. Paramore, of Preswylfa, Neath, and the latter by Mr. William Dalton, of Cardiff, one of a pen of five ewes bred by Mr. William Wilson, of Kington. These two exhibitors, along with Mr. Edward Farr, of Pilleth, Knighton, were the only prize-takers in the class for Radnor sheep at the show. Some of the rams exhibited had horns, but others of improved strains were, like the one here represented, hornless. Indeed, what is called the improved Radnor has not been reduced to a fixed type, the different breeders producing flocks each after his own fashion and ideal. The general rule is to cross with the Shropshire Down for quality of flesh, and with the Leicesters or other long-woolled sheep for weight of fleece—in both cases adding to the size and weight of the original breed. When, therefore, one uses the term "improved" Radnors, it cannot be said that one invariable type is indicated. The Radnor sheep are evidently in a transition state. It remains for some breeder to advance so far before his fellows in developing the breed that a fixed type shall be established, having a uniform and distinct character. The material to work on forms a good foundation; and the further improvement of the Radnor sheep is a praiseworthy effort worthy the attention of local breeders.

CHAPTER XXV.

HERDWICK SHEEP.

By H. A. SPEDDING, Mirehouse, Keswick.

OF the original introduction of this breed into Cumberland nothing certain is known, though there is a vague tradition that the original parents came out of a Spanish ship wrecked on the coast near Dudden Sands; but, however this may be, it is certain that the breed existed in the immediate neighbourhood of Muncaster long before it emerged from thence and spread all through the lake district of Cumberland, Westmoreland, and Lancashire. There it now reigns supreme, having entirely superseded the old Fell breed, which, tradition says, was white-faced and horned, larger than the Herdwick, but neither so hardy nor such quick feeders. Once established, however, it is no wonder that this sheep holds its own against all competition, as no other breed can do so well on the same ground; for, though horned black-faces can endure the cold and wet of the winter nearly as well, yet they want a longer bite, and cannot exist on the " slape " fells—which sometimes consist only of rocks, and grass so short that it seems as if nothing but a razor could get anything off them—and the ewes especially fail in milking. In comparing them with other mountain breeds it must always be remembered that the Herdwicks labour under one very great disadvantage, viz., that the fells they go on are for the most part " common," and the common rights are, as a rule, very imperfectly enforced; thus it generally happens that a great many more sheep are turned on to the fell than the ground ought to carry; and this again, by causing jealousy among the different flockmasters, leads to much unnecessary dogging and driving, which of course takes a great deal out of the poor sheep. However, we hope matters are mending in this respect.

Herdwicks are active, sprightly animals, with a good deal of wild nature about them; thus, ewes which have been sold have been known to jump the most impracticable fences near lambing time, and to travel almost incredible distances, in order to lamb in their old haunts; and if one should happen to lamb on her way home it does not delay her long, as the lamb is soon on its feet, and once on them, can travel nearly as fast and as far as its mother. They are very good managers, too, in hard weather or snow, as they will, if they have warning, make for the exposed places, where they are not likely to be drifted up, and afterwards scratch down till they reach the grass; but sometimes a thick, soft, heavy snow comes on so suddenly that they are drifted up almost unawares, and in this state of being literally buried alive they have been known to live for three weeks, and, after being taken out, recover and do well, their only subsistence having been what they could reach from where they lay.

A Herdwick is very much attached to its own "heaf," or that part of the fell where it generally goes; and in a large stock a shepherd depends chiefly upon this peculiarity for knowing that all his charge is right, mentally dividing the fell into certain tracts, within which he expects to find certain sheep. In spite, however, of this general tendency to stick to the ground, a certain amount of straying does of course take place, and, as there are no fences, often to long distances; on this account meetings are held about twice a year on fixed days at certain places, where men who have lost and men who have found sheep attend, and there is a general clearing. A "shepherd's guide" is published, which contains the sheep marks of all the different farms, with illustrative woodcuts. The "marks" consist of "ear marks," a tar letter or letters on one side, and another mark generally made in red, but sometimes in tar, on some other place; in addition also a mark is occasionally burned on the face. At rutting time the tups fight a great deal, and instances are not unknown where the shock of meeting has been fatal to both combatants.

In hard weather hay is taken up the fells in sheets on men's backs, but, unless accustomed to it as hogs, the sheep will almost starve before eating it.

There are numerous local shows held in the district, but the principal one is the Felldales Association, held in Eskdale. Here all the best known breeders come with a large number of sheep to show in the different classes, and for sweepstakes; keen is the rivalry, and deep and profound the sheep talk on these occasions. Sometimes, too, prizes are given for the best working dogs, when it is a pretty sight to see these sagacious animals turning, gathering, and driving in turn. The dog language used by the shepherds is utterly incomprehensible to any but a native.

The jumping powers of a Herdwick require to be seen to be believed. They run up dry walls like cats, and if they cannot run up, they will leap immense heights.

The legs and faces of this breed, as lambs, are black, or black with a few white flecks; but they soon begin to "brighten out," till by the time they are two years old, all that was black has become a frosty or silver grey, darkening slightly towards the forehead, except a blue-black mark or patch at the back of the neck; any brown tinge is a defect, it being considered a less hardy colour. The eye should be bright and good, and the forehead broad with a tuft on it; the ears should be white and sharp, and the wool should come well up to them, and, in the case of tups, form a kind of mane or heckling of a dark colour. They should be wide between the fore legs, with the breast well forward, and be well ribbed up to endure hardships; the hind legs should be straight, and well muttoned down to the hocks; the knees and feet should be large, the latter white, and the bone between them fine; above all, a Herdwick should stand square and walk well. The ewes are not horned, but the tups generally are, though not always; when they are the horns should be white and waxy, and, rising well out of the back of the head, curl once or twice. Of course most sheep fall off more or less from this ideal, but the commonest fault is a slackness behind the shoulder. There is also in every flock a certain proportion of darker coloured sheep, and these "breuked" ones, as they are called, are often of the best blood; they are not by any means disliked as long as they are of a black and not a brown tinge, for many people deem them hardier, and also consider them useful in keeping up the distinctive markings of the breed;

in fact, some very dark coloured tups have been great prize winners.

As to the wool, which is of a short staple, it is as inferior to that of most low country sheep as the mutton is superior; but it has improved a great deal of late years, and, though still greyish, is nearly free from those "kemps" or grey hairs that used to disfigure it. It varies in price, according to markets, but is always rather above the wool of the horned black-faces.

"Sickness" and the fly are the chief occasions of loss to the flockmaster. The former, which resembles black quarter in calves, principally attacks the hogs in autumn and winter; but some fells have an evil reputation for it, and on these the farmer must lay his account with a considerable annual loss of sheep of all ages, and at all times of the year. The only remedy or rather prevention seems to be change of pasture, and when, as is usual, the hogs are wintered out, the nearer the sea they are the less risk of their sickness. The nuisance of the fly varies very much in different localities, but it abounds on some fells from June till September; and as the sheep on being struck generally seek some cover, such as underwood or deep beds of bracken, it may easily be imagined that on such extensive ranges many are never found till the horrible death of being eaten alive by maggots has overtaken them. Dipping and diligent shepherding will, however, do a good deal. Foot rot is almost unknown, and scab, under more judicious treatment, has ceased to be the terror it was—in fact any serious loss from this cause may safely be set down to neglect. Sturdy also claims a certain number of victims, but not many.

The ordinary routine of a fell farm is as follows: Early in October all the sheep are gathered, and either salved according to the old-fashioned plan, or else dipped, with grease dissolved in the dipping, according to the new and improved plan; the hogs are then sent off to their winter quarters, and the old and broken-mouthed ewes drafted and sold to some low-country farmer—the price varying from 15s. to 25s.—who, after taking one or two crops of lambs by a Leicester or Down tup, feeds them off readily enough on seeds. The rest of the stock then go back to the fell till towards the end of November, when all the ewes are gathered, and the shearling or "twinter" gimmers sorted out and put by themselves in the intakes; then the rest are put to the tup on some inside land, and afterwards go back to the fell, where they range with the wethers till near lambing time, when they are brought inside again, and remain till the beginning of June, or until the grass has grown enough on the fell to enable them to keep their milk. Shearing, or "clipping," as it is called, takes place about the middle of July, some ten days after the washing, and most stocks will average about 3½lb. of wool all round, though where attention has been paid, and there has been no overstinting, they will do 4lb.; and we know of a stock of about 1500 which averaged nearly 4¾lb.

Clipping time is the dalesman's gay season, as the neighbours go to each others' houses in turn to help first in the work and then in the supper, and singing and dancing which follow. After being clipped the four-year-old wethers are sold, but not generally delivered till the autumn. It is hard to say what is the average weight of four-year-old wethers "straight off the fell," as they vary so much according to circumstances; but we should say from 12lb. to 15lb. We have known them from some fells where the pasturage is rather better, and there was no overstinting, do 20lb. per quarter after a few weeks' keep on seeds; and we believe that a pen of Herdwick wethers, that won in a class open to all kinds of mountain sheep at the Kendal Christmas fat show, some years ago, averaged 25lb. per quarter.

It is generally the custom for a stock of sheep to be let with each farm, the tenant being bound (on giving up his farm) to leave the same number of sheep, of the same ages and classes; these are then valued, and, according as they have improved or deteriorated during his occupancy, does the outgoing tenant receive or give compensation. A clause is also generally inserted in the agreement, stipulating that the tenant shall, on leaving his farm, give his landlord the refusal of all his surplus stock, at a price to be arrived at by a valuer on each side. The object of this clause is to prevent any of the stock getting into the hands of neighbouring farmers, as in that case, owing to the attachment of Herdwicks to their native "heaf," they would be very apt to profit their new owner at the expense of their old one's successor, by still grazing on their old fell. Of course, however, if the outgoing tenant is migrating to another farm at a distance, and wishes to keep his surplus stock, this clause is not binding.

A great many objections have been urged against this custom, the most serious being that it enables men without sufficient capital to take farms; but this can be met either by the landlord taking care that his proposed tenant is, if not a man of great substance, a man of steady industry, and one who understands his business, or by requiring a bond. On the whole we think the custom works well and is a good one, as in cases where it has not been in force, and the whole stock has been sold off, we have known great advantage taken of the necessity to buy largely which the incoming tenant was known to be under, to run the biddings up unfairly high.

We have omitted to mention one rather peculiar circumstance, which is, that there was originally a distinct strain of this sheep about Seathwaite, that had fourteen ribs. Though we believe that, as a distinct strain it no longer exists, yet many individual sheep possess this peculiarity; and Mr. Nelson, of Gatesgarth, one of the most noted breeders, tells us that instances are not at all uncommon in his flock, and that he does not notice any other peculiarity or difference between them and sheep possessing the normal number of ribs.

The engravings are from photographs of two tups, the property of Mr. C. W. Wilson, of High Park, Kendal, who has a large Herdwick stock at Kentmere Hall; these tups were shown at the Royal Society's Exhibition at Hull in 1873, where one of them was second to a Lonk tup, shown in the same class. And here we must protest against Herdwicks being made to compete with such a breed as Lonks, which go on immensely superior ground. Surely they might be allowed in the same class with the horned black-faces, seeing that they undergo even greater hardships than these.

THE PIGS OF GREAT BRITAIN.

CHAPTER I.

INTRODUCTORY.

IN following up our illustrations of domesticated animals we propose to devote some chapters to the more prominent breeds of pigs, such as the Berkshire, the Large, Small, and Middle-bred White Sorts, the black Suffolk, the Dorsetshire and the Tamworth varieties, which have of late years acquired distinction in the hands of our leading breeders. In approaching this subject we feel that it may be deficient in interest for the general reader; we cannot fall back on an attractive history. The pig is too common an animal to attract much attention, and yet it will be allowed that a fair amount of profit may be derived from these useful quadrupeds, which have a special value as the scavengers of the farm.

The pig has been domesticated from a very early period. We find mention of it both in sacred and profane history, and we know that in the early days of this country no description of live stock were so abundant, and herds of pigs grazing in the woodlands formed no inconsiderable part of the revenues. Our readers will call to mind the character of Gurth, so ably depicted by Sir Walter Scott in "Ivanhoe," who, by the aid of his horn and his dog Fangs—half mastiff, half greyhound, and whole lurcher—collected together his riotous herd after they had filled themselves with acorns and beech masts. Old England was in those days half covered with wood, in which oak predominated, and it is probable that the acorns formed the principal food on which the pigs for winter use were fed, and after they were consumed the slaughtering and curing for winter purposes would commence. And it may be here remarked that the pig is the only animal that can safely consume an unlimited quantity of this somewhat astringent food, which has proved the source of much loss in cattle and sheep.

The pig has been associated with man from very early days, and it is impossible to arrive at anything definite as to the period of his first subjection. He is the only section of the many-toed division of pachydermatous mammalia that has been thus reclaimed; for though his ally the elephant has been made useful in individual cases the race continues wild, and it is said that, when subjected to captivity, fertility almost ceases. In our own, and in most civilised countries, the wild boar has long since completely given place to the cultivated animal; in Africa and India these animals still form exciting objects of the chase. Whilst the origin of some of our domestic animals is involved in obscurity, so greatly have they become altered by cultivation, the hog retains so much of his original character that we have no difficulty in tracing the resemblance even in the varieties that have been most extensively altered.

According to high geological authority, the boar was coeval with extinct species of the mastodon and dinotherium, and hence must be regarded as the most ancient of our domesticated animals. Much discussion has ensued as to whether the diverse types we see in different parts have sprung from one or more sources. We consider the effect of varying conditions quite sufficient to account for even greater differences. Take, for illustration, one feature which is a very characteristic one, viz., the snout. The use of this organ, in addition to assisting in respiration and containing the smelling apparatus, is to enable the animal to uproot the soil in search for food; therefore its development would depend on the necessity for such work. The British swine, feeding so largely upon acorns and beech masts, has not the same necessity for routing as the African pig, hence we find the snout of the latter much more extended. Then, again, in the domestic pig, the Irish "rint-payer" of fifty years ago was a long-legged, roach-backed, coarse-boned, and long-nosed specimen, that had to rout deep and travel far for its food. The improved animal of to-day is not allowed to rout, has at most a pleasant ramble through the stubbles after harvest, and lives for the rest of the year in well-sheltered yards or sheds; the snout loses most of its utility, and disappears accordingly, until in some of the choicest specimens of the small white varieties the prominent forehead and enormously developed chop almost conceals it from view, and what remains is so ridiculously of the *nes-retroussé* type that its grubbing abilities are reduced to a minimum. Great as these changes are, we can account for them by cultivation; hence we see no reason why pigs should not have sprung from a common origin.

A curious study presents itself in an attempt to explain the degrees in which the pig was estimated in different countries. Take the case of Jews and Mahommedans, amongst whom pork is held in detestation, in obedience to the Mosaic law, which declared these animals unclean. The Egyptians had even a stronger antipathy. Herodotus tells us that " Swine are accounted such impure beasts by the Egyptians, that if a man touches one even by accident he presently hastens to the river and in all his clothes plunges himself into the water. For this reason swineherds alone of the Egyptians are not suffered to enter any of their temples; neither will any man give his daughter in marriage to one of that profession, nor take a wife born of such parents, so that they are necessitated to marry among themselves. The Egyptians are forbidden to sacrifice swine to any other deity than Bacchus, and to the moon when completely at the full, at which time they may eat of the flesh. When they offer this sacrifice to the moon and have killed the victim, they put the end of the tail, with the spleen and fat, into a caul in the belly of the animal, all of which they burn on the sacred fire, and eat the rest of the flesh on the day of the full moon, though at any other time they would not taste it."

Pork is not considered a wholesome food in very hot climates; hence, perhaps, one reason why it should have been condemned. The Greeks and Romans esteemed it greatly, and a young pig stuffed with beccaficoes and served with wine was much approved of, the dish being distinguished as " Porcus Trojanus." In

China, again, where the temperature is high, pork is largely eaten by the natives, though rarely by the English: hence we can hardly explain the differences we have noticed by the effect alone of climate and temperature.

We can well understand the great value of swine in this country before the art of feeding animals through the winter was understood; and even now the acorns are carefully gathered by the children for the cottager's pig. In olden times, when the forests were principally in the hands of the Crown, the copyholders of surrounding land had the right, under certain restrictions, of fattening their swine in the woodlands. The usual time for depasturing extended from fifteen days before Michaelmas to forty days afterwards. In our own time, or rather since the development of agricultural practice, and especially since the formation of the Royal Agricultural Society, wonderful improvements have been made—quite as marked as those in cattle and sheep. The improved pig of to-day is quite a different animal to its half-wild and wholly neglected ancestor, and occupies a most useful position in farm economy.

The store pig, which includes the brood sow, is a scavenger, consuming materials that would otherwise be wasted. Pigs probably convert certain kinds of food into meat as economically as any other animal. They are capable of a good return when properly managed. But, with all these advantages, prices do not rise rapidly in the market. In some exceptional cases particular strains command a high figure, but we have not the competition which prevails, especially for choice cattle or sheep. The reason is not far to seek. The limit to the demand for pork is easily reached, and the fecundity of the pig is so great, that any deficiency is soon made up. Whatever may be the explanation, the fact is undoubted that fresh pork is not sufficiently wholesome to be consumed to the same extent as either beef or mutton; it can only be used occasionally, when it makes an agreeable change. By salting and drying, the too rich properties of the flesh are qualified, and we have a valuable article of diet. Here it is that the value of the animal is best seen. What would the hard-working cottager, with a large family, do without his pig? Wherewith could he otherwise utilise his waste materials, the bad potatoes from his allotment, the odds and ends from the house? With the bedding he is able to collect, valuable manure is made, and eventually the sides of bacon embellish the interior of his cottage, and provide very often the only flesh he can afford during the winter.

The pig has acquired a bad name as a dirty, unwholesome animal, whereas he is simply the victim of circumstances. He is particularly clean in his habits, and has a more sensitive nature than the other animals with which he is domesticated. Confined in a small stye with an open court, the latter unpaved or so badly floored that water accumulates, with little or no litter, is it to be wondered at that the place is dirty, and that a foul smell proceeds from it? The stye is often close to the house, not unfrequently in dangerous proximity to the well, and the innocent and otherwise useful friend of the family may thus prove the source of deadly disease. Fortunately, the recent sanitary legislation is rapidly effacing a condition of things disgraceful to the age we live in. The pigstye has to be removed and properly drained. Unfortunately there is no legislation against the open court, which we consider at the root of the mischief. Let the pig have a roof over his head, and he is no longer the source of bad smells. The manure is doubly valuable, the animal thrives much faster, being particularly sensitive to cold, and less bedding is required.

When only one or two pigs are kept, a building 8ft. by 8ft. is quite sufficient. On the south side should be placed the door, and the wall adjoining need not to be carried up more than 4ft. high, thus leaving open, along the south side a space of about 2ft. from the top of the wall, and close to the plate—this ensures ventilation. The floor may be laid with a slight fall to one side, and an opening in the brickwork allows of the escape of any liquid; but if carefully bedded, there will be little or no waste. A small tank outside is useful to catch the drainings, which otherwise must be taken away to the nearest outfall. By extending the building another 4ft. in the clear, making the extreme outside dimensions 14ft. 3in. by 9ft. 6in., we can make a convenient privy and covered ashpit, which for cottagers' use is preferable to the earth closet.

Were it only for its value to the poor man we should esteem the pig as a most desirable acquisition to our domesticated animals; but when properly managed pigs are a source of considerable profit to the farmer. However economical and careful his management, there must be on a farm waste materials—for example, the bad potatoes, the material taken from the root crops when the latter are prepared for cattle, the refuse corn, to say nothing of the portion of grain unavoidably left in the straw, all of which will be utilised by store pigs. Again, the manure from the pig is valuable in proportion to the food he is consuming; but store pigs are mechanically good manure makers—constantly routing about in the yard in search of food, they cause the materials to be properly mixed up, and do a great amount of good. They are omniverous, so that nothing comes amiss; and, though the fatting pig requires rich food, stores and breeding animals can be kept in a thriving state at little expense. We have seen sows in summer time in good condition with nothing but mangolds and water.

In these days of improved harvest machinery there is comparatively but little corn left in the field; what there is, is better utilised by the pig than in any other way. The whole strength of the farm, with the exception of the advanced feeding stock, should be turned out, previously well rung if young seeds form a part of their range; and, under charge of a lad, they will soon clear the ground, care being taken to provide water at least once a day, for dry corn is thirsty work. It was formerly the practice to a larger extent than now to purchase stores for this purpose, selling them again when the work was done. During the winter store pigs utilise the waste materials from the house, clear up the leavings from the cattle troughs, and act, as we have before stated, as the scavengers of a farm. Inasmuch as they cannot exercise such important functions without being a good deal exposed to the weather, we must be careful that, in cultivating the feeding and early maturing properties of the pig, we do not sacrifice hair and constitution.

This is not the place to discuss the merits of the different breeds which find favour in their own locality. Of late years the public have shown increasing partiality for animals of medium size, which, whilst possessing much of the feeding properties of the Chinese, are yet of sufficient growth to yield good flitches of bacon. The Berkshire and middle-bred white variety are probably more extensively grown than any others, and of these two the Berkshire preponderates. The taste for very small or very large sorts is dying out, because neither are found generally profitable. Quite recently the Tamworth, a red-skinned breed, have come into favour on account of the excellent quality of their bacon.

The question here arises, do pigs pay? Much has been written on this subject, and the majority of writers are of opinion that it is cheaper to buy than to feed our bacon—a conclusion we by no means indorse, because on a farm a considerable part of the feeding can be done comparatively cheaply by using food which would otherwise be wasted. If a gentleman buys a pig and feeds it entirely on purchased food, he will find his bacon an expensive luxury, and there is no doubt that a point is soon reached beyond which the increase of weight is not sufficient to cover the cost. It would be important to the feeder if we had trustworthy experiments to refer to; but, whilst something has been done to elucidate our practice as regards cattle and sheep, with the exception of Sir J. Lawes's experiments, and a series carried out by Mr. C. G. Roberts, to which we shall refer, the pig has not, as far as we know, been subjected to well-sustained investigation; consequently everyone has his own ideas on the subject. That bacon, to be good, should be fat, and the fat interspersed between the lean, is generally acknowledged; but we do not know where a paying system ends and a losing one begins. Mr. John Tyrrel, of New Court, North Devon, states that he put up twenty pigs to fatten upon roots and a large proportional quantity of oats and barley meal. During the first fortnight the increased weight compared with the food consumed paid 9s. 6d.; the second fortnight 4s. 6d.; in the third period of three weeks the increase only represented half the cost. From this, which after all is only a rough test, it would appear that feeding to only a moderate point is best. But a solitary instance like this requires confirmation; we require to know the amount of grain that can be assimilated. If a pig is allowed to eat at discretion, that is as much as it possibly can, much of the food constituents pass away unassimilated in the manure, adding much to its value, but making the feeding process very costly and unprofitable. Had the animal received only such a quantity of nutriment as it could have utilised, the process would be slower, the manure for a given bulk less concentrated; but the result might be profitable instead of the reverse, and we should find for our offal corn and vegetables a better market than we could obtain outside the farm.

We have heard it stated that the cottager's pig is not profitable to him; that it would be better if he neither had his allotment, his pig, nor, as a few favoured individuals have, a cow. And so far back as 1842 Mr. Edwin Chadwick stated, in his sanatory report, that a large mass of evidence supported the opinion of one of his witnesses that pig-keeping and cow-keeping were injurious to the condition of the labourer; that the labouring man pays more dearly for his bacon than he would if he purchased it ready made, &c. We are glad to know that a different opinion prevails at the present time. Competition for the labourer's service renders it desirable that he should have home attractions; and where shall we find them but in a decent cottage, good garden, and allotment, and, if possible, the opportunity for keeping a cow? Now, with the allotment the pig is a necessity, it is the medium for converting the produce into cash and manure; and, apart from political economy, it is good for the labourer to keep his pig, even supposing that he pays as much for his bacon as though he bought it. The pig and garden are his savings bank, into which he puts the value of his spare time and the savings of his weekly wages, which otherwise might go to the publican; and what a source of interest is the pig!—how it is cared for by the whole family, how little it lives upon in its youthful days, and how carefully is every scrap of waste matter taken care of for its use! Then the small and disused potatoes, boiled up and mixed with barley meal, also the produce of the allotment, finish off the pig, whose sides eventually embellish the cottage, and, as Mr. Sturt so well puts it, form there the very best furniture.

We believe that pigs, under favourable conditions, can be made to pay, and leave us valuable manure; but we must have the right sort of animal to start with, judicious management throughout its early life, and careful feeding to its close. Let us inquire shortly how these conditions can be best secured. First and foremost we must secure a good sort; that is, an improved breed, with the minimum of offal that is consistent with a due amount of constitutional vigour—a combination that is not always easy to procure. In some of the modern sorts high feeding and hothouse training have produced marvels as to early maturity and development; but the animals are coatless, delicate, and quite unfit for the rough outdoor life that the working pig has to lead; and we unhesitatingly say that, wonderful as these animals are in a showyard, they are after all only monstrosities, and not rent payers. The farmer must look for an animal that combines early maturity and aptitude to feed with a vigorous constitution. We will not say in which particular line his requirements will most easily be satisfied—there is merit in many directions; but in making his choice he must select the class of animals best suited to fit in with the conditions which surround them. Again, he will be guided to a certain extent by the market he caters for. Thus, if he has a trade for sucklings and porkers, he will not select the same animal that would specially suit when large bacon was the object. However, in these days the distinctive features of different breeds are far less marked than formerly; and whilst in old times we had large slow feeding sorts as compared with small varieties that fed more rapidly, at the present time extremes meet in an animal of medium qualities that can be used for any purpose with advantage. Still there are local conditions that give the preference. At one time colour was a difficulty. In the district of white pigs, blacks were regarded with disfavour, and in the Midland counties the dark sorts predominated, and were preferred. Like the good horse, which it is said cannot be of a bad colour, the well-fed pig is now generally approved, whatever its complexion.

The animal is chosen, and the young farmer invests in the best he can find, not giving fancy prices, and especially not taking animals that have been forced for show. Such are frequently disappointing, either not breeding at all, or producing small and delicate litters. The best plan is to go to a well-known breeder whose success is established, and select a couple or more hilts when fit to be weaned, obtaining the use of a boar from a different strain. The young hilts must be well fed; that is, they must have food containing flesh-forming elements.

F F

We believe a small quantity of ground beans mixed with the house wash will prove economical and sustaining; better still if we add a handful of Smith's palm-nut meal, which, from its high percentage of fatty matter, is very satisfying. The food should be mixed thin, wash or water preponderating, and given at first three times a day—morning, noon, and night. Weaned pigs will eat, in addition to the wash, $\frac{1}{2}$lb. of each kind of meal daily, which may be increased to $\frac{3}{4}$lb. when the animals are six months old. It is surprising how small a measure suffices to keep the animals in healthy, growing condition. Exercise is very important. If young animals are closely housed, the muscles are not properly developed. We prefer an open yard, with suitable shelter sheds. One, two, or three hilts cannot, of course, be by themselves; they will take no hurt if associated with young cattle. Indeed, we have frequently noticed decided friendships between the porcine and the bovine race thus brought together. The pig will nestle up close to the beast, in order to benefit by its shelter and the warmth imparted from its body. Any refuse from the cattle mangers may be placed in the pig trough, and will be greedily eaten.

Assuming that our young sows are of one of the improved varieties, and have been well done to from birth, they should be ready to receive the boar when eight months old, so as to produce the first litter when a year old. We are surprised to find that in Mr. Martin's excellent treatise on "The Pig," edited and revised by Mr. S. Sidney (Routledge and Co., price 1s.) a book which enjoys a well-deserved popularity, the following opinion of Mr. John Tyrrell's is quoted, and apparently adopted: "Sows should be at least two years old before they are mated. They are not full-grown until five or six." Such might have been true of the unimproved sorts, but it is incorrect now, when a pig attains full size at or under two years. It would make the sow very costly if she were to remain unproductive until twenty-eight months old, by which time, according to the practice we advocate she should have had three litters, and might be going on towards a fourth. The possibility of this early fruitfulness depends upon the early maturing properties of the breed being supported by abundance of wholesome nourishing food. Modern experience is in favour of amended practice in this respect with all our domesticated animals. It was formerly held that to breed from heifers before they were fully developed resulted in a puling offspring and a stoppage of growth in the dam. With generous diet the opposite is the fact. The late Mr. Edward Bowly, in his admirable essay on Cattle Management, published in the 19th volume of the "Journal of the Royal Agricultural Society of England," gives a case in which a heifer calved at fifteen months and two weeks old, the calf being at its full time. This animal gained a first premium as a two-year-old in-calf heifer, and a second premium the following year as a cow in calf, and was afterwards sold at a high price to go abroad. The sire of the calf was a bull calf about six months old, sucking a cow in the same field, and neither was a week over six months. Since Mr. Bowly's essay was written, the practice has become common of bulling heifers at fifteen months, and with proper care no evil results; indeed the animals appear to grow out faster whilst in calf than when empty. With cart horses, again, we have found it decidedly profitable to cover the fillies when two years old; early maturity allows of early breeding.

It is a necessity of the present day to secure a quick return, and we must feed highly in order to get the results. There is, however, one point to be considered in reference to the pig which does not apply so much to other animals, owing to their sensibility to extremes either of heat or cold. It is not desirable to have our litters either in midwinter or the dog days; therefore it is best to have the first litter either in the spring or autumn. With this provision, the nearer a sow is to a year old the better for our profits. The period of gestation varies from 115 to 130 days. During the first two months it is not necessary to increase the quantity or quality of the food; but during the later stages a more generous diet is desirable, and towards the time of farrowing the addition of skimmed milk to the wash will materially tend to the secretion of milk. A few days before the expected advent of the family the hilt should be removed to the farrowing pen, so as to become accustomed to her new residence in good time.

The farrowing pen should be roomy, not less than 10ft. by 8ft., but 10ft. square is preferable, well ventilated, and warm. The floor may be of brick or stone, with a slight fall, so as to allow of the escape of the urine. It should be covered with short litter, sometimes rough cut chaff is employed. Around the walls, at about 8in. from the ground, should be fixed a board or a rail projecting at least 6in. from the wall. This is of great importance, preventing the sow from lying against the wall, and allowing the young an opportunity of protecting themselves from being overlaid, which without this precaution is by no means uncommon, especially with a young sow during labour, which is often protracted for some hours. The approach of farrowing is indicated by the restless movements of the hilt, as well as by the appearance of the teats and the hind parts. She will be observed carrying mouthfuls of the litter to one corner, as though preparing her bed. Whilst carefully watched, she must not be disturbed more than is necessary, and especially whilst farrowing she should be left as much as possible to herself. But it is important that the pigman should be near at hand, as her sufferings sometimes make her savage, and cases have been known in which she has killed her offspring as soon as they were born.

It is a good plan to accustom the hilt to be noticed and handled, in which case she is not disturbed by the attendant, whose services may be of great importance to the litter. It is also well to have a basket at hand lined and covered over with flannel, in which the pigs can be collected as they are born, and, if the night be cold, placed near a fire so as to insure warmth; they should be kept from the sow until the business is concluded, which seldom lasts more than a couple of hours, and is frequently over in a few minutes. It is not desirable that a hilt farrowing at the age indicated should have more than seven or eight offspring; and if there should be more, as is sometimes the case, it is a good plan either to destroy the surplus, selecting of course the weaker ones, or to make use of them as roasters when they are three weeks old, and before they seriously draw upon the maternal support; but destruction at birth is the wiser economy. In case the hilt shows symptoms of feverishness after farrowing, it is well to administer a mild aperient in the food—say 4oz. to 6oz. of Epsom salts, with 1oz. of sulphur. The food, for a couple of days at any rate, should be very thin, and, if possible, given warm. Milk and fine sharps or pollards make an excellent mixture. If all goes well, by the third day the young pigs will be well on their legs, and the sow capable of eating the ordinary food, which must be plentiful and nutritious. On dairy farms where pig-breeding can be carried on with more profit than elsewhere, there is either skim milk or whey, both of which are excellent media for the more solid foods. Of these we believe, from a lengthened experience, that barley, and palm-nut meal, with pollard, form both a cheap and sustaining diet. The sow should have exercise once a day if weather permits. Every sow should have twelve teats. It occasionally happens that there is some trouble with these vessels; they are liable to inflammation from cold, in which case they should be carefully fomented, and well softened with lard, or, what is better, goose grease. As the family grow their wants increase, and the sow will require more food. The dust of good linseed cake boiled into a gelatinous condition is said to be a valuable addition to the food we have named, but, not having used it, we cannot speak from experience. The wash from the house, with all refuse vegetables, can be made use of. At three weeks old the little pigs begin to eat, and it is desirable that the family be removed into a larger compartment, still covered over and well sheltered. A partition at one end separates sufficient space for the youngsters to assemble round a trough suited to their capacities, or the latter may be long-shaped and placed against the wall, with divisions by means of iron rods, so that each pig may find a separate supply. Access to this through the partition may be obtained by a slide door sufficiently large to admit the pigs one at a time, and thus they are supplied gradually with food, and can be shut away for a time from the sow if desirable. As the offspring fend (to use a Yorkshireism) more and more for themselves, the food of the sow may be diminished, so that the weaning process, which is finally accomplished from eight to ten weeks after birth, may be gradual. After the first three or four weeks, on fine days the family should be turned out into a sheltered yard, so as to accustom the youngsters to exposure, which is the more necessary if we intend them for bacon purposes, for there is no doubt the combination of aptitude to feed with hardy character is most desirable.

After weaning, the young sow must be kept moderately well.

Just at first there may be some difficulty with the milk, but it rapidly disappears. Some recommend that the process of weaning, as far as the young pigs are concerned, should be gradual, i.e., that at first the pigs should be allowed to suck once or twice a day before being entirely separated; but we have never found this necessary, and it delays the breeding process, for, as a rule from which exceptions are rare, the sow will not come into use whilst the pigs are with her. If our hilt farrows on the first of March, she will be ready to receive the boar by about the middle of May, and come down with the second litter about Sept. 15, thus giving us two litters per annum, which is as much as we can expect. Of course it often happens that the sow turns more than once, which altogether upsets our calculations. It must be an unusually lucky sow that yields two litters a year on an average. When quite clear of milk the young sow may be turned out during the day in a grass field, and have at night and morning a little wash with a moderate quantity of mangold—by this time, i.e., the summer, rich in sugar and nutritious. We have known breeding sows kept in quite good condition with nothing but mangolds in yards. The sow whilst in pig—at any rate, during the earlier stages—should cost little or nothing, otherwise the business will not pay. In the autumn time the sows in farrow, or from which the pigs are weaned, may pick up a good living in the stubbles provided they get water once a day. The treatment of the young pigs depends very much upon the use they are intended for. If pork is our object, the rotundity of form first acquired from maternal support, and afterwards by careful feeding, must be maintained. It will never pay to lose what may be rightly called the milk flesh. The pigs should be placed in a comfortable stye, with plenty of ventilation, fed three times a day with a mixture comprising skim milk, if it is available, pollard, barley, and palm-nut meal, or other meals equally nutritious. There is no doubt whatever that the digestive system of the pig requires a more highly concentrated food than either the sheep or the ox, and this is due to the different proportions between the stomach and intestines. Thus the ox can eat the greatest amount of non-nutritious matter, because for each 100lb. live weight it has $11\frac{1}{2}$lb. of stomach and only $2\frac{1}{4}$lb. of intestines. The sheep requires more concentrated food, because it has only 7½lb. of stomach to 3¼lb. of intestines; whereas the pig has, for each 100lb. of its live weight, only 1¼lb. stomach to 6·2 of intestines. It is in the stomach that the great work of mastication is really done. The porkers, if pork be our object, should be fit for the butcher at about five months old, when they will average from 70lb. to 90lb. each. If, on the other hand, we intend to keep our pigs for bacon at a future period, we must adopt an entirely different system. In order to develop frame and muscle, a portion of highly nitrogenous food is desirable; hence lentils or foreign beans, according to price, may be bought; these may be supplied in small quantities, and eked out by garden stuff, wash, &c. Stores must have exercise, and not be too delicately reared; indeed, the spring litters will have a run of a month to six weeks in the stubbles, where they should grow considerably and prepare for feeding; during this time, whilst out on stubbles, water must be supplied at least once a day, otherwise they will not thrive.

It becomes a highly interesting and important question to determine at what period pigs pay best—whether if sold as stores when weaned, at which age they will, if the sow has been well cared for, bring on an average about 1l. each; or as porkers, when from four and a half to five months old; or as bacon pigs, and if so, how long the stores should be kept—in other words, to what weight should they be fed? Now, in endeavouring to solve these knotty questions, we must not leave out the utility of the pig as a manure maker, nor must we forget that its greatest value in this sense will probably be at an advanced age. As we have stated at the commencement of these articles, the only experiments that have been systematically made are those by Sir J. Lawes, and later—by Mr. C. G. Roberts.

We propose further on to enter at some length into the facts which were elucidated, and to shortly describe the different kinds of food available for pig feeding. It will probably be found most profitable when we keep pigs principally as a source of manure to produce both porkers and bacon pigs, taking care that the former are fit for the market between November and April, after which periods the demand is limited; whereas bacon pigs can be sold perhaps better in summer than winter. The large factories, which are our best customers, go on curing all the year round, and often do their largest business in the summer. Moreover, it has been proved by direct experiment that bacon pigs increase most rapidly upon a given weight of food in summer, which is only to be expected when we consider how extremely sensitive they are to cold. It will be found, when we come to dwell upon these experiments, that feeding *per se* is not profitable; in other words, that the increase of the animal from the time it is put up to its death does not equal the cost of the food required to effect such increase. Hence, as a mere question of profit, it will pay best to keep a lot of prolific sows, making them act chiefly as the scavengers of the farm, and sell the weaners. This we believe would actually pay well; but then we must remember that the value of the manure depends entirely upon the feeding; and sows that work hard for existence are of little use as manure makers. Farming has to be considered as a whole. We shall arrive at very wrong conclusions if we form our judgment only on its separate features; therefore we do believe that judicious pig feeding pays the farmer, because it secures to his crops valuable manure on the spot, which he could not otherwise obtain so economically. But even to reach this moderate position requires careful and economical arrangement.

The elaborate series of experiments carried out by Sir J. Lawes, and described in vol. xiv. of the "Royal Agricultural Society's Journal" of 1853, are of great scientific interest, since they tend to elucidate the relation and value of different kinds of food as sources of meat and manure. Liebig and others had assumed that the value of feeding materials depended upon the proportions of the nitrogenous elements they contained, regarding the starchy or oily ingredients as of secondary importance—a view that was rational enough when it is remembered how important are the functions exercised by the nitrogenised structures and fluids of the animal body, but which is not borne out by actual experiment, or by the relative price of different foods. Sir J. Lawes contrasts the composition, feeding effects, and market value of beans and barley; the former contains nearly double the amount of nitrogenous compounds, yet the cereal produces the better results and commands the higher price. If our object were only the production of meat at the cheapest rate, we should confine ourselves to the use of starchy grain, and materials rich in vegetable fats, especially in dealing with animals fully grown. But, inasmuch as we have a two-fold object in feeding, viz., the production of meat and manure, and as the value of the latter depends so materially upon the proportion of nitrogen in the food, it is quite evident that the most profitable practice will consist in the employment of a mixed diet, which conduces to both ends, and especially will it be desirable to use nitrogenous materials during growth, and substances rich in fatty matters for the later stages of feeding. Both Sir J. Lawes and Mr. C. G. Roberts prove that feeding *per se* is not profitable; our return depends upon the value of the manure obtained. Thus Sir J. Lawes in his opening remarks: "The profit of feeding, indeed, is to be sought *within* the limits of the value of the manure; and that it is therefore much dependent on the quality of the latter, and consequently on the judgment exercised in the selection of the foods and the management of the animals and the manure, is a view which seems to be supported at once by the convergent testimony of current experience, and by a consideration of the laws which regulate the price of all articles in general use. Admitting that the prices of all such articles are regulated by the cost of production, and that they cannot long either be produced at a loss, or be sold at a price which will yield more than a fair profit upon the capital and labour employed in their production; and applying this view to the subject before us, we should certainly decide that the selling price of the meat alone produced upon the farm must be less than that of the food consumed, and that the profit of the feeding process is to be found in the remaining product, namely, in those parts of the food which are rejected by the animal, and which, under the title of manure, gives fresh fertility to the soil, and thus supply a second product for the market."

Mr. C. G. Roberts conducted a definite experiment to test the result of feeding upon a mixture of malt dust, a material tolerably rich in nitrogen, and Smith's palm-nut meal, which contains a high percentage of vegetable oil. The experiment lasted for ninety-four days. At first a small quantity of boiled roots was given, and latterly 1lb. of peas daily per pig was used. Two of the four pigs did not eat the food well during the later stages, which helps to explain the less favourable results as compared with those at Rothamsted. Thus, in Mr. Roberts' case, 561lb. of dry organic matter was needed to produce each 100lb. of increase; whereas 488lb. in one case, and 461lb. in another, gave a similar result in Sir J. Lawes's case. Hence, Mr. Roberts concluded the food used by Sir J. Lawes, and which comprised mixtures of beans, lentils, Indian corn, barley, and bran, was more suitable for pig food than malt dust and palm-nut meal. This was possibly the case, but it is evident that no comparison can be relied upon as conclusive where the conditions as to the animals operated on were different. And it must be noticed, as tending to reduce the difference in the results, that in Sir J. Lawes's experiments 24·84lb. of dry organic matter in one case, and 26·8lb. in the other, was consumed per week for each 100lb. live weight of animal; whereas at Haslemere the average was 18·60lb., or not much more than two-thirds the quantity. Touching this remarkable difference, Mr. Roberts thus comments:

"It will be seen that the rate of consumption at Haslemere was not much more than two-thirds of the consumption at Rothamsted; and when we consider how much of the food is required for respiration and the waste of tissue, I think we must be struck by the comparatively good return of 100lb. increase in live weight for each 561lb. of dry organic matter consumed. It certainly looks as though the fault did not lie in the nutritive power of the food, but rather in its being not sufficiently palatable to induce the pigs to consume it largely. If by the addition of some highly flavoured condiment the pigs could have been induced to eat half as much again of the food, it seems probable that they might have fattened even more rapidly than those at Rothamsted, and the profits in that case would have been very much increased."

Mr. Roberts supplies us with a balance sheet, which shows that, whilst the sale of the pigs was 14s. less than their cost, the value of the manure, according to theoretical standard, left a balance on the four pigs of 1l. 14s.—thus:

COSTS.				RECEIPTS.			
	£	s.	d.		£	s.	d.
Four store pigs	10	11	0	692lb. of pork, at 6d.	17	6	0
Attendance	0	10	0				
Killing	0	4	0				
1327lb. palm-nut meal, at 8s. cwt.	4	14	7	Manurial value at 34s. per ton	1	0	3
668lb. malt dust, at 4s. cwt.	1	3	9	Do. do. at 71s.	1	1	2
116lb. peas, at 1d. per lb.	0	9	8	Do. do. at 62s.	0	3	3
12½ cwt. turnips	0	5	8	Do. do.	0	2	0
Balance, being profit	1	14	0				
	£19	12	8		£19	12	8

Regarding the fact that two out of the four pigs did not eat the food well, and that when killed one was found to have diseased lungs, we think the results were decidedly successful, and proved that the food used was economical.

It would be very desirable if the relative values of different kinds of food could be absolutely determined by experiment, both as regards feeding and manuring properties. Such experiments, however, are difficult to carry out, and therefore we must fall back upon comparative price as a test of comparative value. Bearing in mind the conclusions to be drawn from Mr. Lawes's researches, we should endeavour to combine materials that develope flesh, bone, and fat in such proportions as are best suited to the requirements of the animal at different periods of growth. With the pig, as with our other domesticated animals, early maturity and quick returns are desirable. Whilst growing they should also be feeding. There is no necessity to lose the pork flesh, even though we prolong life until full growth, and this may be done without an extravagant outlay in food; indeed, the best bred pigs of the small or medium sorts cannot be kept thin. In early life our object should be to supply such food as will develope bone and flesh especially, with a proper proportion of heat producing material. We have considerable choice; thus, bran fine or coarse, beans, peas, or lentils, ground cotton-seed cake, all are materials rich in flesh formers, which may be used in conjunction with barley or palm-nut meal. Our own choice would be good sound bean meal, which we should use in equal quantities with a mixture of barley and palm-nut meal, the quantities depending upon the age of the pig. As the animal approached maturity, and as the actual fattening process commenced, we should leave off the leguminous element, and confine the diet to a mixture of barley and palm-nut meal, which we have found admirably adapted to the purpose.

Conflicting opinions exist as to the value of cooking pig food. Where we have vegetables to deal with, as in the case of potatoes or turnips, steaming is decidedly advantageous; and wherever we can cook without a serious increase of expense we believe it answers, because the digestion of the pig is not so complicated as that of the ruminant, and any assistance we can give will effect a saving of force, and an economy of food; but we have grave doubts as to whether the use of warm food is desirable. Let the food be cooked by all means, but give it cold. Where cooking is not convenient, the materials should be soaked in water for at least twelve hours before being used. In the case of barley-meal, and possibly to a less extent with the palm-nut meal, the food is rendered more digestible because the starch cells are dissolved, and this is one great advantage of hot water or steam.

We may here notice a practice which was highly thought of by Arthur Young, viz., giving the food in a sour or acid condition. Thus Mr. H. Evershed, in his prize essay on Warwickshire, tells us that swedes are boiled and mixed with one-third of meal, a couple of cisterns are filled with a week's allowance in each, and the food is used a week old and in a sour state. Again, in Mr. Martin's Book on the Pig, edited by Mr. S. Sydney, "Some recommend that the meal be mixed with cold water in large cisterns, the proportion being five bushels of meal to 100 gallons of water. This mixture must be stirred several times a day for a fortnight or three weeks, until an imperfect fermentation takes place and it becomes acid. In this state its fattening powers are said to be greatly increased." We have not ourselves tested the merits of this plan; everyone knows that the contents of the hog tub—which is the receptacle for all the pot liquor, waste vegetables, &c.—are frequently in a sour state, yet such materials are greedily eaten.

Many writers assume a high feeding value to maize (Indian corn) for pigs. We believe it may be usefully employed in connection with home-grown meal, but if given by itself it is not altogether wholesome; and this is probably owing to deficiency in mineral constituents. Sir J. Lawes found that pigs fed entirely on maize were affected with swellings on the neck, and their breathing was laboured; when, however, supplied with finely-sifted coal-ashes, common salt, and superphosphate of lime, which they soon commenced to eat, these symptoms rapidly lessened, and soon altogether disappeared. It is well known that fatting pigs relish cinders, and, for whatever reason, it is a good plan to give them a shovelful occasionally—possibly the gritty material may assist digestion. In the case of stores, breeding sows, &c., it is not necessary or desirable to be too particular about cleaning the roots with which they are fed; a certain quantity of soil appears desirable.

During the summer time sows that are properly rung may be turned out to grass, and will maintain themselves, especially if supplied night and morning with a few mangolds, which at this season are very sweet and nutritious. Mr. John Tyrrel, in his paper on pig-feeding, quoted largely by Mr. Martin, states that his sows, from fifteen to twenty months old, lived in the fields whilst there was any grass, even up to Christmas, having a well-strawed shed into which they could run at all times; during winter they had roots, and did well. We have seen sows summered in yards on mangolds fresh in condition and healthy; we cannot, however, advocate such treatment. Wash and roots should form the staple food of store pigs; but, as our object should be to mature as rapidly as possible, we do not believe in a low diet either for growing or breeding stock. If the sow is kept low, especially near the time of farrowing, the pigs are weakly, and she has not the proper supply of milk for their sustenance; therefore, a handful of bran, a few brewer's grains if they can be got, or a small quantity of palm-nut meal or home grown corn, added to the wash, will pay, not only in the healthy, thriving condition of the stock, but in the improved quality of the manure.

CHAPTER II.

THE BERKSHIRE PIG.

IN placing the Berkshires at the head of the list, we have been guided by the fact that they are more extensively cultivated than any other distinct breed — a safe indication that their qualifications render them more generally useful. Although classed as a large sort, they occupy in this respect an intermediate position, and are probably smaller than their originals, which are described by Loudon as "being in general of a tawny, white, or reddish colour, spotted with black; large ears hanging over the eyes; thick, close, and well-made in body; legs short; small in the bone; having a disposition to fatten quickly, and when well fed the flesh is fine; feeds to a great weight; is good for either pork or bacon." The tawny or reddish colour has entirely disappeared, although every now and then we have reddish spots occurring, which may be taken as a proof of Loudon's accuracy; and, as in one instance—that of the Tamworth breed—the tawny skin is retained, it is probable that these animals had a common origin with the Berkshire, the present divergence resulting from the effects of crossing. Another evidence of the original colour is found in the pinkish hue of the skin, which still distinguishes the improved Berkshire from other animals—such as the Essex and Suffolk—which owe their prevailing characteristics to a larger infusion of Neapolitan blood. History is silent as to the means by which the improvement was made. We have two distinct varieties—viz., such as are wholly white, of which the Coleshill and Windsor breeds are examples; and such as are principally black, relieved by white down the nose, on the feet, and end of tail. These latter are the more numerous, and represent the recognised type. Mention of the former suffices; our remarks apply to the latter. Some writers attribute the difference to the influence of either Chinese or Neapolitan blood. We are inclined to believe that the Berkshires have been principally, if not entirely, allied with the Chinese, of which there are both white and black varieties. The original type would be easily altered either way; hence the two lines. We are also of opinion that in neither have the original qualities been greatly interfered with, and we question if the Berkshire has been modified by the introduction of foreign blood to the same extent as many other breeds. For instance, we still find occasionally the lop ears spoken of by Loudon. The features are moderately long, and not fine, like the Neapolitan, or short, as the Chinese; but, if the improvement has not been entirely due to crossing, there is no doubt about the changes that care and attention have produced.

The Berkshire pig is, taken in all points, of a thoroughly useful character, fulfilling all the requirements of modern farming. If our object is to feed small for the London market, other varieties can be named more suitable; and again, if we require only large bacon, it is possible that we should find more serviceable animals amongst the large improved white sorts. But if, as is usually the case, we require animals that are suitable either for making pork or bacon, that arrive at maturity early, that are hardy, and capable, if required, of picking up their own living in the straw yard or the stubbles—in short, if we want a rent-paying sort, suitable for practical purposes, not for hothouse pets, then the Berkshire will meet our requirements better than any others. Hence their popularity and distribution, which, originally confined to the county from which they are named, is now extended far and wide throughout the midland and western counties. Nor is their influence confined to this country. Ireland owes more to the Berkshire than to any other, and probably all other breeds combined, for the transformation of the Irish pig, which was noted for ungainliness, length of leg, coarseness of bone and offal, and slowness of growth, and which now is practically of the Berkshire type—

so much so that specimens have come over to our national shows, and competed not unsuccessfully with the stock from which they sprung. A remarkable instance of this was seen at the Royal Agricultural Society's meeting at Worcester in 1863, where Mr. Joyce contributed a choice assortment, and secured several of the premiums. Of course, improved management renders a better class of animals possible, and it is equally or even more true of the pig than of either the sheep or the ox, that half of the breeding goes in at the mouth. So long as the Irish pig had to seek its own livelihood in lanes and villages, a roach back and drooping quarters were suitable for its long journeys and hard, precarious mode of living; indeed, the improved animal could not have maintained its condition under such treatment.

The points of the improved Berkshire are as follows: Head moderately short; forehead wide, nose slightly dished, straight at the end, not *retroussé*, as in the small breeds; chaps full; ears slightly projecting, occasionally pendant and covering the eyes. Prevailing colour black, with white blaze down the nose or white star on forehead: sometimes uniformly dark; but this is the exception, and never the dead black of the Suffolk or Essex. The pink tinge should be always apparent. The eye is not sunk and closed, as in the breeds remarkable for feeding properties, but large, intelligent, and denoting activity. General effect pleasing. The head is well set; the neck, of moderate length, is full and muscular; the shoulders well set— so that we have a perfectly regular outline. There is not the extraordinary wealth of chine seen in the Suffolk, but the forequarters are well proportioned. Occasionally we find a slight deficiency in the girth, caused by the flatness of the fore ribs. The back is fairly level, and the ribs, as a rule, tolerably sprung; a less perfect barrel, however, than is to be found in the Essex and Suffolk blacks. Loins wide and well covered; quarters often rather short and drooping—this is probably the weakest point in the breed. The tail is usually set lower than the hips, which gives a somewhat common character; Mr. Weir has shown this defect in his illustration. The gammon full and deep; under-lines somewhat irregular; the flank is often light. Such are the general features of the improved Berkshire. The carcass stands on short legs, and the bone, whilst stronger than the small sorts, is well proportioned, and by no means stronger than is necessary. The strength and character of the coat varies according to sex and management. The effect of confinement and close breeding is to reduce the hair. We have a great objection to bristles, which indicate a thick skin, coarse offal, and slow feeding; but we also equally dislike the thin, weak, soft hair, which is a sure evidence of delicacy, especially in the boar; here, at least should be plenty of hair, otherwise the offspring will be sadly deficient. In the sow, fine long hair is desirable—too much and too strong hair is indicative of coarseness. But if the pig is required to work for its living and to officiate as the scavenger of the farm, there must be constitution, and we cannot have this without hair. The great merit of the Berkshire over most other breeds consists in the larger proportion of lean meat, and the distribution of fat and lean when properly fed; consequently a given live weight realises a larger proportion of available meat than any other breed, except perhaps the Tamworths. We have stated our opinion that the improved Berkshire comes under the category of middle-sized rather than large breed. It is true that extraordinary weights have been obtained by individual specimens: thus we knew of one instance in Herefordshire where an animal weighed 28 score or 560lb., dead, at a year old; but, as a rule, with ordinary feeding, from 15 to 16 score is the average result of a year's growth, supposing always that the animal is well kept from birth.

The development of the breed into a foremost position reaches back some twenty-five years, when the persevering efforts of certain breeders, and especially Mr. W. Hewer, of Sevenhampton, the Rev. H. Bailey, of Swindon, and others, gained the attention of the Royal Agricultural Society, who in the year 1862 gave a separate class for the first time. Now it is not uncommon to find the Berkshires making up nearly half the total entry. Some really extraordinary collections have been seen of late years. Nor is it at the breeding shows only that the Berkshires have won distinction; some ten years back the gold medal for best animals in the Smithfield Show went to a grand lot of Mr. Biggs's, of Cublington, and the prizes for animals neither white nor black generally go either to pure Berkshires or their crosses. But we must allow that it is amongst the older entries that merit is most noticeable, and it is only fair to state that other sorts are capable of greater development at a given age. Wonderful as are the youngsters shown by Mr. Smith, of Henley-in-Arden, and others, at the Birmingham meeting, some of the white sort are capable of even more marvellous development. As an example, we may notice the improved white Dorset of Mr. T. Homer's at the Smithfield Show of 1876, which at 7 months 3 weeks weighed 17 score, and took the silver cup as best white. It is on the score of general utility and adaptation to the wants of the ordinary farmer that the Berkshire claims consideration. The late Mr. William Hewer was for many years a most successful breeder, having a favourite sow which on one occasion yielded fifteen pigs in the year, the produce of two litters, for which 150*l*. was realised, the pigs being sold when quite young at 10*l*. apiece. The Rev. H. Bailey was for many years a successful exhibitor, but had in his turn to give way to others—first to Capt. Stewart, of Gloucester; then to Mr. Russell Swanwick, of the Royal Agricultural College Farm; and latterly a son of Mr. Wm. Hewer has renewed his father's fame. Nor must we omit from our notice of successful breeders Mr. Heber Humphrey, of Shrivenham, whose persevering efforts towards showyard distinction have been rewarded; Messrs. Wheeler and Son; and Mr. R. K. Fowler, of Aylesbury, whose useful sort want refinement. It is quite evident from the succession of successful breeders that distinction, if easy to reach, is difficult to maintain. The pig improves probably more rapidly than other domesticated animals, but he decidedly goes back readily, and is very difficult to keep up to the mark. Thus we see that names prominent for a time are superseded, possibly to come again with renewed success. Mr. Swanwick had, and probably still has, a large sale for his pigs, and on one occasion made as much as 75*l*. for his magnificent boar Sambo II., which was sold to Canada, and exhibited successfully at their national show; and he has frequently made 25*l*. to 30*l*. of in-pig sows. The late Capt. Stewart and his executors have also sold largely for breeding purposes; indeed, the aim of all leading breeders is to dispose of their animals in this way, as they generally make remunerative prices. It is of course well worth the extra money to obtain animals that possess the qualities produced by careful cultivation; but it is generally a mistake to buy those animals that have been forced forward for exhibition, as they often breed badly and are delicate. The object of exhibiting is to show the public what the sort can do; and it would be better if animals shown in breeding classes which have been unduly forced were consigned at once to the shambles.

We believe one reason for the difficulty of maintaining a breed for long up to the mark is that too close breeding is often pursued. Within certain limits we obtain remarkable specimens from in-and-in breeding, and this is easily explained: If we have brought our stock up to a standard at which we wish to keep them, it may be wise to put them together, in order to solidify as it were the good points they possess; but this must be done with judgment, and the produce must be crossed with a fresh strain of blood, otherwise they lose hair, which means constitution, become small and weak, and are subject to the loss of their tails. Moreover, Mr. Finlay Dun tells us that it is to this cause he attributes many of the diseases to which pigs are liable. He says, in reference to pigs," in his "Essay on the Hereditary Diseases of Sheep and Pigs," published in the 16th vol. of " Royal Agricultural Society of England's Journal: "—"This practice is often pushed to an excessive and injudicious extent in these animals; and from their coming early to maturity, and producing a numerous progeny at one birth, it causes in them a marked deterioration in a comparatively short space of time. In several cases which have come under our observation, it has induced total ruin of the entire stock. At first it merely rendered the animals somewhat smaller and finer than they were before, and improved rather than injured their fattening properties. Very soon, however, it caused a marked diminution in size and vigour, and engendered a disposition to various forms of scrofulous disease and rickets, pulmonary consumption, &c. Many of the boars became sterile, and the sows barren, or liable to abortion. In every succeeding litter the pigs became fewer and fewer in number, and more and more delicate and difficult to rear. Many were born dead, others without tails, ears or eyes, and all kinds of monstrosities were frequent."

This is the opinion of one whose professional experience entitles him to respect. We believe he is correct in his view, and that breeding in and in, though it may be desirable occasionally, is destructive to ultimate success if habitually practised. We have known some of the most successful exhibits thus produced, but it is destruction if made the rule.

The practice of breeders does not vary materially, and the same plan would be followed with the Berkshire as with any other improved sorts. Animals intended for breeding should not be forced when young, although they should be liberally fed on material likely to develope flesh and bone. Bran and pollards are well suited for this purpose. Bean meal in small quantities is also excellent food for young animals when weaned. Above all, breeding pigs should have plenty of exercise; there should be a home field in which they can range; failing this, a roomy foldyard in which they can rout. The litters should, if possible, fall in spring and autumn. Extremes of temperature are objectionable; for, though, when grown, the coaty Berkshire is a hardy animal that will bear exposure well, the high-bred pigs are susceptible to evil influences when young, and especially require to be kept warm. Litters are not particularly numerous: from eight to twelve pigs are about the range; for a first litter six or seven are sufficient. These are often less regular and complete than after families, and it is not considered wise to keep them for breeding. The young hilt may be served when about 8 or 9 months old, and a few days before the time is up she must be removed from her companions to the farrowing stye, which must be covered in, well sheltered, and if necessary further protected by placing a thatched hurdle at such a height above ground, say 3ft. to 4ft., as will allow the sow and her offspring to go under it. This, however, is seldom requisite. The floor should be paved with flat flooring bricks, laid in cement, with sufficient fall to allow the liquid to drain away rapidly. Short straw or 6in. litter may be supplied, so as to cover the floor, and, above all things, there must be, 8in. or 9in. above the ground, a rail projecting about 10in. or a foot from the wall, firmly fixed so that it cannot be displaced.

Shortly before farrowing the food must be improved, milk-making materials being used. As with all stock, gentle treatment is very desirable; the pig man must have patience, and accustom the young animal to be handled. The sow and produce may be removed in a few days to a larger shed, with a division, so that the young pigs may be fed separately as soon as they will take food, which is usually in about a fortnight; this is especially important if we intend to show the pigs when young. It is a great advantage if the pigs have the refuse from the dairy; indeed, skim milk or whey is most useful. This should be given warm, and a mixture of oat and barley meal, with a portion of palm-nut meal, may be used both for the sow and her produce. The particular kind of food will depend upon the market value, but the sow should have such as will make milk—brewer's grains, &c.

It is unnecessary to go into further details, as the general management of the breeding sow has been described in a previous chapter. After weaning, which usually takes place when the pigs are eight weeks old, good food and plenty of exercise are essential. We have seen the most unfortunate results of non-attention to the latter, and shall not readily forget seeing the effects of apoplexy on a pen of youngsters which were being prepared for Birmingham. In a few minutes from the first attack three or four were dead; had they been turned out for an hour or two daily this loss might not have occurred. The pig is a tolerably long-lived animal, and favourite sows may be allowed to go on breeding so long as they produce good litters, but, as a rule, it is best to keep the stock vigorous; and, after four to six litters, the sow, whilst still comparatively young, may be fed. It is hardly necessary to add that the best of the litter should be kept for breeding.

As we have said, the Berkshires are well adapted either for pork or bacon, but it is more especially for the latter purpose that they are valuable; the large proportion of lean meat and the admixture of lean and fat render them superior to any other breed for this object. One peculiarity connected with the bacon-curing process consists in singeing instead of scalding the carcass, a plan which has certain advantages, principally that the contraction of the skin prevents the wasting away of the fat in boiling; yet, though so evident, the practice has not hitherto been introduced with the white sorts, possibly because the appearance would be less attractive. Were a proof required of the superior quality of Berkshire bacon, we should find it in the fact that Wiltshire smoked bacon realises from $\frac{3}{4}d$. to $1d$. a pound more than Yorkshire cured, and the reason is not only on account of its delicate flavour, but because it is found to go further; nothing else is so suitable for rashers. Mr. Thos. Rowlandson describes the process of singeing in his essay on the Breeding and Management of Pigs, which will be found in the 11th volume of the "Royal Agricultural Society of England's Journal." On a large scale, as in the curing establishments of Wiltshire and the south of Ireland, a number of pigs are dealt with at the same time. First knocked on the head, the carcasses are dragged to a spot where the blood can be collected; the throats are then cut, and whilst bleeding a man covers the carcasses with wheat or rye straw; experience soon teaches the quantity, which varies according to the strength of the coat. Then fire is applied in the direction of the wind, and to the uninitiated it would appear as though piggy was being sacrificed; but nothing more than the hair is burnt and the skin scorched. When one side is done, the pigs are turned over and the process repeated. The ashes are next brushed off with a strong brush, and the carcasses are hauled up by ropes and pulleys on to hooks. Here the skin is carefully scraped, warm water being thrown over the carcass, and the viscera are removed, cold water being applied to finish the process. In twenty-four hours the carcass is properly set, and amputation then proceeded with. The description by Mr. Rowlandson is so excellent that we venture to reproduce it.

The heads are first separated; the operator then takes a knife, and makes a clean cut from the tail, along the centre of the backbone to the termination of the neck, baring the whole of the vertebræ; he now takes a sharp cleaver, and, beginning at one side of the vertebræ, commences separating them from the ribs at the points of their attachment. The other side is then cloven in the same manner; by this means the vertebræ are cut clean out. He then cuts the ham about three inches above the knee joint. Thus divided, a side at a time is carried to a table, where another operator is in readiness, who first makes an incision near the neck, where the fore ribs inosculate so largely with the backbone, commonly known as the breast bone; and these, together with four or five pounds weight of the pectoral muscle or breast, are cut out, as it is found by experience that this part does not bleed well, frequently containing several of the larger blood vessels still gorged with blood, especially in large pigs, and, in consequence, it is not well adapted for curing; it is an act of precaution always to be recommended—it has also the effect of baring the scapula or shoulder bone. The latter in ticklish weather is sometimes found a formidable difficulty in the way of good curing, or, as it is technically termed, "striking the meat and taking the salt," the former term applying to saltpetre, and the latter to the common salt used. This portion of the breast being taken out, the ribs are divided with a fine saw at the point named, viz., about the middle. The upper division is cut out with a portion of the muscle attached to them, leaving the lower portion of the ribs attached to the side. In Ireland the piece so cut out is called a strip, and weighs, according to the size of the pig, from two to four pounds weight; in the West of England it is called the griskin, and usually weighs from five to eight pounds. The cutter is provided with a semi-circular saw; with this he dexterously divides the small knuckle of the thigh bone, and detaches it along with the pelvis or haunch bone. The muscle connected with the pelvis and a thin cut from the upper part of the inside of the thigh are taken out with the thigh; this is necessary in order to give free access to the knuckle, which is the most difficult part to cure. These operations completed, the side is taken to another table, where the operator cuts off any straggling pieces of flesh, and, by an ingenious arrangement, embracing a dull iron chisel and a cord passed over his neck, he draws out the scapula or blade bone. The shank of the ham is sawn off, and the side is ready for curing. The salt is injected by a machine, the pressure forcing the fluid into every part of the flesh; and in due time the side, when properly prepared, is hung up in the smoking room and cured. In some of the bacon factories in Ireland and the West of England as many as 1000 pigs per week are slaughtered during the season, and the details are admirably carried out.

Pigs form an important item of the west-country farmer's live stock. It does not always follow that the breeder completes the process. When circumstances are favourable for breeding stock, the farmer frequently contents himself with rearing, and sells his stores as soon as they are strong and ready to go on. A good demand for such usually exists about harvest time, the run of the stubbles supplying much useful food. As soon as all is cleaned up, the pigs are separated into suitable lots, and put up for feeding.

CHAPTER III.

BLACK SUFFOLK PIGS.

E have taken the black Suffolk as our second illustration of the porcine races, because—although not by any means distinguished as an ancient breed—they have latterly assumed a superiority in our show yards over other black breeds not Berkshire, which entitles them to a foremost position. Those who are unacquainted with the breed have frequently and erroneously confounded them with the Essex variety, known as Lord Western's or Mr. Fisher Hobbs' sort, whereas, when contrasted, they are as different as the Shropshire and Southdown sheep. The chief differences are as follows: The Essex are higher on the leg, have fine skins with very little hair, are rather slack round the collar and jowl, and have narrow pointed features, with small upright ears; whereas the black Suffolks are short on the leg, long in the body, well coated with long silky hair, the forehead broad, the nose short and slightly turned up, ears rather short, but broad, with a tendency to droop forward; splendid shoulders, great jowl, body wonderfully symmetrical—the Suffolk more nearly realises the theory of the parallelogram than any other animal we know of; the tail is set quite on a level with the hips, the hams are deep and wide. The presence of such good hair indicates constitution, and this we may safely claim for the Suffolks. Considering the extraordinary aptitude to feed, their fertility is creditable; large litters are not to be expected, but we frequently have eight or nine.

To Thomas Crisp, of Chellesford Lodge, Suffolk, is due the credit of first bringing the breed into prominence, with his very grand sow Black Diamond and her farrow of ten pigs, which took the prize at the Paris Exhibition of 1855. These were by the celebrated Negro, from whom, more or less directly, all the most remarkable pigs at the present day descend. In the following year, 1856, a boar of this litter took the first prize at the Royal Meeting at Chelmsford, beating Mr. R. England's White Yorkshire and others. This boar was sold to an American for the then great price of 50 guineas. The old sow at the same meeting was only highly commended, although there were many who thought her decidedly the best in the class. On this occasion Mr. W. B. Wainman's white sow of the improved Yorkshire sort took the first prize. Three other breeders of Suffolk blacks were exhibitors—the Rev. Martin Shaw, who had been crossing with the Dorsets; Mr. Wolton, who had tried the Devons; and Mr. Badham, with a dash of the Essex. In the following year, at the Salisbury meeting, Mr. Thomas Crisp was again to the fore with Black Diamond, then three years and four months old, and this in a strong class.

In 1860 the society very wisely decided to offer separate prizes for white and black sorts; up to this period they had competed together, the only distinction being as to size. This was a most desirable change, since it was next to impossible to obtain correct judging, as it all depended upon the taste of the judges—as to colour and sort—as to where the prizes would go. It was at the Canterbury Royal in that year that Mr. G. M. Sexton first put in an appearance as an exhibitor, and divided the honours with Mr. Crisp, winning first prize with a three-year-old sow, whilst Mr. Crisp was successful with a boar. Since that year, and for a long period, the Wherstead Hall pigs were successful at every show, but for some years past Mr. Sexton has withdrawn from the show ring though his blood s frequently represented. We were not acquainted with Mr, Crisp's pigs, and therefore have no means of comparing the present with the past, but it is difficult to imagine that they could have surpassed the exceeding beauty of the present Suffolk as illustrated by Mr. Sexton. Up to a recent period they were rendered more glossy even than nature made them by a filthy mixture of lampblack and oil, which was very properly made a subject of disqualification. Such animals require no paint, both in colour, hair, and form they appear to us perfect representatives of a small breed.

The repeated exhibition of animals so uniformly good led to an extensive sale, and their introduction into the northern counties and Scotland, where a few years ago white sorts were universal. We believe they are found more hardy than the latter. The skin never cracks from the sun, and that attention to the toilette is not required which is often necessary in order to prepare white pigs for exhibition. The engraving which accompanies this description is from a drawing, by Mr. Harrison Weir, of Mr. Sexton's boar and sow, the winners of first prizes at the Royal Hull Show in 1873. It will be noticed that in general form, character of head, and smallness of bone, the Suffolk breeders have moulded the blacks into about the same form as the small Yorkshire white sort; the only difference save colour is in the length of the body, which is a remarkable feature of the Suffolks.

One of the most important sales of pigs, if not the most important, ever held in England, was that which took place at Butley Abbey in 1869, after the death of the late Mr. J. Crisp, at which splendid specimens of the large, medium, and small whites, Berkshires, and black Suffolks attracted breeders and the fancy from all parts of England, and even Europe and America were represented. The competition was spirited throughout, and great prices were realised, more especially for the black Suffolks. 50 guineas was given for a favourite old boar, and 27 guineas for his son, a ten-weeks-old pig, Yorkshire and Suffolk competing. The latter, staying the longest, secured these valuable animals for the county, and doubtless the investment proved profitable. Two other Butley sales followed as the executors wound up the business. Thus about 250 breeding pigs were distributed throughout the country, which realised upwards of 2000*l.*

In 1872 Mr. G. M. Sexton had his first sale, at which 100 black boars and sows and fifty small whites realised upwards of 1300*l.* Here again the black sorts were in much demand, in many cases being bought for crossing with the Berkshires. Since this sale two boars, one sow, and three gelts were sold from the same herd for 200*l.* Since 1860 the following prizes have been awarded to the black Suffolks at the Royal Agricultural Society's meetings: Mr. Crisp won eleven prizes; Mr. Greater Stearn, seven prizes (two of the animals were, however, bred by Mr. Herman Biddel); and Mr. Sexton has taken the large number of forty-three prizes.

CHAPTER IV.

THE LARGE WHITE BREED OF PIGS.

THE distribution of white and black pigs offers a study for the curious. Were we dependent on our own resources, a nice little theory might be set up as to the influence of heat and cold, the pigs in the North of England and Scotland being nearly invariably white, whilst those in the middle and southern counties are equally dark. But we conclude that the range of climate in our island is not enough to account for so great a contrast, and that it is probable that the old English pig, varying considerably in detail according to the conditions under which he was reared, and the hardships to which he was exposed, was more or less white in colour, and that our dark breeds owe their colour to the influence of foreign blood. We admit that our argument is incapable of direct proof; but we have this negative evidence, that the large white varieties are uncrossed with foreigners. At the present day we find these animals cultivated principally in the counties of Yorkshire, Lancashire, Lincolnshire, and Leicestershire, and it is probable that they are descended pretty directly from the old English pig. Mr. Rowlandson, in his prize essay on the breeding and management of pigs, says: "There are good grounds for supposing that 'the old English hog' with flop ears was originally the only domestic animal of its kind throughout the kingdom. The genuine old English breed was coarse-boned, long in limb, narrow in the back, and low-shouldered—a form to which they were most probably predisposed from the fact of having to travel far and labour hard for their food, and undergo considerable privations during winter." And he proceeds to say that he has seen surprising improvements resulting from better care, shelter, and food, where no fresh blood was introduced. In these cases the thick flop ears became fine and thin, the bones a moderate size, bristles gave place to hair, and the white skin became fine and ruddy.

Those of our readers who are old enough to have taken an interest in the earlier shows of the Royal Agricultural Society, from its commencement in 1835 to 1850, will remember to have seen enormous specimens of white pigs, often weighing as much as a small bullock—we ourselves remember specimens estimated at 50 imperial stones. These animals were esteemed for the quality of the bacon, for their hardy and prolific character, the litters often reaching sixteen to eighteen; but they were incapable of early maturity, and consumed a quantity of food. Those intended for bacon were kept in a store state until at least eighteen months old, and then if fed for a twelvemonth something remarkable was produced; but, had the cost been calculated, the profits would have been much over the left. Of late years great improvement has been effected by careful selection and by more attention to feeding. It is said that Bakewell was the first to improve the Leicestershire pigs, and this by a process similar to that which proved so successful in the case of the long-horned cattle and the Leicester sheep—viz., by selection, discarding the large, coarse animals, and selecting such as were more symmetrical and finer of bone. It is probable that the first step in the improvement of the Yorkshire was through the improved Leicestershire pigs; certain it is that at one time they were particularly uncultivated, and are described as "of large size, gaunt, greedy, and unthrifty; coarse in the quality of the meat, flat-sided, and huge-boned."

It is somewhat invidious to particularise when merit is due to so many; but we must name the late Mr. Wainman, of Carrheads, as one of the leading improvers of the large Yorkshire. For many years his successes in the showyard were considerable. His factotum, Mr. Fisher, is still often seen as a judge. It was shortly after Mr. Wainman's fame was established that a Lincolnshire breeder began to creep up. The name of Mr. Duckering, of Kirton Lindsey, became familiar in the shows, and few have been on the whole more successful.

Next in point of time we notice the Messrs. Howards, of Bedford, who, selecting the best they could obtain from the two breeders we have named, produced animals remarkable both for form and quality; Mr. James Howard still exhibits, and, though not so successful as either the Lincolnshire or the Yorkshire breeder, always holds a respectable position. It was the produce of a cross between Wainman and Duckering that were disqualified at Birmingham some years ago, on account of variety of dentition and also difference of character. Much satisfactory evidence was produced at the time, showing clearly that the pigs were of one litter, which was to us evidenced by the fact of their remarkable variety. Animals differing so widely in character were likely to produce variable offspring. Later on, Lancashire carried off laurels. Mr. Peter Eden, after having been most successful with pigeons, took up pigs, and did great things both with the large, small, and middle-bred sorts.

Naturally, the result of attention has been to develop an animal with more quality and greater tendency to early maturity. Still the large white pig of the present day retains some traces of the original; we have the large and overhanging ears, and the long large head; the width of the body, though much greater than of old, bears no proportion to the length; and, fairly symmetrical as we now find the matured animals, they are decidedly flat-sided as compared with either the Berkshires or the Suffolks. The back is now level, and the shoulders full and complete, but the hind quarter usually drops a good deal, and the bone is strong, though perhaps not much in excess. The nature and quality of the hair varies a good deal. It is not unusual to find a decided want of coat—a drawback, since the skin unprotected is liable to crack. In the best specimens we have long and moderately fine hair, but never the curly profusion which adds such beauty to the small white sort.

Less prolific than in days of old, we still find the large white varieties good breeders, and it is not unusual even in a showyard to see ten to twelve in a litter. The sows are good mothers, and yield abundance of milk on suitable food; but there is an unfinished, inelegant appearance in the youngsters which is unmistakable, and contrasts forcibly with the admirable proportions of the smaller breeds. Early maturity is not to be looked for; the merits of these animals need time to develop. At four or five months they are overgrown hobbledehoys; at twelve to fifteen months they show their really grand proportions, and if kept to this point will favourably compare as bacon makers with any other breed. Except for particular markets, such as the manufacturing or mining centres, large bacon, however excellent, is not in such demand as the medium or smaller sorts; consequently we do not find the large breed so greatly in request as formerly. Nevertheless a nicely-cured ham cannot well be too big; and we have always heard the bacon well spoken of.

Some few years since, the Earl of Ellesmere purchased largely both the large, medium, and small white sorts from Mr. Peter Eden—the latter retiring from the show ring. We are almost afraid to say how much was paid for one boar, but we believe it was three figures. This, however, was not for a large pig. Under the able management of Capt. Heaton, Lord Ellesmere's agent at Worsley, several important auction sales have been held, and a long array of prizes secured. On the whole, the greatest success has been achieved with specimens of the small breed, but many noble animals of the large sort have been shown. Confining ourselves to the results of the Royal meetings at Bedford and Taunton, Lord Ellesmere took three prizes: first for a young boar bred by Mr. Henry Neild, and second for Cultivator 9th, bred by Mr. R. E. Duckering; and first prize for a valuable sow of his own breeding. At Taunton the Worsley herd was more successful, as the first prizes for breeding sows and pen of three breeding pigs were secured. Amongst other breeders of acknowledged success may be mentioned the names of Mr. Jacob Dove, Hambrook, Gloucestershire; Mr. Clement, R. N. Beswicke-Royds, Littleborough, Lancashire; the Messrs. Howard, of Bedford; Mr. Matthew Walker, of Anslow, Burton-on-Trent; and Mr. Walker Jones and Mr. Strickland, of Thirsk. Our illustration, by Mr. Harrison Weir, represents specimens from the Earl of Ellesmere's herd.

CHAPTER V.

SMALL WHITE PIGS.

THESE offer a marked contrast to the large white sorts last described, not only in the matter of size, but in quality, early maturity, and delicacy of character. It is difficult to imagine that such elegant and complete specimens of porcine development were derived from a common origin with the lop-eared, coarse-skinned, big-boned animals that were the progenitors of the present large variety. The influence of suitable crosses has, we know, a remarkable effect, and so it may be that Chinese blood laid the foundation of the present small white sorts. These are found distributed in several counties, but more especially in parts of Yorkshire—so much so that they are frequently described as small Yorkshires. At the present time, however, they are quite as famous in Lancashire, Suffolk, and Berkshire. These small sorts may be described as "gentlemen's pigs," rather than as being in favour with tenant farmers. The late Lord Ducie cultivated the sort; the late Lord Wenlock's breed was for many years famous in Yorkshire; the late Mr. Samuel Wiley, amongst his other successes, stood high with a breed of small whites remarkable for quality; Sir George Wombwell took prizes, *cum multis aliis*, principally Yorkshiremen. Both Her Majesty and Lord Radnor cultivated white varieties of the Berkshire, but they are neither in size, quality, nor character good types of the small breed of late years. Mr. Sexton has distinguished himself with small white Suffolks, though his fame will rest principally upon his success with the black sort. Mr. Peter Eden, of Manchester, and through him, Lord Ellesmere, of Worsley, have made the small whites famous; and when the latter is to the fore he usually has a good share of the distinctions. From what source derived we know not, but both Mr. Eden's and Lord Ellesmere's animals have been remarkable for an exuberance of long soft curly hair, which, when properly washed, combed, and curled, has a very attractive character. We believe that Mr. Eden's boar Peacock, which won a great number of prizes (amongst others the first at Cardiff in 1872), was peculiarly noticeable in this respect; he was sold to Lord Ellesmere at a very high figure, and did good service at Worsley. As far as we remember this boar, he was somewhat slack in his middle, with a magnificent head, great hams, neat quarters, and curly hair some 7in. or 8in. long. At the present time the Earl of Ellesmere takes a decided lead, as Mr. Peter Eden has retired from the show ring, his principal competitors being Messrs. Duckering, Ashworth, Walker-Jones, and Lords Morton and Radnor, and others, besides those whose fame is not known in the showyard, but who have admirable specimens of the porcine breed.

We have said that the small whites are more adapted for gentlemen and amateurs than for ordinary occupiers; and we make this statement for at least two reasons—the small size they reach, comparative delicacy, and small breeding properties. Whilst the large sort frequently produce really large litters, these seldom exceed seven to nine. Much attention is required in the early stage. Sudden exposure to extremes of temperature is very injurious. The great merit of the breed is its beauty and extraordinary feeding properties. It is impossible to keep them poor; the tendency to lay on fat is remarkable; hence they are well suited for porking purposes, but lack the lean meat so desirable for bacon.

Of late years the middle breed, derived from a cross between the large and small sorts, has come so much into use that it is sometimes difficult to find a pure small breed, and they are valuable when they have been kept intact as improvers of coarser sorts. Thus the pure-bred boar effects a marvellous change in a very short time, and we have found great advantage from this cross upon the Berkshire, whereby we retain the lean meat and increase the fattening properties. It is a curious fact that, generally speaking, the produce of the Berkshire sow and the small-bred boar are white, occasionally, but not often, spotted. In form and character they follow the sow rather than the boar, completely contradicting the theory advanced by Mr. Fowler and others, that form follows the sire and the internal parts take after the dam. We have not gone on with the cross, probably the second generation would be more mongrel. For mere feeding purposes we have found the cross most excellent.

It is rather difficult to write a description of an animal so as to convey an intelligent idea of that which we wish to represent. Mr. Harrison Weir's excellent drawing will assist the reader. The head, although in the matter of money value small, is of the highest importance as giving beauty and character. The snout should be dished, and so small that, when the animal is fat, all we see are the upturned nostrils; these should be small; the forehead, flat and broad. In the fat animals the position of the eyes is indicated by creases of fat; they are invisible. In the store animals the eye should be large and lively. Great importance attaches to the size and form of the ears—by no other mark can we so accurately determine the purity of breeding; they must be small, and not drooping, but slightly inclined forwards, wide set apart, and covered with short, soft hair. In order to complete the short, handsome head, the chops must be full and large; it is this which gives the wonderful side view, which is so admirably rendered in the picture. The neck is very full, and the head well set on, at a somewhat lower level than the line of the back. The shoulders are wide and well covered, sloping back into the carcass, and thus avoiding the hollow and deficient fore flank so often seen and so unsightly. The ribs are full, and the loin sufficiently wide to preserve the uniform *contour*; the tail set on high, though hardly so much in a line with the back as in the black Suffolk, to which breed, it will be seen, the outline bears close resemblance. The hams are deep and square—"meat down to the hocks" is a very correct description of this important part; bone fine and offal light. They are remarkably heavy, according to size, and very complete for age. The admirable specimens shown at Birmingham fat show as under six months prove how capable they are of early maturity. The coat varies as to length and character; we have the thick short staple and the long curly sort, which is not so closely set, but in no case have we strong coarse bristles, which indicate a thick skin and slow growth.

Such are the general features of the small white breeds. They are very handsome, it will be allowed; but, as we said before, they are too small, and perhaps delicate (especially in a young state) for the million, consequently we rarely find them pure, save in a few instances, where their value for crossing is understood. Like the Leicester sheep, they have done good service in improving others; indeed, the middle-bred whites owe their extremely useful character principally to the influence of the small sort.

During a young state, shelter, warmth, and care are required. It is not desirable to commence breeding until the hilt is ten to twelve months old, care being taken that the litters are produced in spring and autumn, so as to avoid extremes of heat and cold. The hilt should be kept well during the later stages of pregnancy; but an over-fat condition, which is so easily produced, should be carefully avoided. There is danger to the progeny if the organs are coated with fat, and the result will be a wretched, puny, and uneven lot, the sow will have difficulty in parturition, and the milk will be deficient. It frequently happens, especially when these precautions are not attended to, that the first litters are very small—four or five on an average, and these somewhat irregular as to size. It is as well not to retain any for breeding, but to select our future dams from a second or third litter, when the maternal powers of the sow are matured. Occasionally we find here, as elsewhere, a much more prolific character than we have described as the rule; but such will be found in herds where forcing for show is not the rule, but only the exception, and where the animals so treated are sold off.

It is most unwise to buy animals for breeding purposes that have been got up for show. If justifiable at any stage, it is when the animals are young, say not over six or seven months, as careful reduction may restore a healthy condition; but it is not always successful, and puny and defective litters—if the animals breed at all—may be looked for. The beginner should form his opinion of the merits of particular blood by the specimens he sees in the show yard, and then visit the animals at home, and select what he wants; such a practice will be satisfactory to all parties.

When fully grown, the small breed maintain their condition upon a minimum of food; indeed, a handful of palm-nut meal in water or house wash suffices for their requirements.

The great value of the breed is for small pork on dairy farms. Nothing can be found so delicate, and for such purposes we can recommend their culture; also as centres from whence improvement of coarser sorts may be safely looked for. For general purposes—that is, to produce both pork and bacon, and especially the latter—the small whites are not so well suited as other breeds.

CHAPTER VI.

MIDDLE-BRED WHITE PIGS.

AS the name implies, these are something betwixt and between the large and small sorts which have been lately described—a cross breed in short, for which the Royal Agricultural Society can as yet find no distinctive title, and it accordingly classes them as pigs not eligible to compete in the above classes, *i.e.*, the large or small division. By-and-bye they will succeed to the dignity of a separate classification, that is, provided breeders keep them pure. At present there is considerable variety, according to the preponderance of either sort. The result, however, is a highly useful breed, which finds increasing favour with the tenant farmer, because it fulfils all the conditions for which pigs are considered desirable. Thus we have size, aptitude to feed, flesh without coarseness, hardy constitution, and productiveness. The middle-bred sorts are essentially tenant-farmers' pigs. At first sight it is difficult, especially in the young stage, to distinguish them from the small sorts. We have no doubt that prizes have been won in consequence of such substitution; indeed, the youngster makes a good small pig, whilst the more mature animal figures in the nondescript division. Up to the present time we have much variety of type, according as the impress of the large or the small sort predominates. Generally speaking, the features resemble most closely the small sort, and this was certainly the case with the specimens which Mr. Harrison Weir has drawn, and which, we believe, represent prize-winning animals bred by the Earl of Ellesmere, who, having such excellent animals of both the pure sorts, is in a good position to found a middle breed of superior merits. Like all other crosses, a considerable interval must elapse, and careful selection be practised, before uniformity of character can be established, although in the case of pigs the operations are much more rapid than with cattle or sheep. The following description applies to a type of which the illustrations are specimens.

The head is longer than the small sort, but more closely resembling them than the large kind, with wide forehead and rather large ears. The nose is not so much turned up; indeed, the features are altogether less curved, and the cheeks are less developed. The head should be well set upon the neck, and the latter, though full, is not so extraordinarily wide and full as in the small breed, where it frequently looks almost out of proportion to the hind-quarters. The back should be long and level, the tail set on rather low, the hams moderately full, and the bone, though small, not noticeably fine; long thin soft hair is desirable; depth of carcass, with well-sprung ribs; the legs being neither so short as the small breed nor so lengthy as some of the large division. We should have a good combination of lean and fat marbled flesh, which is so desirable for rashers. This item is of great importance, and renders the sort so popular at the present time. We have all-the aptitude to feed of the small sort, with a fair development of lean flesh, which renders the animal suitable either for pork or bacon. The large breed, much improved as they have undoubtedly been by such men as the Messrs. Howards, Duckering, Eden, &c., are more suitable for bacon than for pork; indeed, in the young state they are anything but complete, being leggy, lathy, and skinny. The small sorts are always fat, and as round as a dumpling; but then it is all fat. The middle breed, as the name implies, occupy a happy medium, and are equally valuable for either pork or bacon. Then, again, however much we may admire the beauty and precocity of the small sort, limited breeding properties are a serious drawback, they are not rent payers. Middle-bred pigs are certainly more prolific, and usually farrow from nine to twelve offspring. The sows when naturally reared, and not forced on for show purposes, are good mothers; and, above all, they are hardy, in this respect being very superio to the small sort. The future and permanent value of a breed derived from origin so diverse, depends upon the care that is exercised in selection, together with the occasional return to a cross of either of the progenitors, according to the requirements of the case. As a rule, the produce will vary much more than is the case with animals of longer pedigree; if only the best grown are kept for breeding, the quality will be maintained and improved upon. The objects that should be kept in view by the breeder are size and flesh as distinguished from the smaller sort, quality and aptitude to feed as improvements on the large type. The features will vary in size and character according to the degree of mixture; we prefer something between the two. It is quite possible to have the face too short and puggy, as such features indicate want of size and growth. The middle-bred pigs, whilst low on the leg and tolerably fine of bone, should have great length; the three L's should be applicable—long, low, and level. In Yorkshire and Lancashire these pigs are rapidly occupying an important position. Pure herds are found in a few places, but most of the breeders who are famous for show pigs keep two, if not three, sorts; whereas the ordinary farmer goes entirely for the more recent production, which answers his purpose best. Lord Ellesmere, Mr. C. E. Duckering, Mr. F. A. Walker-Jones, Mr. T. Strickland of Thirsk, and Mr. P. Ascroft, are those who have principally distinguished themselves in the show yard of late years as exhibitors of the middle-bred pigs.

CHAPTER VII.

THE BLACK DORSET PIG.

IT has been justly observed by a writer on agricultural subjects that a good pig should have "a small head, short nose, plump cheek, compact body, short neck, thin skin, and short legs." All these characteristics are possessed in an eminent degree by the black Dorsets, as bred and improved by Messrs. John Coate, of Hammoon, near Blandford, and John A. Smith, of Bradford Peverall, Dorchester, distributed by them throughout England, and exported to many parts of the world.

It may be interesting to trace the origin of this class of stock, and it is not difficult to do so, the principal parties to whom is due the credit of having established it being still alive, actively engaged in agriculture, and possessing all the energy and perseverance necessary to preserve, and probably further improve, their favourites.

Mr. John Coate is well known as a most successful exhibitor of black Dorset pigs at the Christmas shows held in London. For twenty-five years, with one exception only, he has carried off first prizes, two silver cups, and on seven occasions gold medals, proving most conclusively the superiority of his animals, as well as evidencing the care and judgment with which they have been bred and fattened. He has informed us that about thirty years ago he obtained two young black sows, whose progenitors were imported from Turkey. These he put to a boar of the Chinese breed; for their offspring he had an opportunity of securing the services of a prize pig of the Neapolitan species, and their produce was in every way satisfactory. About this time he happened to go to the Baker-street show, and, believing his own pigs to be better than any he saw there, the following year he became an exhibitor, and won every prize for which he entered, as well as a gold medal. As before stated, this success has attended him twenty-four years out of twenty-five, up to the last show, when he was awarded first and second prizes in the young class.* Mr. Coate has held his position by purchasing good sows whenever opportunities offered, and mating them with his own boars, endeavouring to obtain size from the strangers, at the same time retaining the beauty of form and quality of his own breed.

What Collings and Bates did for shorthorn cattle, Coate and Smith have done for black Dorset pigs. The latter gentleman (also distinguished as one of the best breeders of Devon cattle) appears to have turned his attention to pigs about the same time as Mr. Coate. In the year 1845 he possessed a good black sow with Chinese blood in her veins, descended from stock the property of the late Earl Poulett. With her he used a boar bred by the late Fisher Hobbs, and from this source he has produced pigs that have made his name famous, and won him numerous prizes at Royal as well as local shows. He has imparted fresh vigour to his herd by occasionally availing himself of the best blood to be obtained from other breeders, including Mr. Coate, and produces animals unsurpassed for symmetry, quality, aptitude to fatten, and fecundity.

The improved black Dorset pigs have many admirers, both in and out of the county; they are hardy, greatly incline to fatten, and in fact can be grazed at any age. They come to a good size, having been made 15 score (dead weight) at 9 months old, and more than double that weight at 18 months. Few pigs of other varieties would maintain the condition they do on similar fare, and they are generally productive, and bring good litters.

A few hints on their breeding and management may not be out of place. The young sow, not less than twelve months old, should receive the boar about the end of March, or beginning of April, and produce but one litter the first year; after that she

* This was written in 1872. Since then many laurels have been added to the long list at Hammoon.

should farrow twice, in January or February, and June or July; the earlier the better. The first pigs, arriving when the weather is probably very cold, require great care and attention, and the dam should be fed on warm food; the young ones are fit to wean in six or eight weeks, and can either be grazed at once, or let run for a year (being kept growing at a small cost) and then fattened, the latter probably being the more profitable course, especially when a dairy is kept.

We doubt if they are as capable of bearing exposure or sudden changes of temperature as those breeds provided with more protection in the way of coat. The absence of hair is certainly a drawback when the pigs have to get their own living in a straw-yard, but does not interfere when the animals are kept under cover and always warm. We have been particularly struck with the absence of hair in the pigs shown at Islington, many of these being from the Hammoon herd. They are capable of producing extraordinary weights for age, but in these overfed animals the flesh is soft and blubbery.

CHAPTER VIII.

THE TAMWORTH PIG.

THERE is no variety of pig known in the British islands which has been improved so rapidly as the Tamworth, which, during the past eight years, has emerged from comparative obscurity as a breed peculiar only to a small locality into one common to the whole country, and which is recognised as a national variety. The Tamworth is no new pig, for we are acquainted with breeders who remember it for the past sixty years, and who declare that it was then, and had been during the lifetime of their predecessors, a whole coloured deep red pig, long in the snout, a prolific breeder, a kind mother, an admirable forager, which could almost be left to find its own living, and at the same time a pig which provided bacon of the highest quality. We have, however, come across at least two opinions equally based upon personal experience, and we arrive at the conclusion that there were at the beginning of the century either two strains of the Tamworth, or one strain which had been much more improved than the other, for where both are described as regards outward characteristics in a similar manner, yet one strain is stated to have been difficult to fatten, and the other one of the quickest to make a return to the farmer. That the Tamworths were prolific there is no doubt, and this valuable point they retain to the present time in perhaps a higher degree than any other of our recognised breeds. Some of the best breeders of to-day declare that young animals of the variety do not fatten until they have reached a certain age, and that the more highly they are fed the more they grow. On the other hand, others have shown by their exhibits, at the leading exhibitions, that young Tamworths will grow and fatten at the same time. The Tamworth is really a large pig, and, remembering the great improvements which have been recently made by such breeders as Mr. Allender, we shall not be surprised to find, in a few years, sows which will exceed a weight of 800lb., which the Tamworths have reached in the early days. At the same time we would rather see them grow in size than manifest a greater capacity to fatten, for they are above all things a bacon pig, and their flesh is equalled by that of no other variety in the even distribution of the fat and lean. Tamworth breeders have to bear in mind that their favourite speciality is in connection with this point—hence great care will need to be exercised both in modifying the type and the size of the pig. The Tamworth has a long body for its size, together with great depth of side, but it does not equal the white pigs in either chops or ham, if regarded from the point of view of size. At the same time the quality of the ham is undeniably good. Mr. Allender has made a step in the right direction in improving the hams, for he has obtained them of the weight of 20lb. cured and dried from pigs not seven months old. With regard to the chops, which are decidedly weak, it may be remarked that they undoubtedly accompany the highest type of bacon pig, and we should regard any attempt to improve them, or to materially shorten the snout, as detrimental to the variety, for full chops should be accompanied, to some extent, by a greater amount of fat with which the lean would be less perfectly interlarded. Tamworths of high type can be made to reach a weight of 300lb. at six months of age with ease, and 500lb. at twelve months, and this alone should show that although they are far from being common out of their special locality, they are a breed which cannot be despised by anyone. The Berkshire has long been claimed as the bacon curer's special variety, but there is plenty of evidence from men who have abandoned the black for the red pig, showing that the latter is superior for conversion into bacon; and a letter which was written by an important bacon curing firm in the West of England at the end of 1886, offering to their customers a premium upon pigs sent to them with a minimum quantity of fat, is sufficiently suggestive of the fact that the

Berkshire is not all it is stated to be even in its own favourite locality. There is, perhaps, no variety of British swine which is ready for breeding at so early a period as the Tamworth, nor one which produces a larger number of pigs at a litter. It is at the same time one of the most vigorous of breeds, and although we do not regard a grazing pig as specially advantageous, yet the Tamworth is a pig adapted for living upon grass or clover—a custom which largely prevails in America. There is no doubt some connection between the grazing of pigs and their qualification as bacon makers just as confinement and sleep with high feeding are connected with the production of fat. With regard to the colour of the Tamworth it appears to be conclusive that the best pigs are now of a lighter sandy red than those of past generations. This has been attributed to a cross with the white pigs; on the other hand, we are of opinion that the Berkshire has been used for crossing, and that the very considerable amount of black which is seen in many reputable herds of Tamworths is derived from that source, although experienced breeders declare that the Tamworth and the Berkshire crossed do not produce a really good pig. This cannot be said of the cross with the white, for we have seen many of the finest specimens of fat and breeding swine produced from this cross as could be desired. The Tamworth is a pig which carries a comparatively small percentage of offal, and this fact, combined with its great prolificacy and quick growth, makes it one of the most desirable breeds for the use of the farmer in this or any other country.

INDEX.

A.

Abortion, probable causes of ...*page*	17
Advice as to feeding cattle	5
Analysis of turnips	254
Of oat straw	254
Anglesea Cattle, by Morgan Evans	206
Early history of	206
Important export trade in	208
General resemblance to Pembrokes	208
Dairy properties neglected in the Island	209
Excellent for beef making	210
Modern improvements in	211
Herd Book of	211
Col. Platt's experience in feeding	211
Angus-Aberdeen Cattle, by the Editor	158
Rapid spread of	158
Judge Goodwin's experience of	158
Origin of, unknown	159
Early history of, by Mr McCombie	160
The Ballindalloch herd of	161
Principal present herds of	162
Mr. Stevenson's herd of	163
Description of	164
Value for meat production	166
Management of	166
Artificial food for cattle	47
For ewes and lambs	256
Ayrshire Cattle, by Gilbert Murray	168
Original character of	169
Early history of, by Aiton	169
Value for milking purposes	170
Milking competition at Ayr	171
Points of	172

Ayrshire Cattle (*continued*)—
    Great care as to dairy properties ... ...*page* 172
    Management of 174

B.

Barley, analysis and value 56
Beans, peas, and lentils, analysis and feeding value of 57
Berkshire Pigs 446
    Change in appearance from originals 446
    Process of improvement not known 446
    Useful character of 447
    Points of 448
    History of development of 449
    Difficulty in maintaining quality of 451
    Mr. Finlay Dun on 451
    General management of 452
    Well adapted for bacon 453
    Killing, singeing, and cutting up 454
Black-faced Scotch Mountain Sheep 376
    Ancient origin of 376
    Distribution of 377
    Effect of improved treatment of 379
    Great intelligence of 379
    General management of 381
    Enlargement of holdings, cause of improvement of ... 382
    Principal breeders of 383
    Mr. Archibald's stock of 384
    Quality of wool not improved 385
    Points of 385
Black Dorset Pigs 473
    Points of 473
    Mr. J. Coate successful exhibitor of 474
    Hints on breeding and management of 474
Black Suffolk Pigs 457
    Sometimes confounded with Essex 457
    Difference in points of Essex 457
    History of 458
    Sale of, at Butley Abbey 459
    Prizes awarded to, since 1860 460
Border Leicester Sheep, by John Usher 289
    Descended from Bakewell 289
    Eminent breeders of 290
    Characteristics of 291
    Influence on mountain breeds 292

Border Leicester Sheep (*continued*)—
    Kelso sales of ... ...*page* 293
    Lord Colworth's flock of 294
Breeding and general management of shorthorns 8
Building materials 72
    Variety of plans 73
    Details of 74
Bull for dairy purposes 4

C.

Cabbage, value of, for sheep 274
Calves, management of 19
    Summer feeding of 21
    Winter management of 25
    Condimental food unsuitable for 25
    Mr. H. Ruck's management of 27
Cattle, early history of 1
Cheese factory in Derbyshire 99
    Manufacture of 98
    The Cheddar process 100
Cheviot Sheep, by John Usher 387
    Early history of 387
    Improvement of, by Mr. Robson 387
    Modern improvement 387
    Management of a flock of 390
    Present breeders of 391
    Value of, for crossing 392
    Value of mutton compared with that of black faces 392
Colour, importance of, in cattle breeding 6
Comparison of Shorthorns and Herefords 8
Condimental food, recipe for, and value 58
Condition of ewes influences crop of lambs 251
Cotswold Sheep, by the Editor 297
    Origin of name 297
    Early value of wool of 298
    Modern breeders of 299
    Influence of, on other breeds 300
    Characteristics of 301
    Grey faces not impure 301
    Weight of wool 301
    General management of 302
Cotton cake, inferior, injury from 51
    Analysis of 50, 51
    Decorticated, should be ground 50

Covered yards, when desirable ...*page* 59
    Increased value of manure in 64
    Economised litter 65
    Economy of food 65
    Details of construction 66
    Cost of, and return from 67

D.

Dairy cow, food required for 82
    Management of 84
    Treatment before calving 85
    Treatment after calving 86
    Food rations for 87
    Cooked v. raw food 88
    Average produce of 90
    Profits from 91
    Advantage of artificial food for 91
Dairy, temperature and regulation 92
    Construction of 92
    Sanitation of 93
    Value of cream separator 94
    Description of Danish and Swedish inventions 95
    Swartz system of setting cream 96
    Churning from milk 97
    Devonshire system of raising cream 97
    Churning from cream 98
Devon Cattle, by Lieut.-Col. J. T. Davy 125
    Common origin with Hereford and Sussex 125
    Compared with Sussex 126
    Influence of climate and food on form and character of 126
    Description of 127
    Economical meat makers 128
    Value of, in America 129
    Herd Book of 130
    Value for draught purposes 130
Devon Long-woolled Sheep, by Joseph Darby 310
    Mixed origin of 310
    Best flock derived from Leicester and Bampton 311
    Ancient history of 312
    Description of 314
    Management of 315
    Present breeders of 316
    In good demand for Somerset marshes 319
    Recognised by R.A.S.E. at Taunton 320

Devon Long-woolled Sheep (*continued*)—
    Col. Luttrell's statement as to increase of value ...*page* 319
    Success with, by S. J. Heathcote Amory ... 320
Dorset Horned Sheep, by Joseph Darby... 395
    Value for early land ... 395
    Claridge on ... 395
    History of... 396
    Present breeders of... 400
    Competition of, with South Downs ... 401
    Mr. J. Ensor on ... 403
    Experience of different breeders with... 404

E.

Early maturity of lambs ... 270
    Capable of cultivation ... 250
    Mr. De Morny's experiences ... 250
Economy of well-bred cattle ... 7
Evils of over-feeding for show ... 13
Ewes, system of dropping ... 249
    Quantity of turnips desirable for ... 252
    Breeding importance of dry food... 253
    Preparation for and attention during lambing ... 259
    Pen, situation and nature of ... 260
    Pen, method of construction ... 260
Exmoor Sheep, by Joseph Dreaby ... 369
    Locality of ... 366
    Character of ... 369
    Origin of ... 370
    Management of, by T. B. Acland... 371
    Arthur Young on ... 370
    Principal breeders ... 372
    Management of, by R. Stranger ... 372
    Management of, by Mr. Birmingham ... 373
    Management of, by Mr. T. M. King ... 373

F.

Feeding, management in summer ... 31
    Management in winter ... 33
    Irrational mode of ... 36
    Cost and return from ... 58

G.

Galloway Cattle, by Gilbert Murray ... 151
    Native home of ... 151

Galloway Cattle (*continued*)—
 Characteristic features of ...*page* 152
 Great importance of, in early times ... 152
 Not injured by Irish cross ... 154
 Bad milkers due to mismanagement of ... 155
 Improved modern management of ... 156
 Crossed with Ayrshires ... 157

Glamorgan Cattle, by Morgan Evans ... 188
 Now rapidly being superseded ... 188
 Improved by Normandy cattle, 12th century ... 189
 Devon influence of ... 189
 Great value in 18th century ... 190
 Bred at Windsor by George III. ... 190
 Decline of, ascribed to breaking up pasture land ... 190
 Modern breeders of ... 191
 Description of the Treguff breed ... 191
 Description of, by Ewart and Martin ... 192
 Formerly bred in Monmouth and Gloucestershire ... 193
 Description of Badminton herd, by Mr. J. Thompson ... 193

Guernsey Cattle, by a native ... 226
 Increasing reputation of ... 226
 Legislation to prevent introduction of foreign cattle ... 227
 Weights of show cattle ... 231
 General management with a view to milk production ... 232
 Large American demand for ... 233
 Importance of pedigree ... 233
 Herd Book rules ... 233
 The general Herd Book started in opposition, great evil therefrom ... 234
 American Herd Book ... 234
 Bye-laws and forms of same ... 235
 Milk tests ... 237
 English Society scale of points ... 238

H.

Hampshire Sheep, by E. P. Squarey ... 335
 Origin of ... 335
 Great variety in early breed of ... 335
 Early breeders of ... 336
 Mr. Humphrey's improvement of ... 336
 Wilts and Dorset flock merged into ... 337
 Management of ... 338
 Statistics of flock ... 339
 Wool of ... 339

Herdwick Sheep, by H. A. Spedding ...page	419
Obscure origin of	419
Hardihood and instinct of	420
Characteristics and habits of	421
General management of	422
Loss of, from sickness	422
Injury of, from fly	422
Flocks let with the land	423
Hereford Cattle, by T. Duckham	112
Rapidity of feeding and fine quality of flesh	112
Early breeders	113
General character of	114
Herd Book of	114
Localities of	115
Spread of, at home and abroad	116
Value of, in America	119
American Herd Book of	120
Success of, in Australia	122
Moderate milking properties of	123

I.

Importance of progress in cattle breeding	3
Indian corn, analysis of	54
Pig feeding, Lawes on	55

J.

Jersey Cattle, by John M. Hall	219
Early Improvement of	219
Export to England, by Mr. N. Fowler	220
English breeders of	220
Character of	220
Selection of bulls	221
Great value for butter production	221
Late Mr. Dauncey's herd	222
Average yield of milk of	223
Management of calves	223
Management of cows and heifers in milk	224
Liability to milk fever and preventive treatment	225

K.

Kerry Cattle, by the late R. O. Pringle	213
Early history of, by Sir R. W. Wilde	213
Points of	214

Kerry Cattle (*continued*)—
    Dexter variety of ... *page* 214
    Points of Dexter cattle ... 215
    Crossed with West Highland ... 215
    Neglect of breeders ... 215
    Suitable for suburban farming ... 216
    Dimensions of Alderman Purdon's ... 216
    The poor man's sort ... 216
    Yield of butter and milk ... 216
    Quality of flesh ... 217
    Modern breeders ... 217
    Prize bull at Dublin show ... 217
    English breeders of ... 218
    Mr. Bogue, salesman of ... 218

L.

Leicester Sheep, by the Editor ... 283
    Improvement of, by Bakewell ... 283
    Prices realised by Bakewell ... 284
    Prices realised by the Dishley Society ... 284
    Early breeders of ... 285
    Influence on other breeds ... 287
    Characteristics of ... 287
Limestone healthy for sheep breeding ... 247
Linseed oil cake, analysis, variety, value of ... 48
    Objections to use of ... 49
Longhorned Cattle, by Gilbert Murray ... 132
    Past and present value of ... 132
    Original form of ... 132
    Early Improvers of ... 133
    Bakewell's herd of ... 133
    Description of the bull Shakspere ... 134
    Description of, by Mr. Marshall ... 134
    Ousted by Shorthorns ... 135
    Valuable milkers ... 136
    In Derbyshire ... 137
    Present herds of ... 138
Longwoolled Lincoln Sheep ... 304
    Description of, by Melbourne ... 304
    Improved by Leicesters ... 305
    Early improvers of ... 305
    At R.A. shows ... 305
    High prices realised for ... 306
    At Lincoln Fair ... 307

INDEX. 487

Long-woolled Lincoln Sheep (*continued*)—
    Weight of wool... ... ...*page* 307
    Average prices of 307
    Experiments on by the Parlington Club 308

M.

Macdougal's disinfecting powder 79
Management of ewes up to lambing... 247
    During lambing 261
    During lambing, Mr Cleeve on 262
    From birth to weaning 265
Manure, management of 77
    In open yards 80
Method of improving a herd 7

N.

Nigrette Merino Sheep, by the Editor 363
    Peculiarities of... 363
    Native of Spain 363
    And George the Third 363
    Society formed... 364
    Messrs. Sturgeon, breeders of 364
    General management of 365
    High value of Sturgeon ewes at the Cape and in Australia 366
Norfolk and Suffolk Polled Cattle, by T. Fulcher... 144
    Mr. Marshall's description of 144
    Difference in type of and Scotch Polled Cattle... 145
    Mr. George's herd at Eton 145
    Principal herds in Norfolk and Suffolk 146
    Characteristics and qualities of 147
    Recent improvement of 148
    Herd Book of 148
    Mr. Hausman's report on, 1884 149
    At the Norwich show 150

O.

Oat straw, feeding value of 255
Oxfordshire Down Sheep, by Messrs. Druce and Hogg 351
    History of 351
    Early breeders of 351
    First appearance as a breed at Battersby in 1862 352
    Champion prize at Smithfield, 1872 353
    Characteristics of 353

Oxfordshire Down Sheep (*continued*)—
    Weight of fat sheep... ...*page* 353
    Present breeders of... ... 354
    General management of 354
    Mr Druce's history of Freelands ... 355

P.

Palm-nut meat (Smith's), analysis and value of ... 53
Paterson, Andrew, Lord Polwarth's shepherd ... 295
Pedigree, true value of ... 31
Pembrokeshire Cattle, by Morgan Evans ... 195
    Antiquity of the breed ... 195
    Early colour uncertain ... 195
    Suitable for exposed situations ... 197
    More suitable to their district than Herefords or Shorthorns 197
    Character of ... 200
    Meat and milk excellent... 201
    Tradition of improvement by Devon bulls ... 201
    Experience of the writer opposed to crossing ... 202
    Herd Book of ... 203
    Errors in popular descriptions of ... 203
    Localities in which they are bred ... 205
Pigs of Great Britain... 425
    Early history of ... 425
    Superstition of foreigners with regard to ... 427
    A source of profit and economy ... 428
    Management in breeding ... 433
    Question of profit ... 431
    At what period they pay best ... 439
    Experiments in feeding, by Mr. Laws... 440
    Experiments in feeding, by C. G. Roberts... 441
    Different practices in feeding ... 444
Pulping machines, Messrs. Hornsby and Son's ... 38

R.

Radnor Sheep, by Morgan Evans ... 414
    Characteristics of ... 414
    Value of mutton ... 415
    Influence of Shropshire cross ... 417
    Variability of type in modern flock ... 418
Rape cake, analysis of ... 52
Romney Marsh Sheep, by the Editor ... 322
    Its peculiarities... 322
    Characteristics of, as described by Mr. Price ... 322

INDEX.

Romney Marsh Sheep (*continued*)—
    Professor Wilson's description of... ...*page* 323
    Hardiness of ... 323
    Improved by crossing with Leicesters... 323
    Mutton superior to other long-woolled breeds ... 324
    Treatment of lambs ... 325
    Breeders of ... 326
Roscommon Sheep, by the late R. A. Pringle ... 357
    Early history of ... 358
    Improved by crossing ... 358
    Rams smuggled from England ... 359
    Recent improvement of ... 360
    Principal food grass and hay ... 360
    Wool of ... 361
    Breeders of ... 362

S.

Science and practice of breeding ... 14
Sheep, difference of character due to cultivation ... 243
    Early history of, obscure ... 244
    Qualities of different breeds ... 245
    Food after weaning ... 269
    Food in summer ... 271
    Tick ... 280
    Louse ... 281
    Washes ... 281
    Farming in Ireland ... 357
Shorthorns, by J. Thornton ... 103
    Universality of ... 103
    Early history of ... 103
    Points of ... 104
    C. Colling's sale, 1810 ... 104
    System of breeding ... 106
    Herd Book of ... 107
    Comparative past and present value of ... 107
    Prices of, at New York mills ... 108
    Preponderance of, at royal shows ... 108, 109
    Value for milking purposes ... 109
    Value for crossing ... 109
    Progress of, in America and colonies ... 110
Shropshire Sheep, by the Editor ... 341
    History of ... 341
    Past and present breeders of ... 342
    Characteristics of and value of the breed ... 348

Silage, history of ... page 42
    Methods of preserving ... 43
    Stacked, Johnson's Press ... 45
Southdown Sheep, by the Editor ... 327
    History of ... 327
    Arthur Young's description of ... 328
    Improvement of, by Thomas Ellman ... 329
    Prices realised by Ellman for ... 330
    Success of Jonas Webb as a breeder ... 330
    Distinguished breeders of ... 331
    Value for crossing ... 332
    Characteristics of ... 333
    Feeding and management of ... 333
Statistics of cattle ... 5
Statistics of sheep from 1868 ... 242
Straw, feeding value of ... 40
Substitutes for straw ... 69
Suggestions for economical feeding of ewes ... 257
Sussex Cattle, by A. Heasman ... 139
    Original value for draught purposes ... 139
    Contrast of modern and ancient type ... 140
    Improved by Mr. John Ellmann ... 140
    Dimensions of the Burton ox ... 141
    General features of ... 141
    Improvement of, by Mr. E. Cane ... 142
    Description of Fry's steer at Smithfield, 1867 ... 142
    Herd Book of ... 142

T.

Tamworth Pigs ... 476
    Recent improvement of ... 476
    Variety of colour in old breed ... 476
    Improvement of, by Mr. Allender ... 477
    Special value for bacon purposes ... 477
    Hardy and prolific ... 478
    Present colour of ... 478

W.

Water, value of, in sheep feeding ... 258, 271
Weaning, season for ... 267
Welsh Mountain Sheep, by Morgan Evans ... 408
    Characteristics of ... 409
    Wool of ... 409
    General management of ... 411

INDEX.

West Highland Cattle, by Robertson ...*page* 176
    Original breed of the West ... 177
    Originally from the mountain districts ... 178
    Characteristics of ... 180
    At Falkirk Trist ... 181
    Anecdote of the "Baron" ... 181
    In the Western Islands ... 182
    Herd of Messrs. Stewart ... 182
    Potalloch herd of ... 183
    Duke of Athol's herd of ... 184
    General management of ... 185
White Pigs, large breed ... 461
    Uncrossed with foreigners ... 461
    Mr. Rowlandson's account of ... 461
    Great size of ... 462
    Different breeders of ... 462
    Points of ... 463
    Success of different breeders ... 464
White Pigs, small breed ... 465
    Different breeders of ... 465
    Described as gentlemen's pigs ... 466
    Defects of crossing ... 466
    Points of ... 467
    General management of ... 468
    Bad effects of forcing for shows ... 468
White Pigs, middle breed ... 470
    Increasing favour of ... 470
    Variety in appearance of ... 470
    Points of, as compared with large and small breeds ... 471
    Different breeders of ... 472
Winter feeding of sheep ... 273
Wool, imports and exports of ... 276
    Growth of ... 277
    Washing of ... 278
    Influenced by soil ... 277

Y.

Young lambs, food of ... 265
    Hampshire and Wiltshire management of ... 266

The Tamworth Pig.

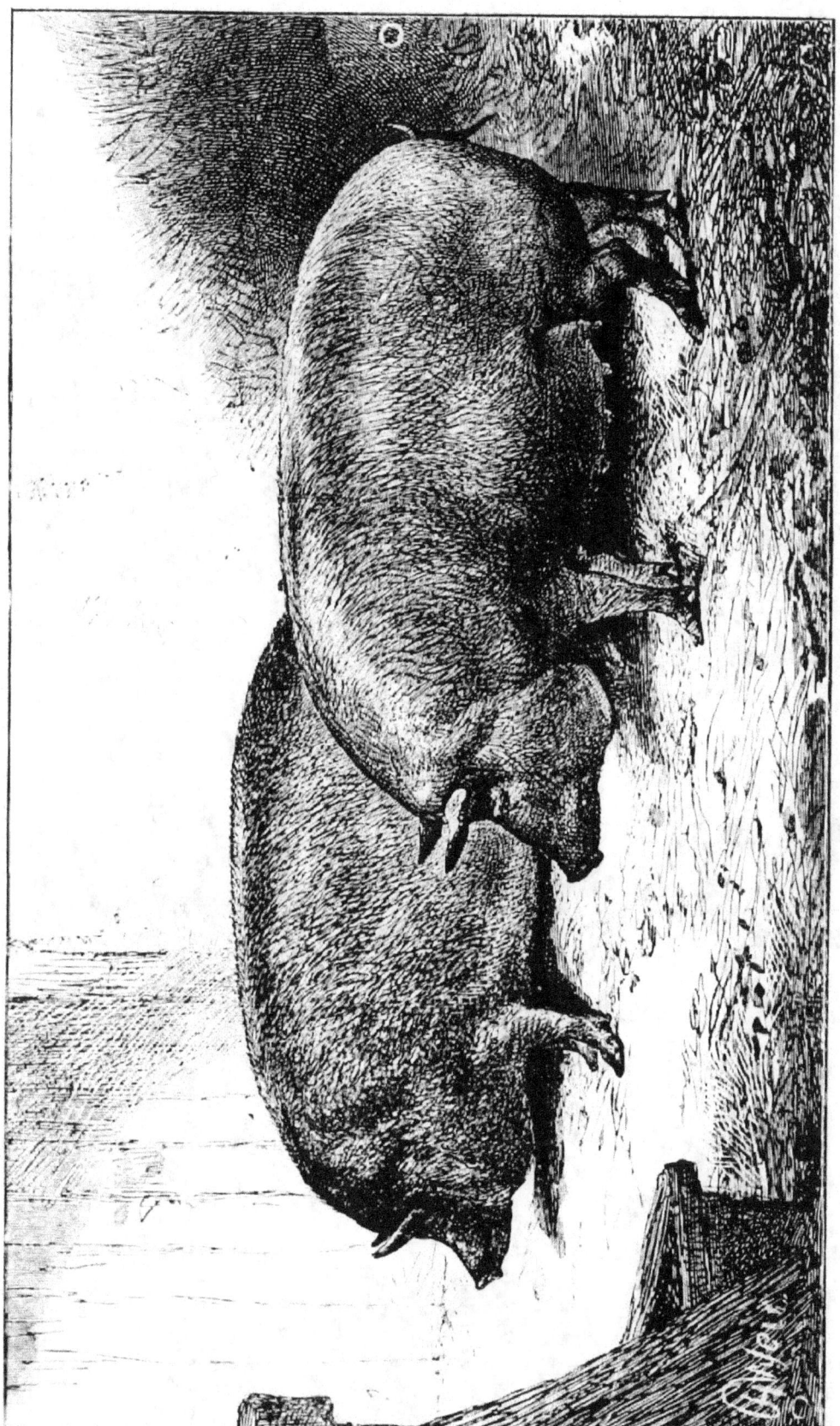
Black Dorset Pigs.

MIDDLE-BRED WHITE PIGS.

Small White Breed of Pigs.

LARGE WHITE BREED OF PIGS.

BLACK SUFFOLK PIGS.

The Berkshire Pig.

Hardwick Sheep.

RADNOR SHEEP.

Welsh Mountain Sheep.

Dorset Horned Sheep.

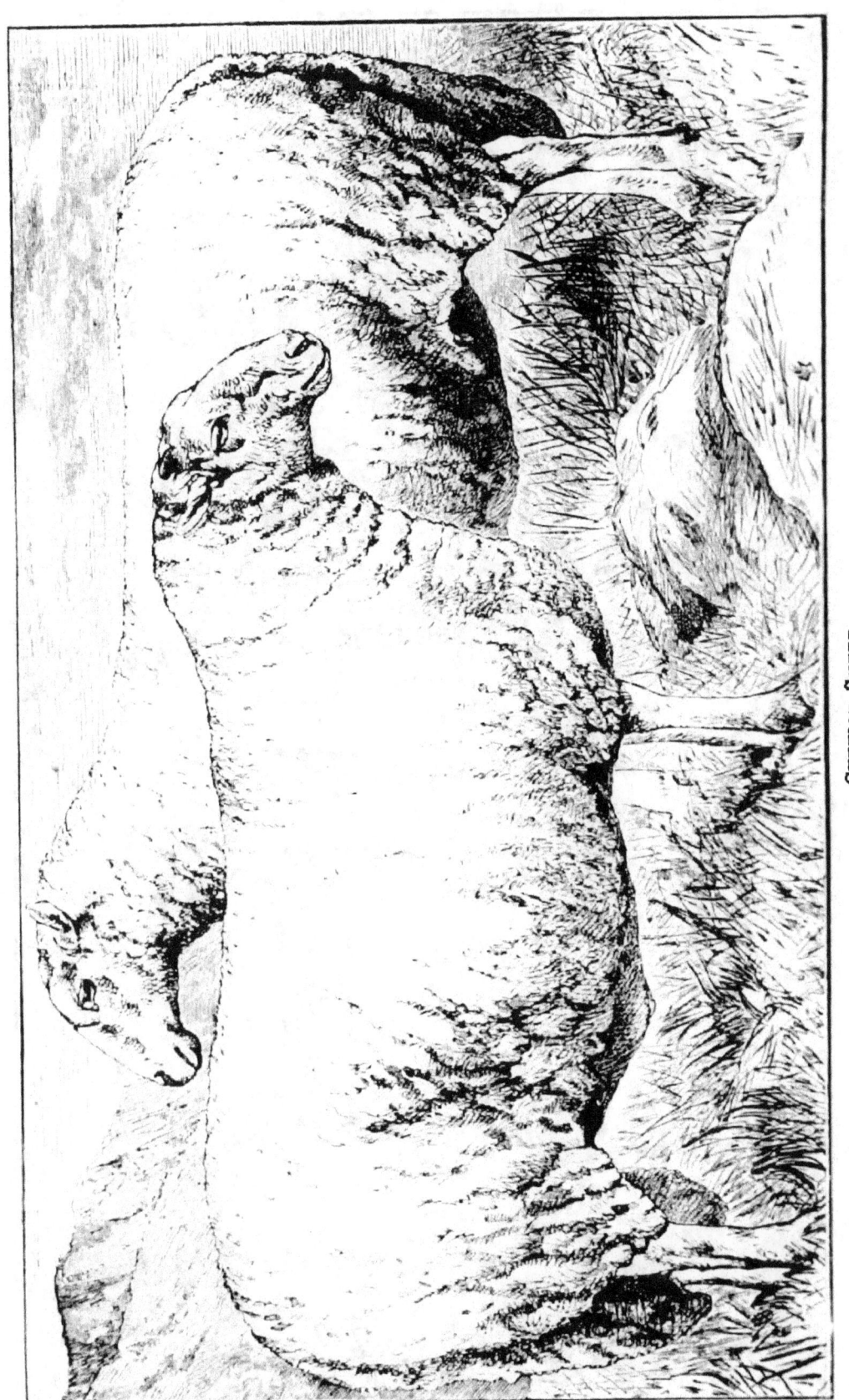

CHEVIOT SHEEP.

Black-faced or Scotch Mountain Sheep.

Exmoor Sheep.

NEGRETTE MERINO SHEEP.

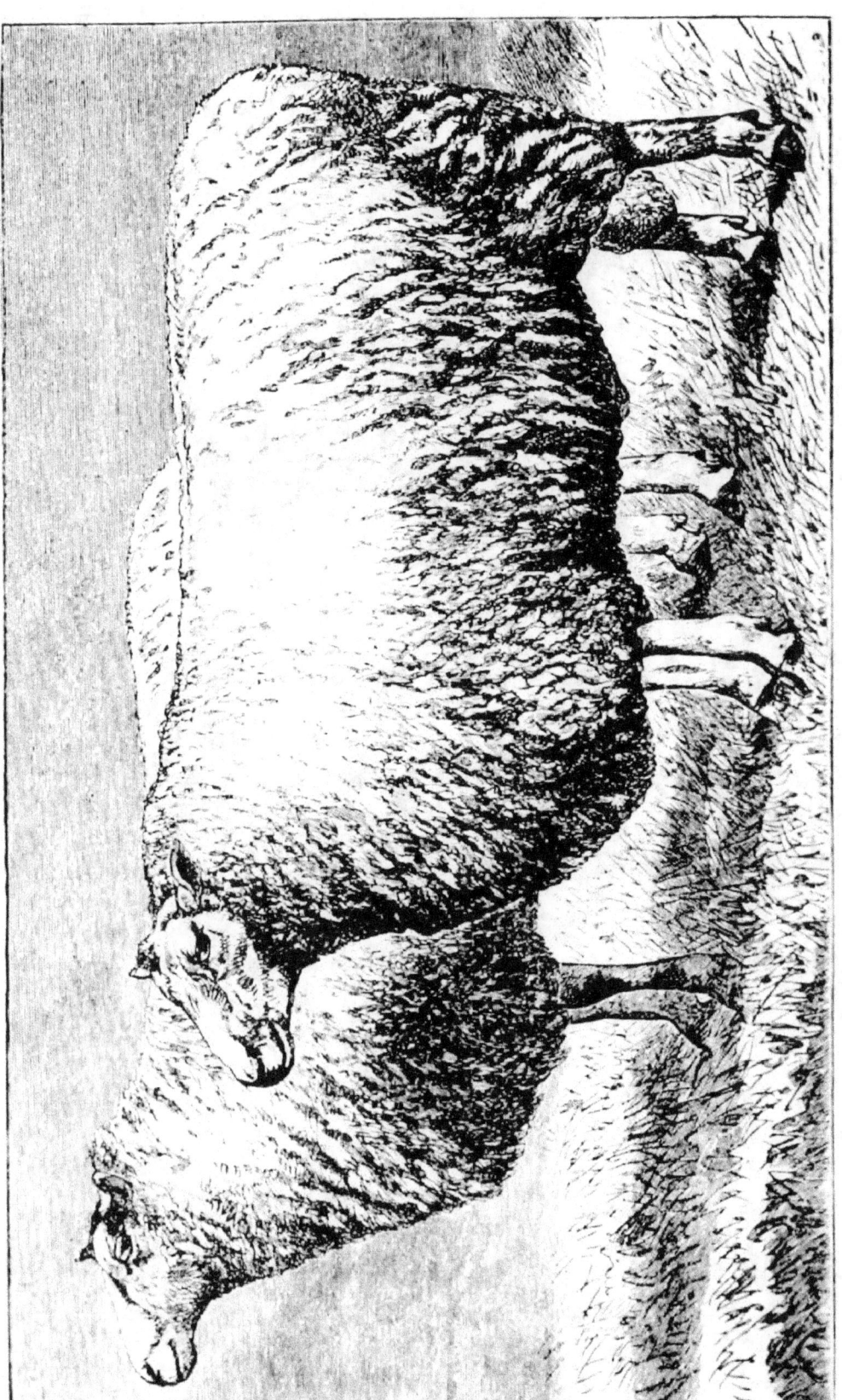

Roscommon Sheep.

OXFORDSHIRE DOWN SHEEP.

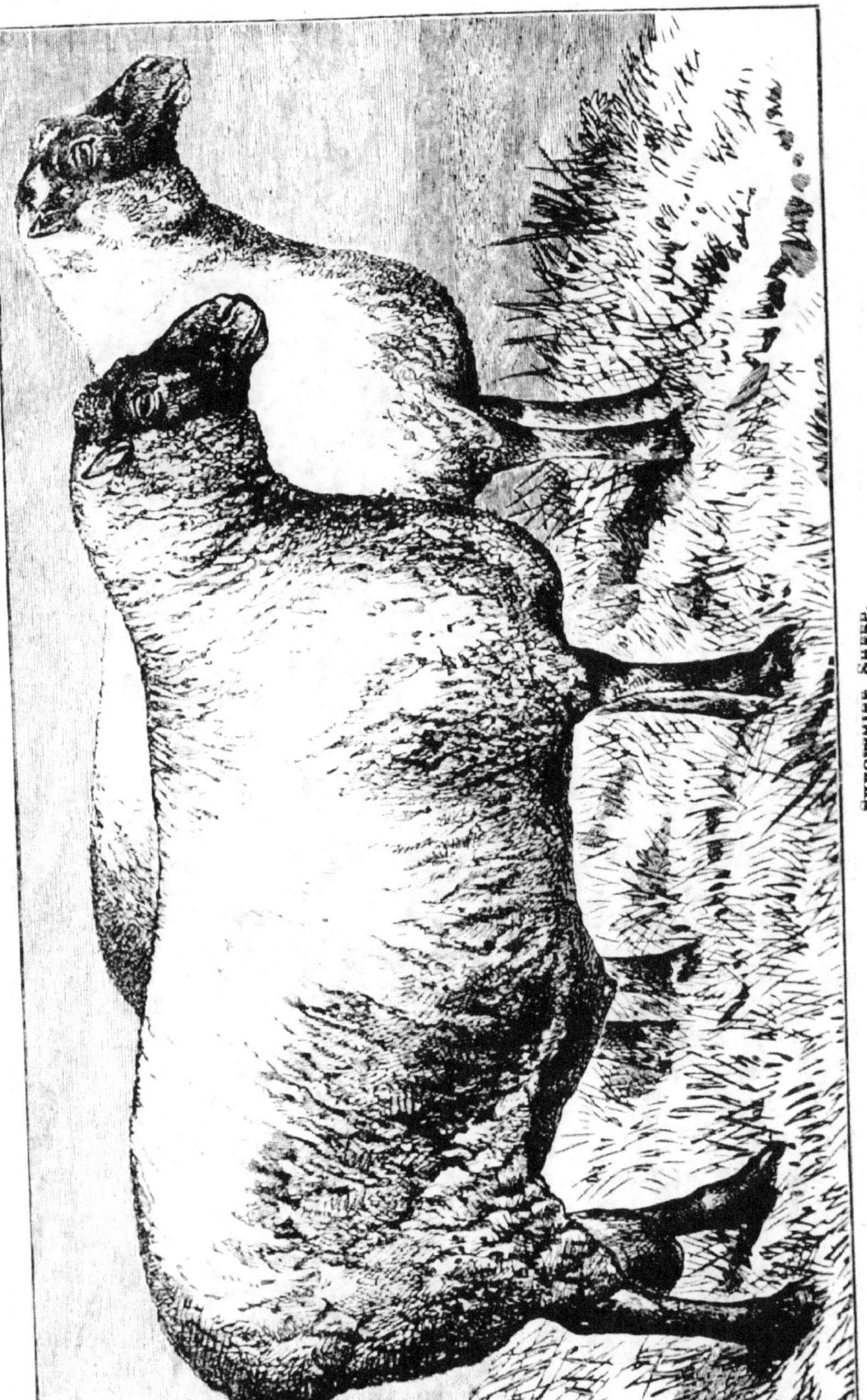
Shropshire Sheep.

Hampshire Down Sheep.

Southdown Sheep.

Romney Marsh Sheep.

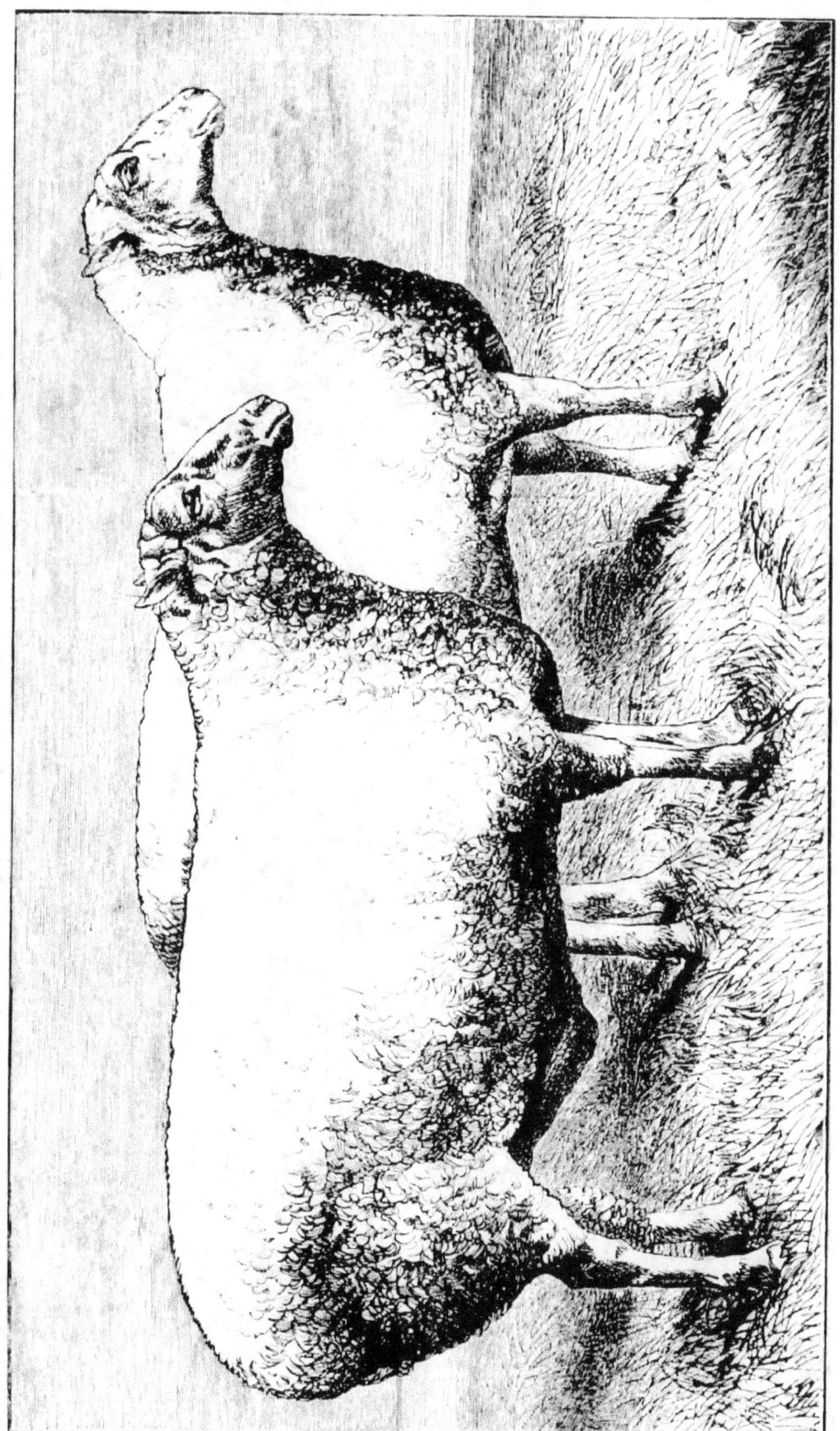

Long-woolled Devon Sheep.

LONG-WOOLED LINCOLN SHEEP.

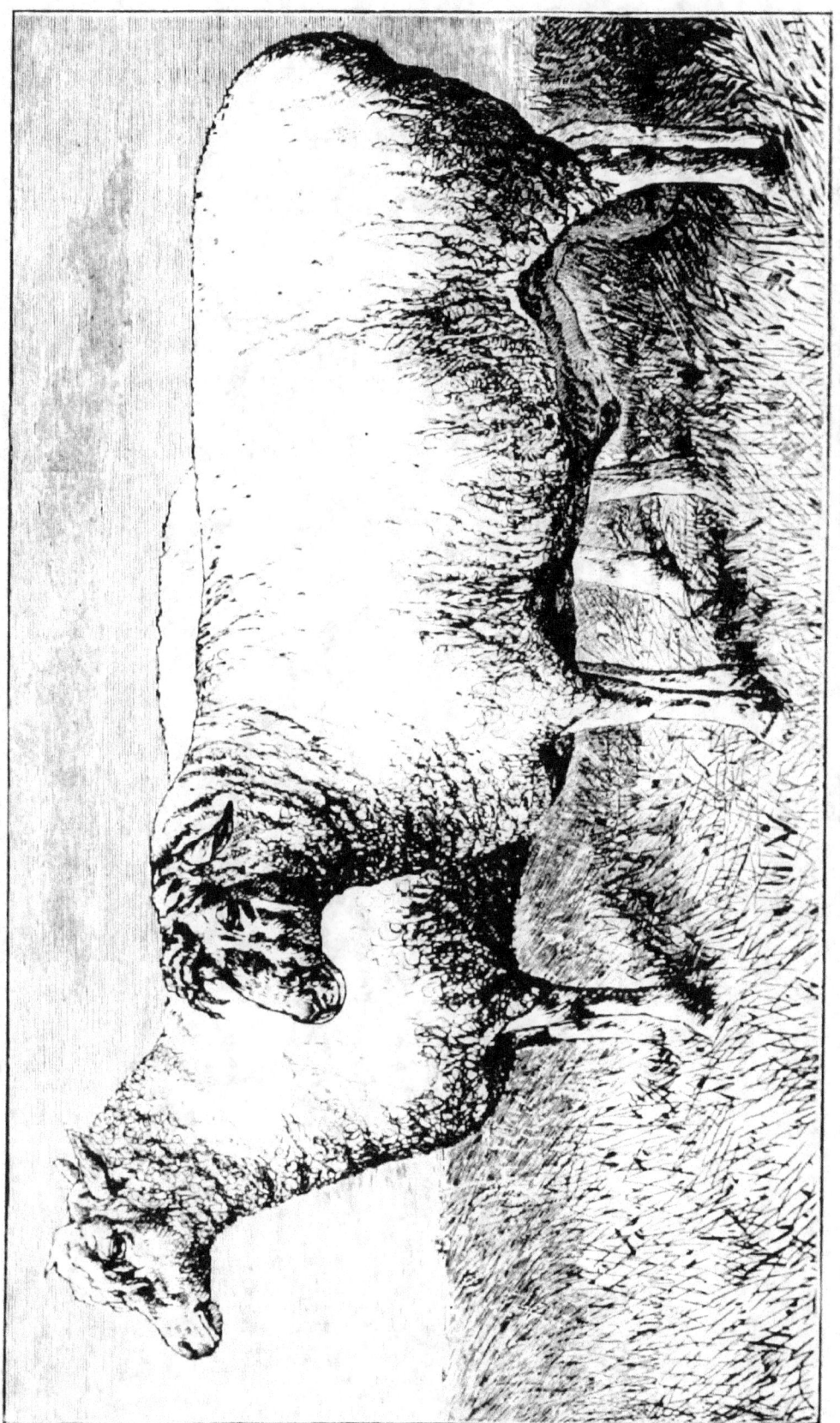

COTSWOLD SHEEP.

BORDER LEICESTER SHEEP.

LEICESTER SHEEP.

Guernsey Cattle, the Property of the Rev. T. B. Watson.

Jersey Cattle, the Property of E. Marjoribanks, Esq.

KERRY CATTLE, THE PROPERTY OF J. H. MURCHISON, ESQ.

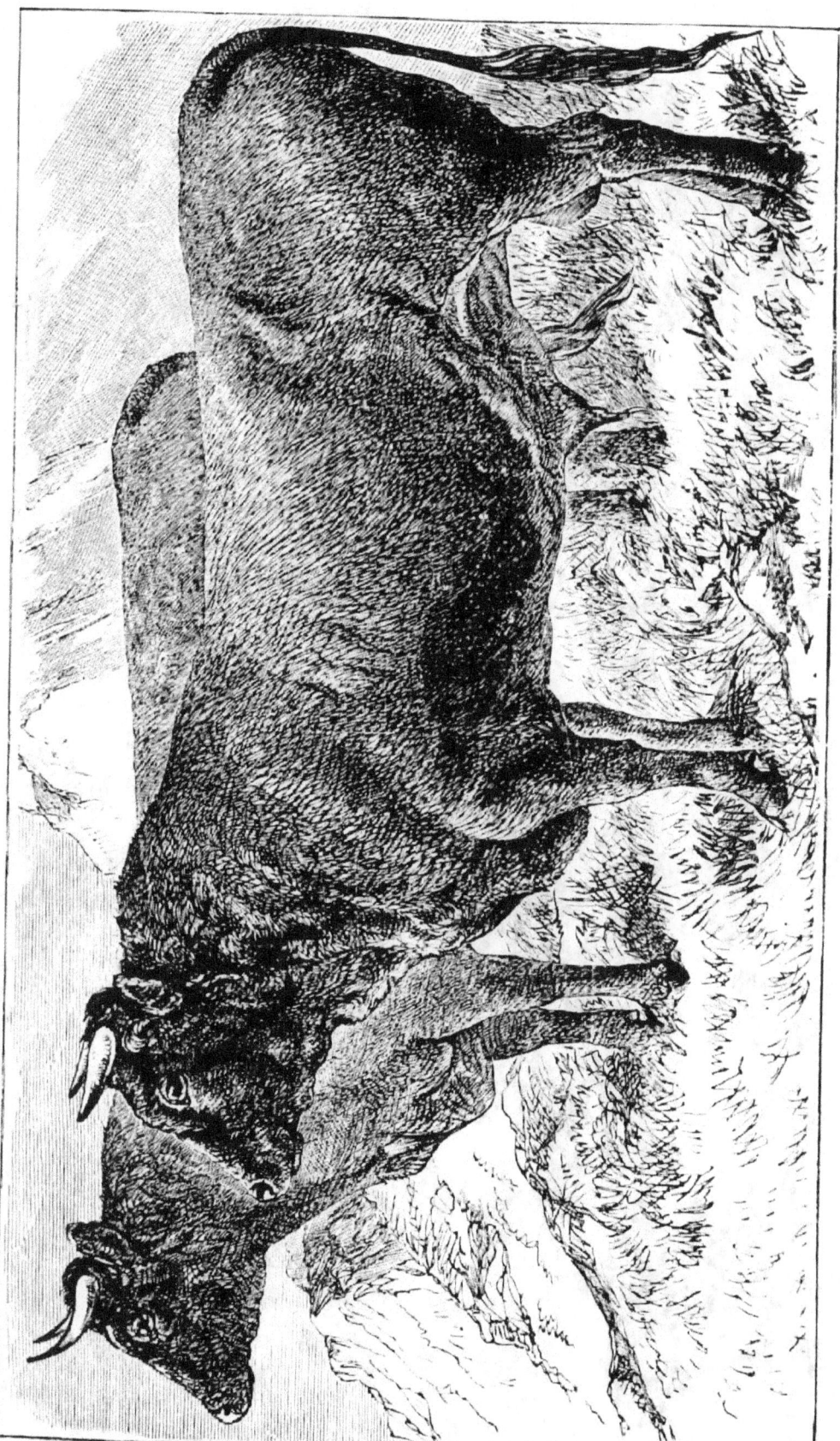
ANGLESEA CATTLE.

Pembrokeshire or Castlemartin Cattle.

Glamorgan Cattle.

WEST HIGHLAND CATTLE.

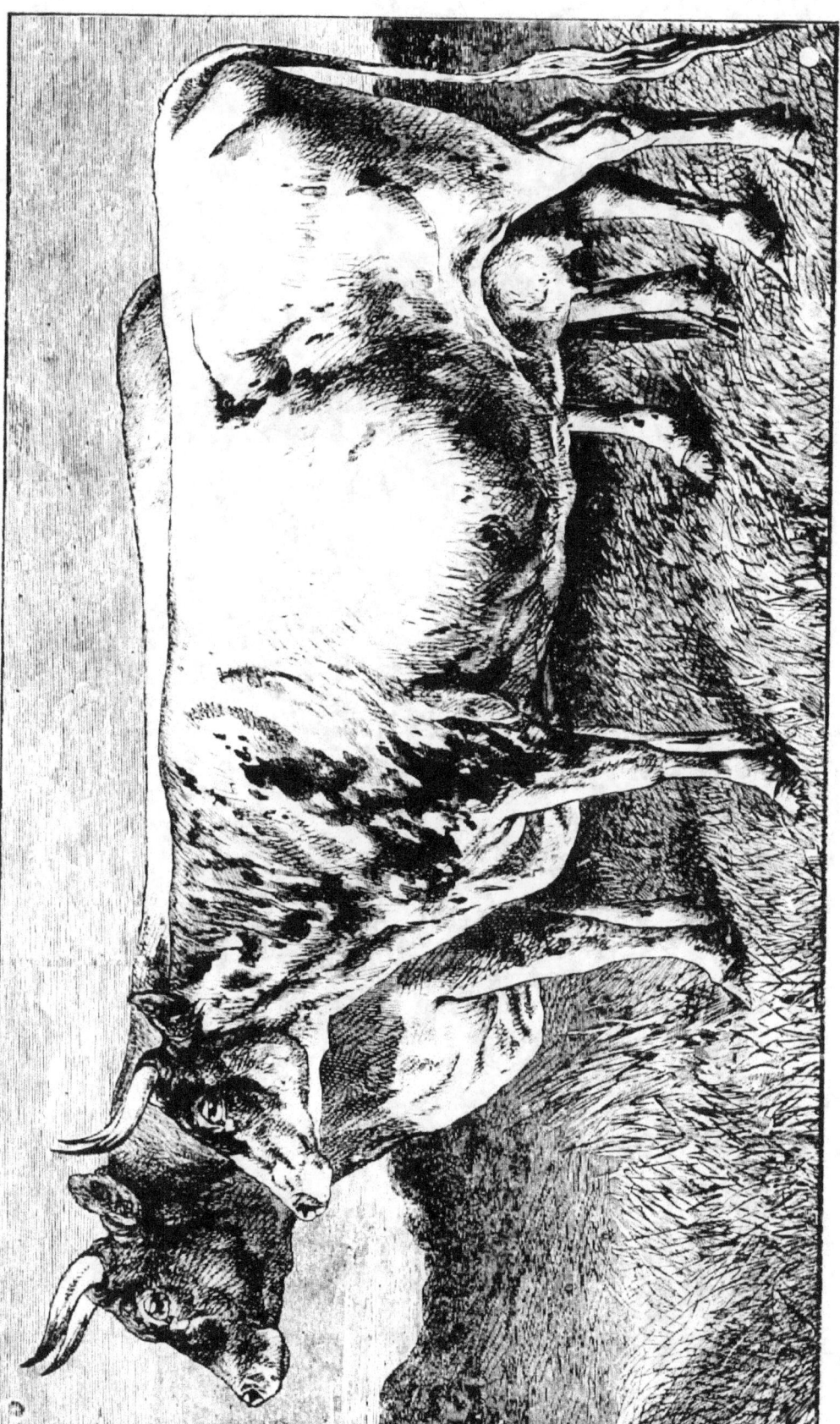

Ayrshire Cattle.

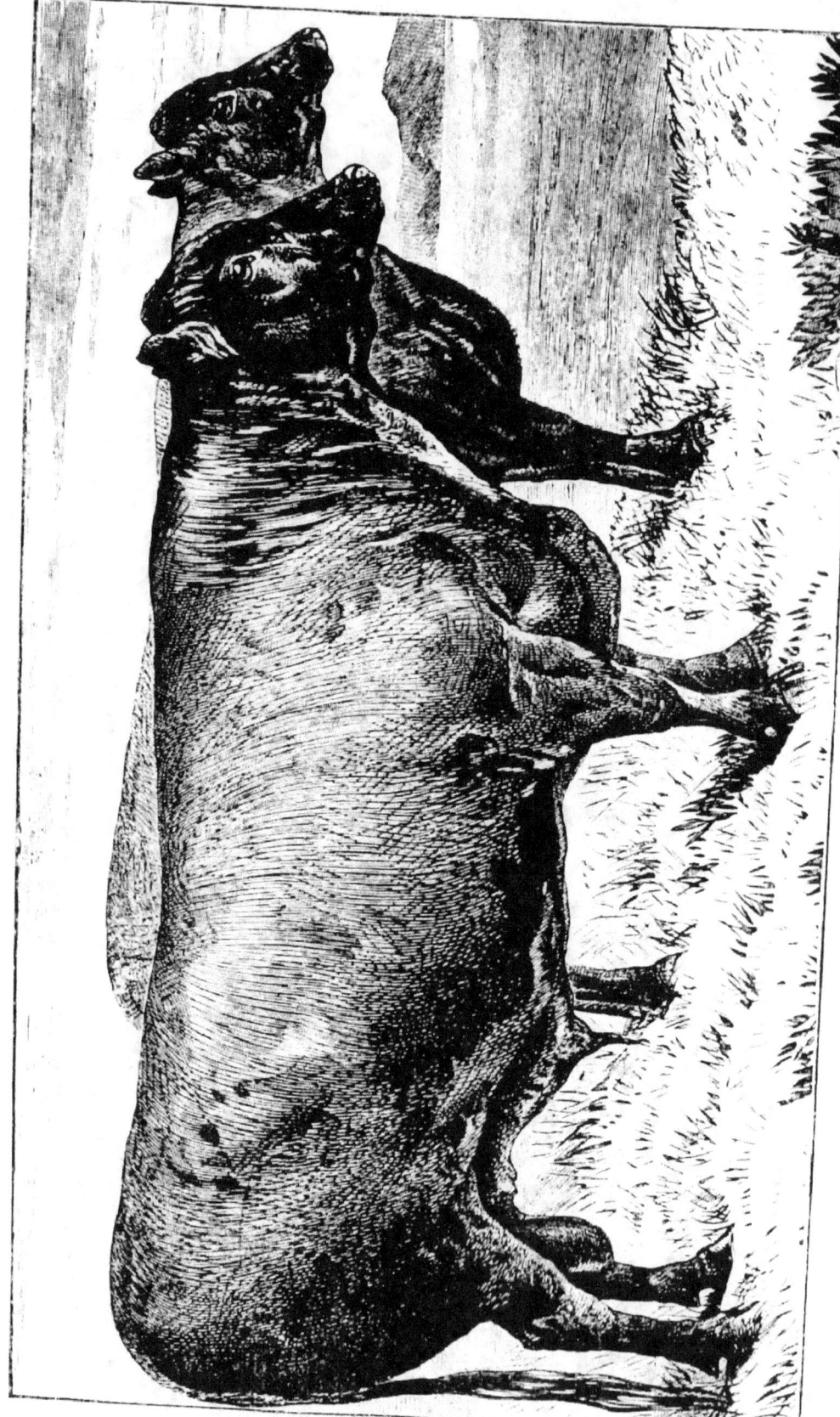
Polled Angus or Aberdeenshire Cattle.

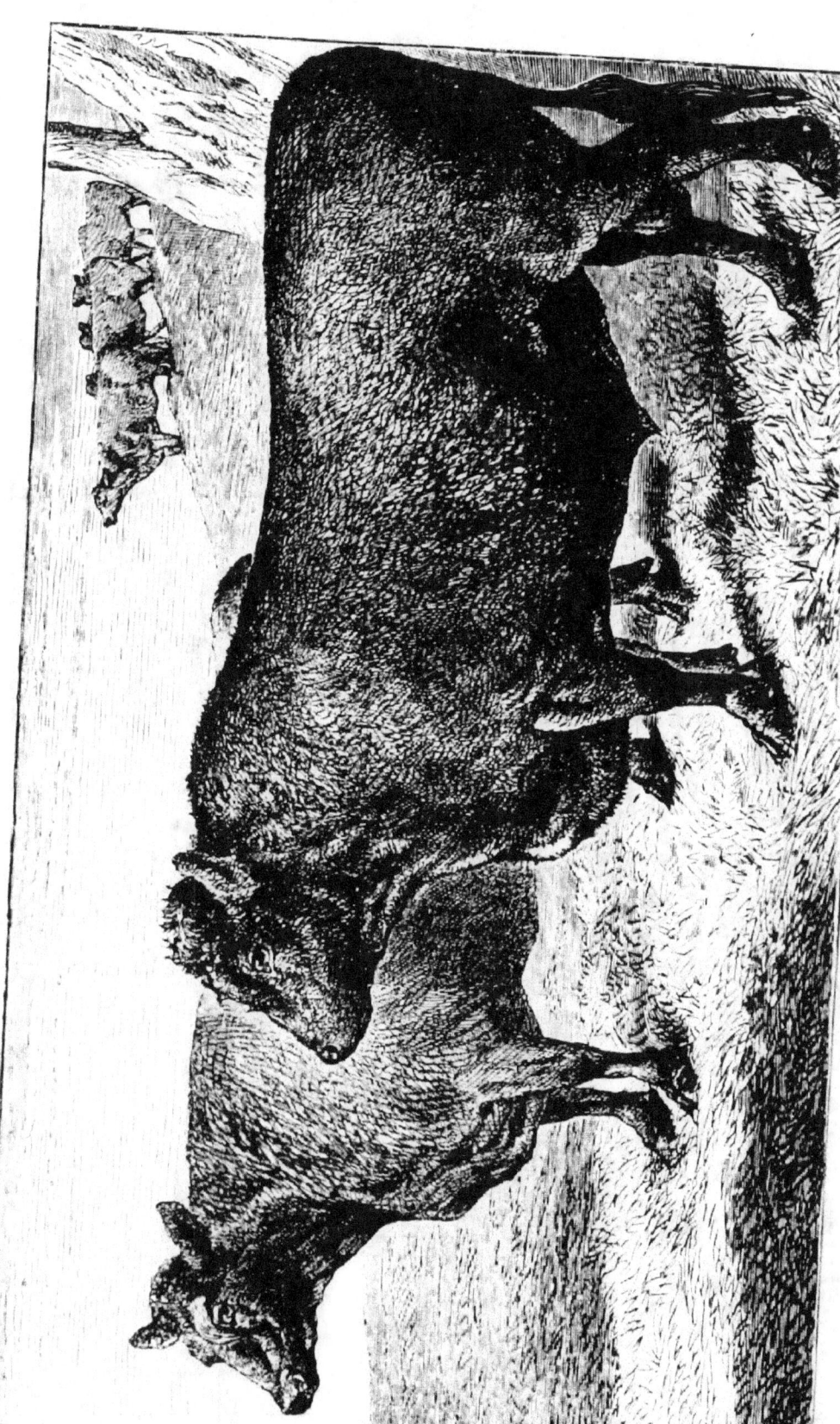

Galloway Cattle.

Norfolk and Suffolk Red Polled Cow "Dolly."

Norfolk and Suffolk Red Polled Bull "Falstaff."

SUSSEX CATTLE.

Longhorn Cattle.

Devon Cattle.

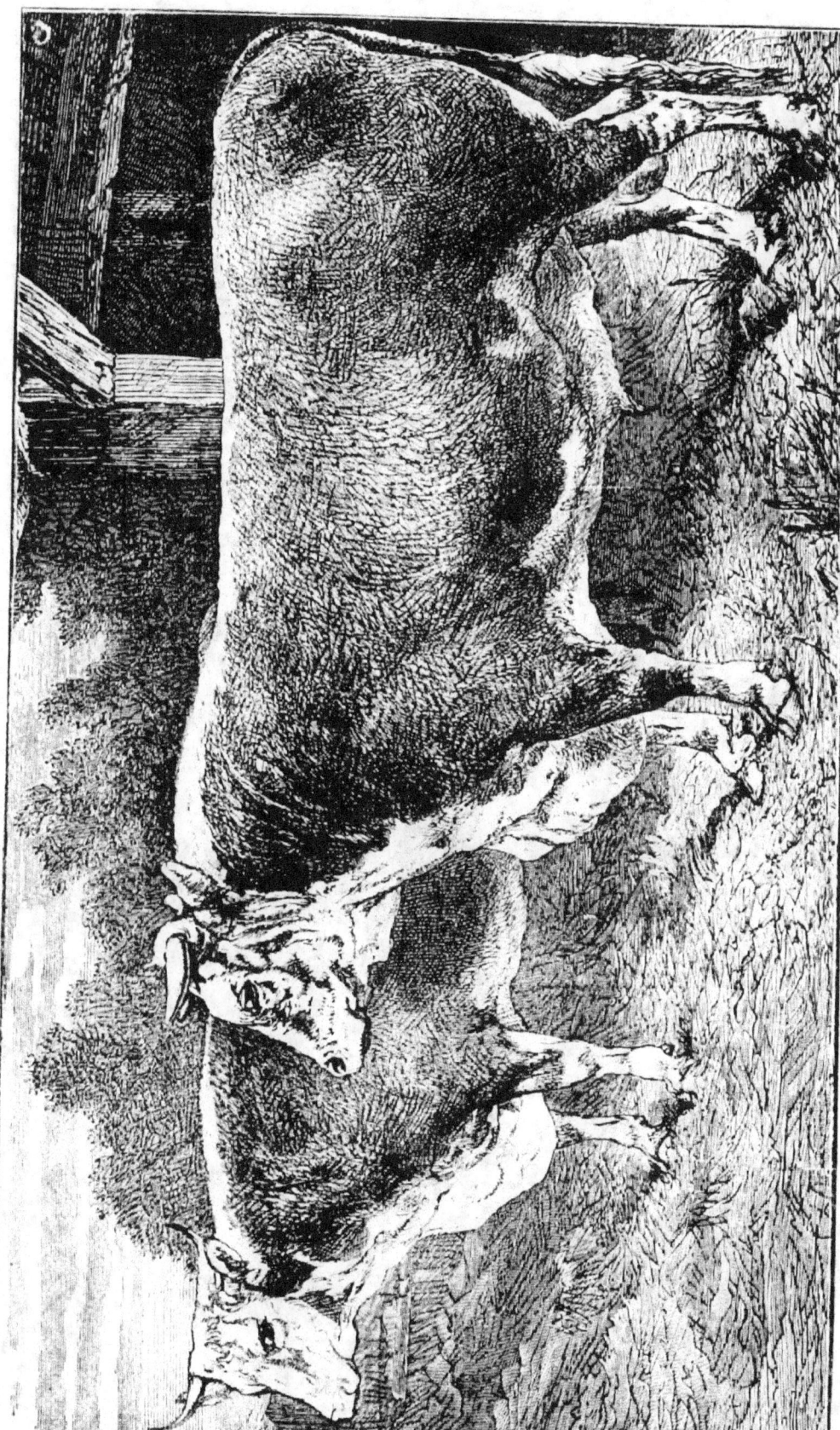

HEREFORD CATTLE.

SHORTHORN CATTLE.

www.ingramcontent.com/pod-product-compliance
Lightning Source LLC
Chambersburg PA
CBHW081017240526
45471CB00017B/3151